S0-BWW-818

Biological Information Processing

Current Theory and Computer Simulation

JEFFREY R. SAMPSON

Department of Computing Science
The University of Alberta
Edmonton, Canada

A Wiley-Interscience Publication

JOHN WILEY & SONS

New York • Chichester • Brisbane • Toronto • Singapore

Library of Congress Cataloging in Publication Data:

Sampson, Jeffrey R., 1942 -
Biological information processing.

"A Wiley-Interscience publication."
Includes references and index.
1. Information theory in biology—Mathematical models.
2. Information theory in biology—Data processing.
I. Title.

QH507.S27 1984 599′.0188′01 83-26058
ISBN 0-471-06281-2

Printed in the United States of America

10 9 8 7 6 5 4 3 2 1

The mysteries of development and of the central nervous system will ultimately be explained in terms of processes, data structures, virtual machines, methods, algorithms and the particularities of their implementation, control structures, and types and styles of representation of knowledge together with detailed specifications of the knowledge required for different tasks. A novel feature of the contemporary scientific scene is that the computer allows one to try out information-processing theories on real-world data. One can argue that a clever enough scientist might not need direct computational experience to formulate the appropriate methods and prove that they will work; but the intuitions needed for understanding biological information processing are not easily available. Only by wresting them from actual experience does one gain a feel for what questions need to be asked, and develop a language in which to ask them. Even with this help, progress is slow, and only small advances have so far been made. But without it, larger ones never will be.

David Marr,
Approaches to biological
information processing.
Science, 190, 875-876, 1975.

Preface

This book grew out of my research and teaching experience during the past decade at the University of Alberta. There have been several reasons for writing it. First, colleagues engaged in simulation of biological systems may find the book of use; many simulation studies are described, including some that previously have not been easily accessible. Second, I hope to promote the study of biological information processing, with particular emphasis on the use of computer simulation. Many biologists, and even computer scientists, see the computer in biological science as merely a processor of experimental data or a numerical integrator of differential equations. Using illustrations from their own fields, I expect to convince some of these scientists that the computer can play an active role in model development and that simulation can benefit their research programs.

A third motive for this book is to make available a unified resource for teaching biological information processing and its study by computer simulation. As a text, the book is primarily designed for computer science students at senior and graduate levels. Little background in biology is required, since chapters preceding the simulation discussions present intensive introductions to relevant biological systems. A course based on this book has a place in the computer science curriculum because biological information processing is of intrinsic interest to many students of artificial information processing. Biological computation is also of potential relevance to future developments in hardware and software, notably in areas like fault tolerance, parallel processing, man-machine communication, and artificial intelligence. Although most of the book can be covered in a one semester course, Parts Two, Three, and Four are relatively independent, allowing the instructor to emphasize biological systems of particular interest.

The book can also serve students of biological science, provided they have at least a rudimentary knowledge of mathematics and programming. Selected chapters can introduce the student of genetics, development, or neurophysiology to the use of modelling and simulation in these fields. Other chapters can provide an information-processing oriented approach to related subjects.

As already suggested, the book has a modular organization. Relevant background material and recent theories of biological information coding, storage, transmission, and retrieval are presented prior to the discussion of specific simulation projects. Successive parts of the book treat distinct levels of biological systems: cellular, developmental, and neural. The reader may thus proceed from the essential preliminaries of Part One to any of the other parts of the book, with minimal loss of continuity.

Many people helped with the preparation and production of this book. My former student and present colleague, Dr. Catherine Descheneau, coauthored Part Three, drawing on her considerable experience with modelling and simulation of developmental systems in order to provide most of the initial draft for Chapters 8 and 9. Dr. Jeff Pelletier, of the Departments of Computing Science and Philosophy at the University of Alberta, furnished similar material for Section 14.2.

The superb illustrations were all drawn by Diane Hollingdale, whose creative insight often enhanced the synergy of pictures and words. The manuscript, written using the TEXTFORM system developed by the University of Alberta Department of Computing Services, was meticulously prepared for printing by Erik Vink, using the phototypesetting facilities of the University's Department of Printing Services. Both the artwork and manuscript processing were supported in part by grants from the National Sciences and Engineering Research Council of Canada.

Valuable critical appraisals of various manuscript segments were provided by several colleagues, including Dr. Ken Morgan of the University of Alberta Genetics Department, Drs. Bernie Zeigler and Michael Conrad of the Wayne State University Computer Science Department, and Dr. John Holland of the Logic of Computers Group, Department of Computer and Communication Sciences, University of Michigan. My graduate students Anne Brindle and Mireille Dubreuil carefully read nearly all of the initial draft and supplied vital improvements in both style and substance. Lisa Higham and Dr. Sven Hurum made similar contributions to certain chapters.

Finally, I am deeply indebted to all the individuals at John Wiley and Sons who were associated with this project. My editors, first George Novotny and then Jim Gaughan, provided remarkably patient guidance and support throughout the long periods of manuscript development; and Rose Ann Campise supplied highly sympathetic assistance and direction during production.

Jeffrey R. Sampson

Edmonton, Canada
May 1984

Contents

PART ONE

Preliminaries

Chapter One

Biological Information Processing and Computer Science

Current advances in molecular biology rival those seen recently in information technology. Following the dramatic deciphering of the genetic code have come vastly improved theories about processes like the control of genetic information expression, hard data like the nucleotide sequences of whole chromosomes, and revolutionary techniques like gene splicing. Computer scientists are thus aware that biological systems contain an impressive array of information processing machinery. We can be easily persuaded that cells and systems of cells frequently function by applying sophisticated algorithms to complex data structures.

For many, this facile application of computer terminology is all there is to say about the interface between natural and artificial information processing. Most students and practitioners of the science of computation do not realize that application of their unique perspectives and skills to the biological counterpart can redound to the benefit of both disciplines. This book has been written to support that claim. It offers students and colleagues thorough descriptions of current theories of biological informatics along with a wealth of examples, some drawn from my own recent research, of how computer simulation is used to advance both the understanding of such natural systems and the art of simulation itself.

This introductory chapter first imposes some essential boundaries on these excursions into natural systems by adopting a narrow view of biological information processing. The systems that survive this filter seem to offer the most fertile ground for both computational models and discrete simulation studies. After an inventory of reasons why computer scientists should study biological information processing (Section 1.2), we then view the obverse of the coin to see how simulation forwards biological research (Section 1.3). The last section of this chapter offers a guide to the plan of the rest of the book, so that the selective reader can choose a path through its modular organization.

1.1 BIOLOGICAL INFORMATION PROCESSING

Information processing is ubiquitous in natural systems. Even seemingly blind mechanical operations, like amoeboid movement or plant shoot elongation, are based on complex signalling systems responsive to a variety of environmental inputs. But we must focus on only a few systems, ones in which information processing is the major function. Nearly every introductory computer science text describes a general purpose digital computer in terms of a few basic logical components: input-output, memory, control. Perhaps these features can be used to select biological systems in which information processing is fundamental.

Since most biological systems have some sort of interface with their immediate environments, input-output capabilities are not a useful measure of information processing. Memory is another matter. Systems which can store large amounts of information for long periods of time usually seem to have information processing as their primary task. One clear example is in genetics, where sequences of millions of nucleotides constitute a repository of information governing all aspects of cell behavior and enduring through countless generations. The central nervous systems of higher animals also have enormous memory capabilities; human experience is probably recorded by modifications to trillions of synaptic connections.

An example of an informational system which does not meet the memory criterion is found in endocrinology. The complex array of glandular control circuitry responds to environmental conditions on an essentially moment to moment basis; there is no historical record or even any underlying behavioral program (other than that specified in the genetic system itself). This book therefore treats genetics and neurophysiology extensively, but endocrine systems not at all.

A general purpose computer is much more than an information repository. The control unit provides a repertoire of basic operations which can be combined into programs for arbitrary manipulation of data. Further, these programs reside in memory and are thus subject to self-modification. This concept of program, in particular of stored program, seems to be another defining characteristic of biological systems whose major function is information processing. Thus the enzymes which process DNA for replication and transcription of information have their primary structures encoded in the DNA itself. The developmental programs which determine what parts of a cell's genetic information will influence its behavior are also recorded in the same genetic medium. And the routines that access and process stored neural information arise from the same networks of neurons.

An example of an informational system having memory but no stored programs is found in immunology. The immune response systems of higher animals can recognize huge numbers of previously encountered antigens for as long as the organism lives. But the system is "programmed" almost entirely by its environment and lacks the autonomy associated with execution of stored programs by a control unit. If the organism never encounters an antigen there will never be any antibodies produced; the system has a purely reactive role. Such information processing is not uninteresting. Given the need to be selective, however, we deal exclusively with biological systems capable of autonomous execution of stored programs.

1.2 WHAT BIOLOGY OFFERS THE COMPUTER SCIENTIST

There are half a dozen reasons why computer scientists study biological information processing. This section reviews these motivations and gives some examples of each. Perhaps the basic attraction of these natural systems for the student of information science is simply that they do process information. There is no apparent reason to confine study of that phenomenon to its manifestation in human artifacts. A more fundamental understanding of architecture, data structures, and algorithms can be derived from viewing them in both their natural and artificial settings. Nature has, after all, devised some fascinating solutions to quite challenging information processing problems, including automatic error correction in DNA replication, representation of 20 amino acids with a 4-symbol coding alphabet, identification of a cell's position in a growing array of cells, transmission of neural signals over noisy channels, and extraction of relevant cues from visual inputs. Since the remaining reasons for studying biological information processing all have an inherently pragmatic flavor, we should happily acknowledge that we are sometimes drawn to information processing systems by their sheer elegance of interacting form and function.

Computer design. Current prospects for innovations in computational hardware have in most cases been anticipated by biological systems. It is possible that the day is not far off when the computer designer will benefit from an understanding of natural systems architecture. Advances may be anticipated in at least four areas. (1) Reliability: although electronic components have become extremely failure resistant, it may be noted that the human central nervous system has somehow been configured to have an expected time to failure of more than 60 years. (2) Compactness: as the speed of light already limits processor volume, packaging at the molecular level may soon be essential to further progress. (3) Parallelism: while current parallel computers are still quite crude, bacterial ribosomes have long known how to have several copies of a translated message in various stages of completion before the template is even fully transcribed. (4) Associativity: in order to construct large associative, content addressable hardware memories it may be necessary to learn more about how these functions are implemented in the brain. Finally, it is tempting to speculate that some day hardware will be adaptive in the way that some biological systems reorganize themselves on the basis of experience.

Man-machine communication. Study of the information processing characteristics of human sensorimotor channels may contribute to an improved interface between people and their mechanical contrivances, especially computers. There is much more to this issue than such simple questions as how fast to present information on a display. Computers are on the verge of employing multimodal (visual, audio, tactile) input-output channels. We need to know more about how the brain deals with modalities if we are to determine optimal assignments of message categories to media. Such research is clearly not exclusively biological and must be integrated with psychological studies.

Prosthesis. The ultimate in man-machine interaction is the direct hookup. Prosthetic visual devices will be able to channel information directly into the optic nerve or visual cortex of blind persons. To make this interface effective a better understanding of the neurophysiological organization of the human visual system is required. Artificial limbs can be equipped with microprocessors capable of directing rapid and complex patterns of movement, as well as providing elaborate sensory feedback to the user. Connection of such devices requires thorough understanding of the types of signals that flow (in both directions) along peripheral nerves.

Artificial intelligence. In this important and rapidly expanding domain of computer science, the goal is to write programs which cause computers to carry out activities that would be considered intelligent if done by people. Recent work has emphasized natural language communication, acquisition and representation of knowledge, problem solving and search, automatic theorem proving, computer vision, and robotics. Many investigators have forgotten, if indeed they ever knew, that one of the roots of artificial intelligence lay in the study of neural networks. Experience has shown that it is unrealistic to approach computer intelligence by trying to build an analog of brain circuitry. Nevertheless, a better understanding of organization and process in the seat of human intelligence may yet have something to contribute.

A related issue is the role of learning in artificially intelligent systems. No one denies that learning is an inextricable component of human intelligence. But early attempts to build intelligent programs that improved performance on the basis of experience were mostly (though not entirely) unsuccessful. In the human brain learning most probably occurs through some form of modification at synaptic connections between neurons. Again, the specific physiological details may not contribute much to machine intelligence (since computers do not have any reasonable analogs of synapses). But researchers in the artificial intelligence community may be ready to take a new look at learning as a process, a process which reaches its current epitome in neural tissue.

Simulation. One view advanced throughout this book is that simulation, especially discrete time simulation on a digital computer, can make important contributions to biological research; more about this in the next section. In addition, computer scientists can benefit from the study of biological information processing systems because they offer an ideal testbed for studying and advancing the art of simulation.

One reason is that biological systems are usually so richly structured that it is possible to model them at different levels of resolution, using a variety of formal techniques at each level. Information processing in a single cell, for example, can be studied as a network of chemical reactions modelled with differential or difference equations, as a flow of materials among a small number of chemical pools, or as a sequence of states modelled as a finite automaton. Neural information processing has been modelled using primitive components at the level of the synapse, the neuron, the neural network, and the brain region. When models of the same system at different levels, and perhaps in different formalisms, are simulated in parallel, each program serves as a check on the others. New simulation techniques for new levels or formalisms can be developed with greater assurance of validity, within the context of established methodologies applied to the same system.

Another feature of biological information processing that provides a challenge to current simulation methodologies is the mixed discrete-continuous nature of such systems. In many other applications exclusively discrete or continuous models (and simulation languages when they are used) suit the phenomenon being studied. There have been few reasons to develop simulation techniques (or languages) for mixed systems. But a neuron is perhaps best modelled as a discrete signalling device attached to a continuous integrator of inputs. And the discrete cells in a developing tissue are embedded in a multitude of continuous gradients. The use of simulation to study mixed models of such systems can considerably advance simulation techniques.

1.3 WHAT SIMULATION OFFERS THE BIOLOGIST

As simulation methodology improves through application to biological information processing systems, we might expect there to be some direct benefits to biological science as well. Five chapters of this book are devoted to analysis of specific simulation studies and their contributions to biology. This section briefly summarizes the types of help the biologist can expect from computer science. The contributions of simulation are presented in order of their increasing potential value to biology. Regrettably, this sequence is inversely related to the frequency with which the benefits have thus far been seen in practice.

First, simulation confers rigor on the modelling exercise. Computers do not tolerate vagueness or undefined variables. A working program always represents a fully specified and unambiguous model. Of course the model may not be a particularly good one; it may not even be the one the investigator had in mind. But these situations can often be uncovered rather quickly by running the simulation for some simple test cases. So, if the required programming time is not too long, simulation can also provide a relatively rapid form of preliminary model checkout. Full model validation is obviously another, much longer process.

Simulation can also guide the commitment of scarce resources. The amount of money, personnel, and equipment available to a laboratory project usually limits the number of questions that can be asked. Even when it cannot replace any portion of the actual laboratory work, simulation can often identify the most promising questions and channel research onto paths most likely to succeed. For example, various competing theories about the nature of synapse modification during learning could be tested in simulations of an otherwise realistic neural network model. Those synaptic mechanisms which could not be shown to support learning under any reasonable combination of simulated circumstances would be assigned much lower priority for laboratory investigation. Simulation probably cannot identify the learning mechanism; but it might considerably shorten the search.

Experiments which are impossible, too dangerous, or too costly to perform in the laboratory can often be studied through simulation. As one example, the consequences of the loss of a particular enzyme system could be studied by simulation of a suitable model of bacterial biochemistry, even if no such mutant were naturally available. Or, potentially hazardous types of recombinant DNA research could be carried out, at least initially, on the computer.

Simulation gives investigators control over the inputs to the system that may not be possible in the laboratory situation. Sources of variation can be precisely specified, sometimes leading to an understanding of system interactions otherwise obscured by fluctuations in actual data and errors of measurement. Inputs and parameters can sometimes be varied more smoothly and over broader ranges than in nature. Time scales can be expanded or contracted by simulation, slowing down phenomena which occur too quickly for adequate measurement, or speeding up processes which would take too long to observe. Simulated experiments can also be repeated under identical or systematically varied conditions. The simulated system can always be reset to any particular state. No information is lost or material destroyed when a simulation experiment is aborted.

Ideally, simulation and laboratory research should cooperate intimately in the exploration of a model. Simulation studies should make predictions testable in the laboratory; the results of these tests should suggest refinements of the model and further simulation studies, and so on. There are two major reasons why this situation is seldom realized. First, many biologists do not appreciate the utility of computer simulation since they have never encountered computers in contexts other than statistical data processing. Second, a sound understanding of simulation and its role in the design and testing of complex models is relatively new to computer science itself. The hardware, software, and formal prerequisites to the existence of a science of large scale simulation have just begun to coalesce. One rather rigorous view of modelling and simulation is offered in the next chapter.

1.4 PREVIEW

This book is organized to suit the needs of a variety of readers. After the two remaining chapters in this opening part, which provide essential background and terminology in modelling and simulation (Chapter 2) and biological systems (Chapter 3), the reader may proceed directly to any of the subsequent three parts. Each treats a successively higher level of biological organization: cellular system, developmental system, neural system. Within each of these modules, alternate chapters present current biological theories about aspects of information processing in the particular biological system and, then, a variety of simulation studies of those aspects.

A final introductory observation concerns the selection of simulation studies. There is no attempt to provide a comprehensive review of the research in any of the areas. Such a compendium would fill several books. The goal is rather to present studies which effectively illustrate the simulation methodology, with an emphasis on the discrete time approach. The selection obviously reflects personal preferences and biases. In some cases studies carried out by myself or my students receive relatively detailed treatment, especially if the work has not been fully described in readily available literature.

Chapter Two

Modelling and Simulation

This chapter serves several purposes. First, we expect reader background and experience with modelling and simulation to vary considerably in approach, application, and depth. We therefore try to provide a unifying view of these fields with respect to terminology, scope, and formalization. Second, we forward, rather forcefully at times, a personal view of modelling and simulation activities, a view based directly on Zeigler's (1976) recent treatment.

A third motivation for this chapter is related to both of the above reasons. In the course of the book we wish to present, compare, and evaluate simulation projects which use varying methodologies to study models in diverse fields of biological science. These discussions will be greatly facilitated by a uniform context, a way of viewing individual models and their simulation which permits direct comparison and comparable evaluation.

The first section below looks at the modelling process, in a discussion mostly independent of simulation considerations. We discuss types of systems which may be modelled, the nature of system-model correspondences, and some reasons for and risks of modelling. A taxonomy of modelling is also presented, with several biological examples. Finally, various aids to modelling are considered, ranging from informal description through simulation.

Section 2 examines the use of digital computers to study models by means of simulation. We first describe the three standard simulation paradigms (continuous, discrete event, and discrete time) and review some strategies for simulation within each of these paradigms. A discussion of costs and hazards of computer simulation is followed by a special look at interactive simulation. The final section of this chapter outlines Zeigler's formal framework as presented in his *Theory of Modelling and Simulation* (1976).

In addition to personal experiences with modelling and simulation, and Zeigler's definitive work, this chapter has been influenced by the writings of other authorities in the field. In particular, the discussion of modelling owes much to the study of cybernetic modelling by Klir and Valach (1967) and Olinick's (1978) recent book on mathematical models. Fishman's (1973) text on discrete event simulation contributed to the views of both modelling and simulation. Finally, most of the discussion of interactive simulation (Section 2.2.3) is based directly on papers by Sohnle, Tartar, and Sampson (1973) and Sampson and Dubreuil (1979).

2.1 MODELLING

2.1.1 Systems and Models

Systems may be classified into static and dynamic types. Static systems are more limited, in that time is not a factor. Although we will mention some models of static systems in the next section, the concern here is exclusively with dynamic systems, which manifest changes with time.

A *system* may be defined as a set of interacting *components*. For example, the brain is a system with nerve cells as components and events at synaptic interfaces as the basis for interaction among components. A computer is a system with electronic components interacting through wires. These are two *physical* systems, one *natural* and one *artificial*. Contrasting with physical systems are *abstract* ones, such as mathematical systems in which interactions among component variables may be specified by equations. Natural, artificial, and abstract systems share the following five fundamental characteristics.

(1) *Time base*. In a system characterized by a *continuous* time base, events may occur at any point in time. The brain has a continuous time base. In a system with a *discrete* time base, events may occur only at regularly spaced points in time, called time steps. The cycle of a digital computer is an example of a time step.

(2) *State set*. Each component of a system will have certain properties, or variables, whose values fully define the component. At any point in the time base the system will be in some *state* defined by the current values of all variables for all components. The set of all possible combinations of such values is the state set of the system. Like the time base the state set may be continuous, if the variables are continuous, or discrete, if they assume values at (not necessarily regular) intervals. A state set in which some variables are continuous and others discrete is termed *mixed*. Whereas a continuous or mixed state set is necessarily infinite, a discrete state set may contain either a finite or

infinite number of possible states. The brain is a system with a continuous state set comprised of possible combinations of parameter values (threshold, firing rate, and so on) of all the nerve cells. A digital computer has a finite (though very large) discrete state set, in which a state is defined by the contents of all the switches, registers, and memory cells.

Note that a system may have a continuous time base and a discrete (or mixed) state set. A simple example is a waiting line in which a state is the number of people in line, a discrete set, but people may enter or leave the line at any point in time. A system may also have a discrete time base and a continuous (or mixed) state set. This situation can be illustrated by a computer control system which collects temperature readings (a continuous variable) at discrete time intervals.

(3) *Environment*. A system exists in some context. The brain resides in the body. A computer functions in a community of users. The boundaries between a system and the rest of the world are determined by the interests and requirements of the investigator. The environment can usually be characterized as a system too. Just as the system being studied exists within larger systems, so does it contain subsystems, which may themselves be the objects of interest on another occasion. This situation allows study of a phenomenon at various levels of subsystem depth, called *resolution levels*. One hierarchy of resolution levels in biology is that of molecule, biochemical pathway, cellular component, cell, tissue, organ, organ system, organism, and population. A system at one level in this hierarchy normally has a system at the next higher level as its environment.

(4) *Input-output*. A system interacts with its environment by means of inputs and outputs. The brain receives signals about body conditions from sensory organs and alters the body through signals sent to muscles and glands. A computer communicates with its users through an assortment of I-O devices. These systems are termed *open*, to distinguish them from systems in which there is no input-output (or, more precisely, where the system and environment cannot affect each other's states). Such *closed* systems, of little relevance to biological information processing, do not play much of a role in this book.

(5) *Behavior*. When a system in some initial state is presented with a sequence of environmental inputs, it will exhibit a sequence of states and outputs called its *behavior*. The course of such behavior is a consequence of the rules by which the system components interact, respond to inputs, and generate outputs. The transition functions of finite automata (discrete-time finite-state machines) can be generalized to all types of systems. There is a *next state function* and a *next output function* which use the current state of the system and the current input from the environment to determine, respectively, the successor state and the next output of the system.

It is often possible to evaluate the next state and output functions without examining the value of every variable associated with every system component. A smallest set of variables which determine a system's behavior is called a *state variable* set. Although there may be more than one such smallest set, we henceforth assume that one has been selected and refer to its members as state variables. Knowing the values of the state variables (and inputs) at any point in time allows computation of subsequent values of all system variables, in particular future state variable values.

If there is just one state-output pair for every combination of state and input, the system is *deterministic*. If more than one future event is possible under a given set of circumstances, the system is *nondeterministic*; if there is a probability distribution assigned to the alternatives, the system is *stochastic*. A system can have a deterministic next state function and a stochastic next output function, or vice versa.

A system with just one state exhibits a special kind of behavior. Such a system is purely reactive, in that the output is a consequence of only the current input. In order to have behavior which is influenced by past events (a form of memory), a system must have at least two states. As defined in the previous chapter, biological information processing systems do not have this memoryless character, although they may have subsystems which do.

This completes the specification of fundamental system characteristics. We turn now to the idea of a model of a system. Whenever two systems correspond in some way, it may be possible to treat either one of them as a model of the other. Although modelling is frequently thought of exclusively in the context of abstract models of physical systems, there are artificial models of both abstract and physical systems, and even natural models. The important point is that modelling is a two way relationship between systems. It is only the investigator's viewpoint and intentions that identify one system as the "model." Henceforth when we use the word *model*, it should be taken to mean "system intended as model."

To illustrate the modelling correspondence between systems, we return to the two systems used as examples throughout this section, the brain and the digital computer. We might wish to regard the computer as a model of the brain, since both use large numbers of extensively interconnected basic components to carry out complex information processing tasks. The fact that one system is continuous in time and state, and the other discrete in both respects, should present no special problems if we are prepared to accept a discretized view of the neural activity in the form of a succession of "snapshots." Yet upon closer examination this putative modelling relation seems inappropriate. A nerve cell is really nothing at all like a transistor or any other component of a computer. Storage and retrieval of information are so fundamentally different in the two systems that it is difficult to say what might correspond to an address in the nervous system. Clearly, we need some stronger notion of similarity between systems than superficial resemblance of structure or function. As a first step, we may attempt to formalize the notion of *similarity of behavior*.

The mathematical concept of *morphism* can help define the idea of system-model correspondence. A morphism is a mapping (or function) between two sets that preserves desired properties of the domain set in the range set.

The kind of morphism that results from a mapping depends on the properties which are preserved. In what follows we describe morphisms which are successively stronger in that they require increasing numbers of properties to be preserved.

There are two general classes of morphisms, depending on the nature of the underlying mapping. If the mapping is one-to-one, so that there is a unique element in one set corresponding to each element in the other, the relation is an *isomorphism*. In this case the sets must have the same cardinality. If the mapping is many to one, the range set may have lesser cardinality than the domain set; the relation is a *homomorphism*. There exist homomorphisms from infinite sets (e.g., the real numbers between 0 and 100) to finite sets (e.g., the integers from 0 to 100 obtained by rounding the reals).

We may use morphisms between system characteristics to build or assess system-model correspondences. With respect to the time base, we should require that the model time base be at least a homomorphic image of the system time base. We thus allow the common practice in modelling of approximating continuous time systems by discrete time ones, as well as the contraction of time scale so that a model time step corresponds tc several successive steps in the system. In either case the property of *chronology* is preserved, since model events occur in the same (partially ordered) sequence as system events. Similar considerations suggest that a homomorphic relation is a sensible requirement for sets of inputs and outputs. Mapping of subsets of system inputs to single inputs in the model may provide an essential simplification of the modelled system. If a homomorphic contraction of time scale is in effect, we can establish a similar relation between sequences of system inputs and single model inputs.

We have thus arrived at the weakest form of behavioral equivalence between system and model, at the level of observed inputs and outputs. The requirement can be strengthened by also requiring a homomorphic mapping from the states of the system to the states of the model. Once again, we may choose to model an infinite state set with a finite one, or a larger one with a smaller one. And sequences of states in the system may be mapped to single states in the model.

Homomorphic preservation of the system state transition function may be required of a model. This condition can be explained, and tested, in terms of a commutative property. There is a homomorphic mapping h from system states to model states. When the system is in some state s which maps to state s' in the model, the next state in the model should always be the same whether we (1) determine the successor of s using the system transition function and map this successor to a model state using h, or (2) compute the successor of s' directly, using the model state transition function. The order in which we apply h and the appropriate transition function is immaterial if there is homomorphic transition-function preservation.

Beyond behavioral equivalence, we may require or impose morphisms governing the relations between *structures* in the two systems. At whatever level correspondence is sought, as long as the morphisms are made explicit and adhered to in the design and analysis of the model, we know exactly how the system and model are intended to correspond. Whether this is a productive correspondence depends on the reasons for modelling and is often partly a matter of subjective judgement.

This section concludes with some remarks about motivations for modelling. A major reason for building a model of a system is frequently just to increase understanding of the system by organizing knowledge about it in a systematic way. As Zeigler (1976) has observed, models embody hypotheses about reality. Models also enable manipulation of a substitute for reality, whether by simulation or other modelling techniques (see the discussion of aids to modelling in Section 2.1.3). Manipulation of the model may be cheaper, faster, and/or easier than working with the modelled system.

Models designed for the best of reasons can be counterproductive, however. One of the primary risks of modelling is that the correspondences established between system and model may be inadequate or inappropriate to the investigator's goals, and may actually obscure system characteristics intended for study. Morphisms are no help here, since the most rigorous similarity between wrong things still leads to wrong answers. A related problem occurs when an initially valid and productive modelling relation deteriorates with time. The investigator may have become so involved with the model that he has lost sight of the original system and the reasons for modelling it. He may even be ignoring new data which tend to invalidate aspects of the model. He may become so rigid in his thinking that he can consider no alternatives to the hypotheses about reality that the model embodies. Clearly the system-model relationship does not end with the creation of the model, but must be continually reviewed.

Lastly, any model can be misapplied beyond the scope of its original purposes and validity. If morphisms have been established only at the input/output level, for example, inferences about structural correspondences between system and model cannot be supported. Models which imitate or even predict behavior do not necessarily explain it.

2.1.2 Varieties of Models

Since a model is just a system viewed in a special relation to some other system, it is not surprising that the taxonomy of model types will closely parallel the types of systems uncovered in the previous section. We begin with a few quick examples of *static* models, by way of contrast to the dynamic ones that are of main interest here, and to suggest the pervasiveness of modelling in everyday life as well as in scientific endeavors.

Physical static models include: scale models, like a miniature airplane designed for wind tunnel experiments, but also sculpture and other art forms; pictorial models, such as photographs and X-rays; and schematic models, like maps and circuit diagrams. *Abstract* static models include: symbolic models, like logic systems or structural descriptions of languages by linguists; and mathematical models, such as systems of algebraic equations or representations in another domain effected by various kind of transforms. Like the dynamic models to be discussed next, abstract static models may have a nondeterministic character, such as a mathematical model of some natural phenomenon in terms of a probability distribution.

Dynamic models fall naturally into six classes according to whether the time base is discrete or continuous and, in either case, whether the state set is discrete, mixed, or continuous. Terms like discrete-mixed will be used to describe these classes, with the understanding that the time base is always described by the first adjective in the pair. It should be kept in mind that, although we characterize the model with such terms, the system being modelled may have a different type of time base and/or state set. In most of the examples the model is abstract, although physical dynamic models are common enough; an example is an electrical circuit equivalent to some natural system (like the nerve cell membrane, as discussed in Section 11.2.1).

Several types of dynamic models are now illustrated with thumbnail sketches of models to be considered in full later in the book. For this reason, the reader with little background knowledge about the biological systems being modelled may find some of these discussions less than fully intelligible. But it is more important at this stage to notice the character of the modelling process in the various categories than to attend to biological detail. Before these models are examined again, all of the necessary biological background will have been provided.

Discrete-discrete models can in most cases be regarded as finite or infinite state automata, although this is not always the representation most convenient for manipulating the model. Probabilistic transition functions may be used to express stochastic versions of such models.

In Section 9.1.3 we will look at one of several discrete-discrete models of cell sorting. In one version of this laboratory phenomenon, cells of two different tissue types which have been uniformly intermixed move in such a way as to "sort out" into clusters of one cell type (foreground) within a background of cells of the other type. In the model, the environment is represented as an unbounded two dimensional grid each position of which is occupied by a hexagonal cell of one of the two types. A state in this model thus corresponds to an assignment of cell types to grid positions. The initial state can be set up to represent the starting situation in the laboratory by dispersing the desired number of foreground type cells randomly through a region of background type cells. New states occur as a consequence of motion of foreground cells. Each such cell undergoes a random walk, moving with equal probability to one of seven positions (six neighboring grid positions plus its present position) during each time step. This simple transition rule is complicated by special provisions to deal with situations where foreground cells become adjacent and form clusters. Further details are not important here. Notice that this model treats the cell sorting phenomenon as occurring in a closed system, since there is no external influence on the model state. The realism of this assumption is of course open to question.

A *discrete-mixed* model of a neuron (nerve cell) is considered in Section 12.1.2. A time step in this model corresponds to about one millisecond of real time. During each time step the neuron can either fire (emit an output) or not. Although this aspect of its state is obviously discrete, the firing decision is based on the comparison of two continuous variables: a threshold which decays exponentially from an effectively infinite value after each firing, and a potential which is computed as a running sum of exponentially decayed values derived from the effects of (discrete) environmental inputs through synapses onto the model neuron.

Several of these model neurons can be interconnected to form a model of a small neural network. Some of the inputs to neurons in this network may come from the outputs of other model neurons, whereas other inputs may be external to the model. The time base is unchanged in the aggregate model, but provision must be made for updating the states of the individual neurons simultaneously during each time step. The state set of the network model is the composite of the state sets of its individual neurons. Network behavior is a direct consequence of single cell behavior rules and interconnection patterns.

Discrete-continuous models are frequently ones in which the time base of a continuous time system has been approximated with difference equations rather than directly modelled with differential equations. Such an approximation can be quite accurate and computationally more convenient.

Although discrete-continuous models of biological information processing systems are not especially common, this method was employed at one point in the development of the bacterial cell model of Section 7.1.2. A state in this model is a collection of chemical concentrations, one for each of the 31 chemical pools in the modelled bacterium, plus a few additional variables like the volume of the cell. The way in which each concentration changes as a function of other variables was originally expressed as a differential equation, producing a continuous-continuous type model. But in preparation for computer simulation a difference equation version of the model was specified with a time step corresponding to about 60 seconds of real time. (Actually, it is a moot point whether the discretizing step in this project was part of the modelling phase or the simulation phase; for purposes of this exposition we have adopted the view that the difference equation formulation was a distinct version of the model.)

Turning to continuous time dynamic models, we first mention the *continuous-discrete* case. When a model is characterized by discrete state transitions occurring at arbitrary time intervals, it is often natural to view the situation in terms of the scheduling of events. This *discrete event* approach has become an important simulation paradigm because of its wide applicability to the modelling of queueing systems, such as computer operating systems. Event scheduling models have been less common in biology, and are not often encountered in this book, although there are examples in Sections 7.1.1, 9.2.2, and 12.1.1. Event scheduling logic can also be applied to *continuous-mixed* models, the only difference being that some of the variables specifying the state are continuous.

The last case is that of *continuous-continuous* models, in which both time and state change smoothly rather than in discrete jumps. Such models are usually represented as systems of differential equations. A model implemented in any of the simulation systems for studying biochemical reaction networks, discussed in Section 5.1, would be of the continuous-continuous variety. A system of differential equations specifies the interaction of continuous variables, representing reactant concentrations, in continuous time.

2.1.3 Aids to Modelling

We turn now to a brief consideration of some techniques which can be useful in developing and using models. Zeigler's proposed notation for informal description of models is described first, followed by an inventory of more formal tools. Finally we look at simulation, in the broadest sense of the term, and discover a number of commonly employed modelling aids which simulate without the use of a digital computer.

Informal model description. This discussion is based directly on Section 1.3 of Zeigler (1976). He argues that the informal description of a model in some reasonably standard format is an essential part of model development, particularly in the early phases when such a description serves to maintain perspective. Informal descriptions can also be used to introduce fully developed and formalized models to users and colleagues who may lack the background, time, or motivation to cope with formal and detailed specifications.

The first of the three parts of Zeigler's informal description format is a list of model *components*. We have already encountered this term in connection with the composition of systems in general. The components are the elementary building blocks out of which the model is constructed. In the description each component should be clearly related to some aspect of the system being modelled. The component list can be very simple, as is the case for the cell sorting model described in the previous section, where the only component is the CELL. The neural network generalization of the neuron model would have the components NEURON, SYNAPSE, and (external) INPUT. The correspondence between the components and real world entities is self-evident in both these cases. On the other hand, the list of the 31 chemical pool components of the bacterial cell model would be accompanied by a description of which groups of molecules in the real cell were intended to be modelled by each pool.

Following the list of components is a specification, for each component type, of its *descriptive variables*. The values of these variables will fully specify the component's contribution to the state of the model at any point in the time base. Both the names and ranges of the variables should be given. (The following lists have been simplified somewhat, for compactness of presentation.) In the cell sorting model each CELL has a location, described by an XCOORDINATE and YCOORDINATE, both of which range over the integers. In the neural network, a NEURON is described by its THRESHOLD (range, the positive reals), POTENTIAL (positive reals), and OUTPUT (0 or 1); a SYNAPSE has GAIN (integral) and PERSISTENCE (real); and an INPUT has a VALUE (0 or 1). In the bacterial cell model, each chemical pool has a single descriptive variable, its CONCENTRATION (real).

Appended to the list of descriptive variables should be a list of the roles and ranges of any model *parameters*, characteristics which remain constant for any given use of the model but which may change from one use to another. The usually challenging task of establishing proper values for parameters in a given context is called the *parameter identification problem* and is an essential step in validating model-system morphisms. Parameters in the cell sorting model would include probabilities of cell movement in the alternative directions. One parameter in the neuron model is the time constant of threshold recovery.

The third part of Zeigler's format for informal model description is a statement of *component interactions*. These are the rules according to which components affect each other and thereby generate the model's behavior. This part of the description is frequently the least precise since the elaborate equations and functions that may underlie complex models are difficult to summarize verbally. Zeigler proposes the use of *influence diagrams* as part of most descriptions of component interactions. These block diagrams show components affecting each other by means of directed links which can be labelled to indicate the nature of the interaction. We will not attempt to specify here the component interactions for the examples we have been following, since the detail required to do so would involve repetition of large parts of the descriptions of these models that appear later in the book. Zeigler supplies complete informal descriptions of five simple models in his Section 1.3.1.

Formal tools. In the course of this chapter we have mentioned a variety of mathematical and other formal techniques which can be used for model construction and manipulation. These methods are here drawn together in summary fashion. No attempt is made to describe any of the techniques; and it is not necessary for the reader to be deeply familiar with any of them.

Block and flow diagrams, just mentioned in connection with informal model description, can also formally specify aspects of models, particularly those which are at least partly algorithmic in character. An alternative way to specify algorithms rigorously (short of actual programming) is to use pseudocode like pidgin ALGOL.

Equations of all types are used in defining models and parts of models. Many purely mathematical models use no other descriptive medium. We have seen how difference and differential equations fit various categories of models. Integral and integrodifferential equations are also used extensively in modelling dynamic systems. Probability and statistics play special and important roles in modelling, especially when computer simulation is involved. The theory of random variables underlies the use of probability distributions to generate pseudorandom numbers, a prerequisite to simulation of stochastic models. Statistical theory supplies tests which can be used to evaluate data from experiments with models in the same way as data from experiments with real systems. The statistical behavior of queues is the basis for many types of models involving event scheduling.

Automata theory offers a body of knowledge about the behavior of various classes of machines, from finite state sequential machines to cellular automaton arrays. Many of the models treated later in this book are couched at least partly in automata theoretic terms. Lastly, systems theory (with which this chapter began) can be especially useful when models are being evaluated or compared. In a sense, much of this chapter is an exercise in applied elementary systems theory.

Simulation. This term refers to a broad class of modelling tools. Those activities employing digital computers will be covered in the next section. Some authors regard the manipulation of a system directly, without benefit of a model, as a form of simulation, sometimes called identity simulation. Although such activity may not be too productive, parts of the real system are often usefully coupled with simulated model components. An example would be the use of dummies for people in the destructive testing of automobile safety systems. In other cases people may play their normal roles and have their environments supplied by simulation. Spacecraft simulators and war games (where a computer has the role of the enemy) are examples of such man-machine simulation enterprises. In yet another variation on this theme, a physical model is tested in a real or simulated environment, such as a wind tunnel or vacuum chamber.

When a model has been expressed algorithmically with the intention of using computer simulation, it is not uncommon for the investigator to work by hand through the program or parts of the program. Such hand simulation can often reveal oversights in design.

Finally, there is simulation by analog computer. Although much slower and smaller than general purpose digital machines, these computers are useful for simulating certain classes of models, especially where small numbers of continuous variables interact in a parallel fashion. And, although they are not much so used, hybrid digital-analog computers could provide ideal environments for simulating some kinds of mixed state models. Analog and hybrid simulation are, however, beyond the scope of this book.

2.2 SIMULATION

2.2.1 Paradigms and Strategies

Although dynamic models were grouped into six categories in the previous section, one of three general simulation paradigms is usually appropriate for exploring a model with the assistance of a digital computer. General and special purpose programming languages have been developed in recent years to facilitate simulation within each of these three paradigms. Adopting one of these languages usually commits the investigator to one of the two or three alternative strategies for carrying out simulation within the paradigm. A similar choice of strategies confronts the programmer who implements a model in some language not specifically oriented toward simulation. But there is some trade off between ease of implementation and flexibility of alternatives. Although use of a language like FORTRAN or PL/1 (or even an assembler language) may increase the programming task when an available simulation language fits the model well, the user may thereby be able to get closer to the problem and exert more direct control over the simulation strategy.

Continuous simulation. The first paradigm is appropriate for models with continuous time and state, but also encompasses many continuous-mixed types of models. The model is usually specified by one or more differential equations in which time is the independent variable. Although there are several strategies for numerical approximation of solutions of differential equations, many are not well suited to computer implementation.

Since the digital computer is a discrete device, the standard strategy for using it to solve differential equations is numerical integration by discrete variable approximation. The computer can deal with a continuous variable only as a sequence of closely spaced values, where the minimum interval is determined by the number of bits used to store its value. A differential equation like $dx/dt = f(t)$ specifies the change in x at any point in time. We can use this information to estimate the value of x at $t+i$ based on its value at some t by assuming dx/dt is constant over the interval, so that $x(t+i) = x(t)+f(t)i$. To simulate the "model" through any desired point in time, only a starting value for x is required. This approach is the basis for Euler's classic numerical integration method.

As just described, the Euler method is impractical. An appropriate step size is so small that huge amounts of computation would be required to cover even a small fragment of model time. But when the step size is increased to a practical value, the assumption of a constant derivative becomes unreliable. Fortunately there are ways to improve the approximation by using values of x and $f(t)$ at two or more points in past time. There are also ways of adjusting the step size dynamically so that it becomes smaller when the derivative changes faster. Since many of the better numerical methods are widely available in computing center program libraries, an investigator who elects to program a simulation of a continuous system may find that a large part of the work has been done.

As a logical extension of the numerical integration package, general purpose continuous simulation languages, like DYNAMO, CSMP, and ACSL have been designed. The sources cited in Chapter 5 of Zeigler (1976) provide entry points to the literature on some of these languages, along with some general references on numerical integration techniques. Alternatives to general purpose languages have also been built for those using particular classes of models. For example, in Section 5.1 we will look at systems designed to facilitate the implementation of models of biochemical reaction systems.

Discrete event simulation. The discrete event paradigm is applicable to continuous time models with discrete or mixed state sets. The paradigm is employed when the state variables of interest change at times separated by variable but predictable intervals. There are three basic strategies in discrete event simulation: event scheduling, activity scanning, and process interaction. Since there are very few discrete event simulation studies in this book, only the simplest strategy (event scheduling) is outlined here. Thorough descriptions of all three may be found in Fishman (1973) and Zeigler (1976). These authors also treat a variety of discrete-event simulation languages, including GASP (actually a package of FORTRAN subroutines), SIMSCRIPT, GPSS, and SIMULA.

Event scheduling is the most straightforward of the discrete event strategies. An *event* is a change in the state of a model component. In the usual implementation of this strategy, the primary data structure is a list of events and their (future) times of occurrence. The list is initialized with one or more events to get things started. In the basic simulation cycle, the model clock is first advanced to the time of the next event since nothing significant can happen until then. If events can occur simultaneously, a tie-breaking procedure must be provided. An event is processed by updating the state of any affected component and then entering any consequent events in the event list. Event scheduling strategies applied to models of cell division and neural networks respectively are considered in Sections 7.1.1 and 12.1.1.

Discrete time simulation. This paradigm is the one we will encounter most frequently in the simulation studies discussed in later chapters. Biological machinery has often been modelled in automata theoretic terms. Since, in any given simulation, continuous or unbounded discrete variables can assume only a finite number of values, it is possible to view any discrete time simulation as the behavior of a finite state machine. But sometimes when applying this paradigm to discrete-mixed or discrete-continuous types of models, we wish to maintain the viewpoint that we are actually dealing with continuous variables.

Simulation of a model specified as a finite automaton is straightforward. The model will consist of three finite sets (some of which may be empty) and two functions. We denote the set of possible combinations of external inputs I, the set of states S, and the set of outputs O. The next state function is denoted

F and the next output function G. There is a designated element of S, called the initial state, in which we find the machine at the start of the simulation ($t = 0$). To simulate a time step, we obtain the next element of I and the current state, apply F and G to determine a new state and output, produce the output, place the machine in the new state, and advance time by one unit. The inputs can come from a prespecified array, with sufficient values for the desired time period of the simulation, or be produced by some programmed function (possibly stochastic) at each time step.

The discrete time simulation strategy just outlined is of the *sequential* variety, in which the model is viewed as consisting of a single component (the finite state machine) which undergoes state changes one after another. Of course the description of a state may be so elaborate that it might be more realistic to consider the system as consisting of several components changing state and interacting in *parallel.* An example is the cell sorting model mentioned earlier in this chapter. It is possible to think of the grid as a single machine with a state consisting of an assignment of a foreground or background cell to each position. But the state transition function for this machine would be much too complex to work with computationally. The preferred view of the system is as it was presented; each foreground cell is a component whose state is its position. All components change state together at each time step.

Simulation of a model of parallel interacting components can involve some important strategic alternatives since the conceptually simultaneous component state changes must be executed sequentially on current digital computers. Recall the related situation in the discrete event paradigm, where we required a rule for breaking ties and ordering the execution of simultaneous events. In some discrete time simulations the order in which component states are updated makes no difference. But the interaction rules may impose constraints on certain kinds of simultaneous occurrences.

In the cell sorting model, for example, two foreground cells cannot move into the same position at the same time. We can use this constraint to illustrate the synchronous and asynchronous strategies for discrete time simulation of parallel processes. In the *synchronous* strategy all components are permitted to change state as if no conflicts could occur. The new states must be stored in a supplementary data structure because we cannot throw away the old state information until conflicts are resolved. Thus we would move all foreground cells independently, then scan the new array to see if two cells occupied the same space. If so, some conflict resolution rule would be used, perhaps requiring one of the cells to move somewhere else or not move at all.

In the *asynchronous* strategy some rule is applied to order the components for state change. Conflicts are resolved (again by some supplied rule) as they arise. In the cell sorting model we would not permit the cell being updated to move into a position occupied by another foreground cell. The asynchronous strategy is usually faster but carries the risk of introducing systematic distortions into a conceptually parallel interaction. In the example, if the same cell always moved "first," it could have slightly more freedom and hence slightly different properties than other cells. In the actual simulation this distorting tendency was counteracted by moving the cells in a new random order each time step.

Although there are no widely available general purpose simulation languages devoted to the discrete time paradigm, discrete event languages can often be used to simulate discrete time models. Also, some local special purpose discrete time simulation systems have been implemented because of interest in particular classes of problems, like cell space models or neural networks.

2.2.2 Costs and Hazards

In addition to the potential dangers of using models, mentioned in Section 2.1.1, simulation contributes some of its own. The most obvious cost associated with simulation is that of using a computer. For large scale projects this factor can be prohibitive or can at least severely curtail the amount of simulation that can be done. Even if money is not a consideration (perhaps because the investigator owns the computer), time and space considerations can impose significant limitations. These effects can sometimes be offset by judicious choice of programming language and data structures. But some problems are just too big. At present, for example, it is theoretically possible to use a computer implemented model to predict worldwide weather weeks in advance, but not fast enough to keep pace with real weather.

Another hazard attendant upon using a computer in the modelling effort is the incorporation of excessive detail. Investigators designing models destined for simulation are often too thorough. Inclusion of every possible complicating factor not only adds to the cost of simulation but can also obscure the nature and intentions of the original model. In addition large complex programs are notoriously difficult to debug and maintain. Simulation should not be a reason for abandoning either judicious model simplification or normal restraint in the choice of a resolution level at which to study a model.

A third risk of simulation is that it may be undertaken when it is neither necessary nor appropriate. Computer users tend to forget that there are other ways to study models, ways which in some circumstances may be faster and give better results than simulation. If a model is specified as a system of differential equations, for example, it pays to spend some time considering the possibility of an analytic solution before plunging into programming.

Finally, there is the risk of confusing (or even identifying) model and program, or at least of viewing models intended for study by means of simulation as somehow fundamentally different. The latter type of thinking can lead to the most damaging misconception, that simulation is used only when a model is not sufficiently well defined or understood to be studied by more classical modelling aids. The reality of these hazards is attested to by the increasing use of the popular but confusing misnomers "simulation model" or (worse) "computer model." Since a given model may be implemented by different algorithms, and/or in different programming languages, and/or on different computers, it is vital that modellers maintain clear and constant awareness of the critical distinction between the *validation* relation, between system and model, and the *verification* relation, between model and program. Models are not improved by tinkering with code.

2.2.3 Interactive Simulation

Simulation has traditionally been done most often in a "batch" computer environment. The user submits program and data on cards or tape, waits for a period of minutes to days, collects printed output, and goes off to analyze it in preparation for another run. The advent of time sharing operating systems and fast flexible terminals for communicating with such systems has made interactive simulation possible. In this mode the terminal user initiates and monitors a simulation run. He can interrupt the simulation and restart it after adjusting model parameters, or save and later recall states of the model.

Because a good deal of extra software support is required to do simulation interactively, such implementations are usually not limited to one particular model. Rather an interactive simulation *system* is designed, in order to handle a broad class of models. Some of the simulation studies reported later in this book were carried out on such interactive systems. The importance of interaction, especially in the early phases of model development and preliminary testing, is further underscored by the comparative analysis of two groups of such systems, in Sections 5.1 and 12.1. All of those systems eliminate the use of a programming language, substituting a more or less natural model description language (see below) appropriate to the class of models which can be studied.

Another approach to interactive simulation makes available a "front end" which is to be combined with a user's program. One example is SIMCON (Hilborn 1973) which can be used for interactive control of any discrete time simulation program coded in slightly constrained FORTRAN. SIMCON has features equivalent to those outlined below, at least with respect to graphical output, simulation monitoring, and modification, storage, and retrieval of models. Vehicles like SIMCON may considerably expand the classes of models that can be handled, while correspondingly limiting the community of users to those who can implement (or can have implemented for them) models in the appropriate programming language.

The rest of this section highlights important features of good interactive simulation systems. Some of this material was presented by Sohnle, Tartar, and Sampson (1973). Sampson and Dubreuil (1979) reviewed methodological aspects of interactive simulation system design in biological modelling, drawing examples from their work with systems for models of biochemical reaction networks, neural networks, and ecosystems. As set forth by Sampson and

Dubreuil, interactive simulation has three major components: model definition, simulation monitoring, and data collection. These are now discussed in turn.

Model definition. For effective model implementation and manipulation, an interactive simulation system should offer the user a convenient model description language, based on a command syntax suited to the typical user's level of experience with computers and programming. Appropriate and lucid diagnostic messages should be produced during the dynamic process of implementing a model. An implemented model should be easy to modify, either before or during simulation. An editing facility may either reinvoke portions of the model description format or supply a special syntax for model modification.

Simulation monitoring. The monitor in an interactive simulation system manages the running simulation program and acts as a communications device through which the user has access to the model. A concise and flexible command language should allow the user to communicate desired actions to the system. Among the user's options should be: specification of duration of simulation runs and partial runs; the ability to restart the simulation, possibly after having modified the model; and facilities to save the state of the simulated model at any point and later recover such states for further simulation. In many kinds of modelling endeavors, the user will also want a facility for specifying conditions under which the simulation is to be automatically interrupted before its nominal termination. A simulation might cease, for example, if a particular variable acquires a value outside some preset range.

Dynamic state display is an important aspect of simulation monitoring. The user needs to know what is going on, in order to abort a run that has gone totally off the track, or just suspend the simulation to make a small adjustment of some parameter. Among the user's options should be determination of both the frequency with which the state is displayed and the focus of such displays on selected components of the state.

Data collection. An interactive simulation system should have comprehensive facilities for collecting measures of the performance of the simulation. In the preliminary testing phase the user may want to have the complete state recorded at every time step (or equivalent). Later, selected summary statistics may be more appropriate. Of great importance in assessing the results of a simulation run is some sort of graphics capability, which allows the user to see global relationships in the output data at a glance.

2.3 ZEIGLER'S FRAMEWORK

This section summarizes key aspects of Bernard Zeigler's framework for modelling and simulation, as described in his *Theory of Modelling and Simulation* (1976). In the course of the book, Zeigler presents two or three treatments of some of the main subjects, at increasing levels of formalism. His notation for informal model description has already been presented here in

Section 2.1.3. And there is no attempt in what follows to deal with his most rigorous treatment, which is at the level of theorem and proof. Rather, the present discussion is pitched at an intermediate level of formalism, with what we hope is sufficient rigor to reassure the mathematically inclined reader. All readers are encouraged to supplement this section by reading Zeigler in the original, where much fuller discussions and many extensive examples may be found.

2.3.1 The Five Elements

Zeigler characterizes the activity of modelling and simulation in terms of five basic elements: real system, experimental frame, base model, lumped model, and computer.

The real system. This is the system being studied. It may be of any of the (dynamic) types set out at the beginning of this chapter, and will have the characteristics outlined there. The real system is a source of *behavioral data.* Such data may be collected as the values of some *descriptive variables*, which are of two types. Observable variables are those whose values can be measured by currently available techniques. Nonobservable variables are inaccessible to the investigator. Observable variables may be input or output variables. An input variable is one which gets its values from the system's environment, whereas an output variable gets its values as a result of the system's response to inputs. Some input variables may be under the control of the investigator, in which case it is possible to conduct experiments with the real system.

An input or output *segment* is a sequence of values of all variables of the given type over some period of time. The combination of an input segment and an output segment for a given temporal interval constitutes an input-output (I-O) segment pair. The set of all possible I-O segment pairs is all that can be known about the real system from direct observation.

The experimental frame. Most real systems are too complex to be studied in complete detail. Normally, we wish to focus attention on a subset of observable variables. An experimental frame imposes this constraint by limiting the circumstances under which the real system is observed. Typically there will be many possible experimental frames appropriate to the varying objectives of different modellers of the same system. Models of information processing in the brain, for example, usually employ an experimental frame which ignores the presence of blood vessels, connective tissue, and other components in the real system. And among such models, experimental frames may or may not acknowledge such variables as neuron fatigue or synaptic potentiation.

The base model. A model is a set of instructions for generating behavioral data. The base model can account for all the I-O behavior of the real system. All variables found in any possible experimental frame are acknowledged. In other words, the base model is *valid* in all such frames.

Model validity can be assessed according to three increasingly strong criteria. If the term is not qualified, only *replicative* validity is implied. A replicatively valid model is one whose behavior fits data previously acquired from the real system. A model with *predictive* validity anticipates real system behavior. The predicted behavior may have already been known, but was not employed in model design or parameter estimation. *Structural* validity is the strongest condition. A structurally valid model not only replicates and predicts real system behavior, it generates data in the same way that the real system does. There is a morphism between the modes of operation of the two systems in addition to the one between their I-O segments. Rigorous structural validity is seldom achieved in modelling enterprises.

The base model of an interesting natural system is usually incredibly complex. The base model of the brain, for example, must account not only for the presence of non-neural elements but also for the complete state of around a trillion neurons. For most real systems, the base model can never be fully described, much less simulated. Some rather restrictive experimental frame is required to make investigation practical.

The lumped model. Within a chosen experimental frame the investigator will usually be able to construct a model which is sufficiently simple for computer implementation. Zeigler calls this the lumped model because it is frequently built by lumping components of the base model together and making appropriate changes in the interaction rules. Unlike the base model, the lumped model is completely known to the modeller. All of the models mentioned previously in this chapter have been of the lumped type. Perhaps the best example of lumping is found in the model of the bacterial cell, where the 31 chemical pools are the lumped counterparts of the more than 3000 species of molecules present in the real system.

A lumped model is derived from a base model by the application of one or more *simplification procedures.* This is an operation fraught with hazards, since simplification procedures produce valid and simpler models only when properly applied. Much of Zeigler's formal analysis is concerned with ways to insure valid model simplification. In essence, simplification procedures represent hypotheses about the base model, assumptions that aspects of it may safely be combined or ignored in the context of the particular investigation.

Zeigler discusses four types of simplification procedures. (1) Dropping of one or more components, descriptive variables, and/or interaction rules leads to an approximation of the base model in which supposedly minor factors are neglected. (2) Replacing deterministic variables by random variables is often required after step (1) since deterministic causes of some model phenomena may have been eliminated. (3) Coarsening the ranges of descriptive variables, so that distinct values in the base range become indistinguishable, is a routine simplification procedure. We have already encountered it in the context of approximating continuous variables by discrete ones. (4) Grouping components and aggregating variables is the most powerful technique, and was mentioned above for the case of the bacterial cell model.

The computer. The last of Zeigler's five elements is a computational process (not necessarily a physical machine) capable of employing the model to generate behavioral data suitable for comparison with real system behavior. At this point it is appropriate to recall the above discussions of simulation, its various forms, its advantages, its paradigms and strategies, and especially its limitations.

2.3.2 The Postulates

Zeigler summarizes his formal framework in terms of 12 postulates, which are presented here in a somewhat simplified form. We have dispensed with most of the mathematical symbols and substituted less formal equivalents for terms which can be understood only after reading Zeigler's complete presentation.

Postulate 1. There exists a real system, which is identified as a universe of potentially acquirable data.

Postulate 2. There exists a base model, denoted B, that structurally characterizes the real system and consists of a time base, a set of input segments, a state set, and a state transition function.

Postulate 3. There exists a set of experimental frames restricting access to the real system.

Postulate 4. An experimental frame, denoted E, is composed of a set of input segments applicable in the frame, a set of output values observable in the frame, a function which maps base model states to frame outputs, and a range of output validity determined by experimental control conditions.

Postulate 5. The real system observed within the experimental frame is structurally characterized by a base model in the frame, denoted B/E.

Postulate 6. B/E is characterized by a time base (identical to the one for B), and inputs, input segments, states, outputs, and next state and output functions, all derived from B and suitably restricted or constrained by E.

Postulate 7. The data potentially acquirable by observations of the real system within E are identified with the I-O relation (set of pairs of input and output segments) of B/E.

Postulate 8. The real system is the set of data potentially acquirable by observations within any of the experimental frames in the set of all such frames.

Postulate 9. A lumped model is an iterative system specification (essentially a sequential machine characterization), denoted M.

Postulate 10. A lumped model is valid for the real system in experimental frame E if its I-O relation is equal to the one specified in Postulate 7. Note that this postulate limits empirical tests of validity to comparisons of data collected from B/E and generated by the system specified by M.

Postulate 11. A computer (program) is an iterative system specification, denoted *C*.

Postulate 12. A computer is a valid simulator of a lumped model *M* if there is a specification morphism from *C* to *M*. A specification morphism is a behavior preserving mapping from one iterative system specification to another.

2.3.3 Problems in Modelling and Simulation

The framework just outlined allows modellers to address precisely a number of key problems in modelling and simulation, and in some cases to propose at least partial solutions. One such problem is valid model construction and simplification. How can we choose among candidate base models (as described by Postulate 2) one which is valid according to the criteria of Postulate 10? Zeigler shows how the use of structure morphisms and analysis of systems by various forms of decomposition (parallel, series, slow-fast, feedback) can assist in answering such questions. A related problem is simulation program verification. Given a program as described in Postulate 11, how can we know that it is valid according to the criteria of Postulate 12? Iterative specification morphisms are again a useful tool.

When the equality asserted in Postulate 10 is not exact, the issue of approximative modelling arises. The framework provides perspective on goodness of fit and related considerations. To go beyond Postulate 10 and make inferences about the structure of *B* from that of *M* requires a consideration of structure preserving morphisms and the kinds of justifying conditions under which they may hold.

Finally, Zeigler mentions the problem of model integration. Given a number of experimental frames producing different *B*/*E*s and lumped models of the same real system, how can these various views be coordinated to produce new insights about the real system? Although he does not attempt a comprehensive answer to this question in his book, Zeigler has since illustrated the power of what he calls multi-level multi-formalism modelling with a group of models of an ecosystem. Research in modelling methodology is actually just getting started. The work involves development of formal frameworks, integration of models, and other aspects of computer assisted modelling (see Zeigler 1979).

Chapter Three

Introduction to Cellular Biology

This chapter deals mainly with the milieu in which biological information processing transpires; concepts and terminology fundamental to later parts of the book are introduced. There is brief treatment of ancillary processes which, although not inherently informational, interface closely with mechanisms of primary interest. The last section of this chapter initiates biological information processing proper with a discussion of informational macromolecules.

In this and subsequent chapters that consider current theories of how biological systems process information, to achieve a compact and yet comprehensive presentation it has been necessary to minimize consideration of the laboratory techniques and massive experimental programs which have been essential to the development of the theories. Textbooks of biochemistry, genetics, developmental biology, and neurophysiology thoroughly document the experimental origins of current theory. Such sources also provide more detailed consideration of many topics which can be given only cursory treatment here.

3.1 CELLULAR ARCHITECTURE

3.1.1 Components of Procaryotic and Eucaryotic Cells

The fundamental unit of biological organization is the cell. Cells can be classified in a number of functionally important ways. Animal and plant cells, for example, have complementary roles in ecosystems. Another distinction is that some cells must provide all the functions of a complete organism; in multicellular organisms, by contrast, cells can play a great variety of specialized roles, like the neurons which process information in the brain.

Perhaps the most fundamental classification of cells is into the procaryotic and eucaryotic types. Bacteria and blue-green algae comprise the *procaryotes*, which are smaller and simpler than the *eucaryotes*. This distinction is important to bear in mind, since many major advances in molecular genetics have been based, at least initially, on work with procaryotes exclusively. The results of such work typically cannot be simply extended to eucaryotes, where different types of molecular machinery are likely to have evolved. Also, since multicellular organisms are built exclusively of eucaryotic cells, little can be learned from procaryotes about information processing in cell differentiation and the development of organisms.

Among the defining structural characteristics of all normal biological cells is a membrane which physically limits the cell and provides an interface with its environment. Since the fluid matrix inside the membrane is called the *cytoplasm* ("cell fluid"), the interface is usually referred to as the *plasma membrane*, to distinguish it from membranes within the cell. Many procaryotes and plant cells have a *cell wall* outside the plasma membrane. This rigid or semirigid structure provides protection and support, usually without significantly affecting the transport of material between cell and environment.

A second defining characteristic of cells is the presence of genetic material organized into one or more *chromosomes*. The single chromosome found in most procaryotes consists mainly of DNA. A eucaryotic cell typically has at least a few chromosomes, each a complex of DNA and several other materials. During most of the eucaryote life cycle, chromosomal material is confined to the cell *nucleus* by a double membrane structure called the *nuclear envelope*. Some highly specialized cells, like the human erythrocyte (red blood cell) which does little more than carry oxygen, actually exist without a nucleus. But such cells are direct descendants of nucleated cells and are themselves unable to divide and propagate the cell line. Although the procaryote has no nuclear envelope and hence no true nucleus, it is usually possible to identify a *nucleoid* (nuclear region) in which the chromosome tends to remain localized.

In addition to the nucleus or nucleoid, the cytoplasm of most cells contains a variety of other structures. In eucaryotes some of these constituents are large and complex membrane enclosed structures known as *organelles*. Visually prominent are *mitochondria* (sites of energy production) and, in plant cells, *chloroplasts* (sites of photosynthesis). The relative complexity of procaryotes and eucaryotes is demonstrated by the similarity in size and internal organization of mitochondria and bacteria. There is recurrent and well founded speculation that large eucaryotic organelles are actually evolutionary descendants of procaryotic symbiotes. Many eucaryotes also contain an intricate web of membrane-bound channels called the *endoplasmic reticulum*, which serves a variety of transport and membrane recycling functions.

Lastly, all cells contain *ribosomes*, small particles which are sites of protein synthesis, one of the cell's key informational activities. Procaryotic ribosomes are slightly smaller than eucaryotic ones. Ribosomes are sometimes found aligned along parts of the endoplasmic reticulum, in which case it is termed "rough" (as opposed to "smooth").

Most of the cellular components mentioned above are sketched in Figure 3.1, which shows a possible cross section of an animal cell. This diagram could be misleading in two ways. First, chromosomes are not always seen condensed into distinct strands, as discussed in Section 3.3.2. Second, not all structures in the figure are drawn to the same scale; ribosomes, for example, would be too small to be seen at the suggested level of magnification.

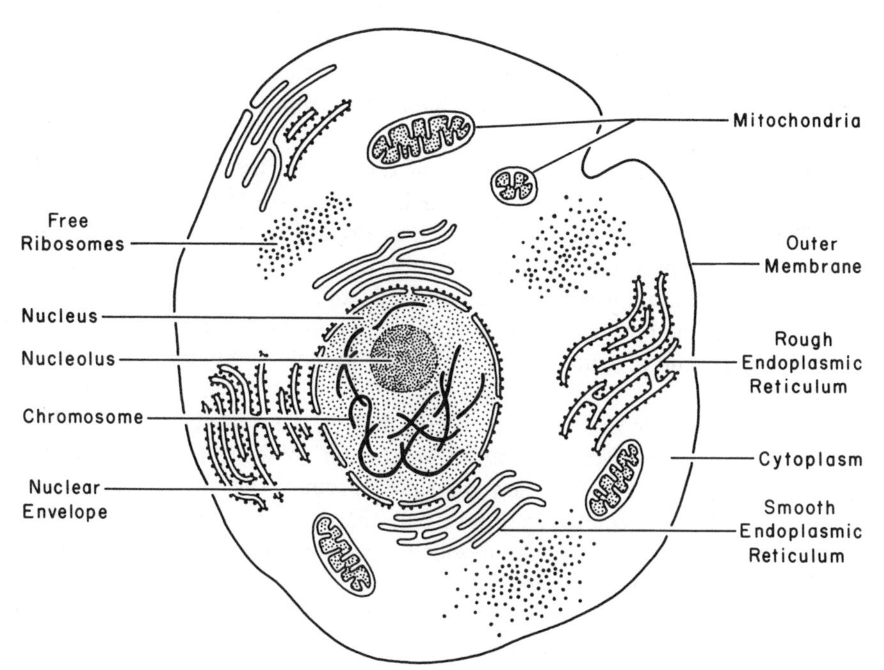

Figure 3.1 Generalized diagram of an animal cell.

3.1.2 Membranes and Transport

Like any open system, a cell survives by interacting effectively with the surrounding environment. The cell-environment interface, the plasma membrane, is critically involved in acquisition of nutrients and elimination of wastes. The membrane must also participate in cell movement, attempt to screen out toxic substances, and regulate the delicate balance of concentrations of inorganic molecules and ions dissolved in the cytoplasm. Not surprisingly, then, the plasma membrane has a complex structure.

Theorizing about the structure of plasma membranes began in the late nineteenth century, when it was first suspected that polar lipids (fat-like molecules with regions of electrical charge) were a major component. Two layers of these lipid molecules can form a stable barrier if they are aligned with their polar "heads" on the outside and their nonpolar "tails" (which tend to avoid water) clustered within. In fact, recent studies with synthetic lipid bilayers have shown them to have many properties similar to those exhibited by natural membranes.

In addition to lipids, however, most membranes contain variable amounts of protein. These large, structurally complex molecules are now thought to be involved in both the membrane's architecture and its dynamic functions, such as transport of materials and adhesion to other cells. Since the nature and distribution of membrane proteins vary significantly, these molecules give membranes their characteristic properties, which differ from cell to cell and even between different regions on the same membrane.

A recent and widely accepted model of the plasma membrane has the descriptive title "fluid mosaic." In this model, proteins are of two types, integral and peripheral. Integral membrane proteins are rather tightly bound to the lipid matrix and in some cases extend all the way through it, facilitating simultaneous interactions with molecules inside and outside the cell. Peripheral proteins are attached more loosely, possibly to the integral proteins rather than the lipid matrix, and in some cases are thought to move freely within the membrane.

Materials move into and out of a cell through its plasma membrane in three major ways. The first of these is *passive diffusion.* Whenever a substance has size and shape suitable for easy passage through the interstices of the membrane's local molecular configuration, the substance will tend to flow across the membrane in a direction which equalizes its concentrations in the cytoplasm and extracellular fluid.

This flow with a *concentration gradient* may be complicated by the existence of a *charge gradient*, if the substance is a charged ion like chloride (Cl^-) or potassium (K^+). As a result of the simultaneous attempts of these passively diffusing ions to satisfy both concentration equality and electrical neutrality, combined with the mediated transport of sodium (Na^+) discussed below, the plasma membrane usually exists in an equilibrium where there is a relative internal abundance of K^+, a relative external abundance of Na^+ and Cl^-, and a voltage drop across the membrane of about 50 to 70 millivolts with the interior of the cell negative. Disturbances of this steady state are the basis for transmission of information in the nervous system, as discussed in Chapter 11.

In an important special type of passive diffusion, called *osmosis*, water may flow across the membrane in a direction tending to equalize internal and external concentrations of substances which cannot diffuse easily. Osmosis can be dramatically demonstrated when a cell like an erythrocyte is removed from its normal saline environment in the blood stream and placed in distilled water. The influx of water to dilute the concentration of the cell's interior substances causes the cell to explode.

The second type of transport across membranes, known as *facilitated diffusion*, resembles passive diffusion in that the transported substance flows with a concentration gradient, but differs in that the substance cannot easily cross the membrane without assistance. Facilitated diffusion is nicely illustrated by the transport of the sugar molecule glucose in the human erythrocyte. This transport system exhibits a number of characteristics that are strikingly similar to those associated with biochemical reactions catalyzed by enzymes, suggesting that the facilitation involves an *allosteric*, or shape dependent, attraction between glucose and some membrane molecule, possibly an integral protein. The carrier-glucose complex is then assumed to change configuration in some way that brings glucose across the membrane.

One strong argument for a limited number of glucose carrier sites on the membrane is that the system displays *saturation kinetics*. The rate of transport approaches an asymptote (saturates) and cannot be further increased regardless of the magnitude of the glucose imbalance. This situation may be contrasted with the kinetics of passive diffusion, where the rate is linear over the entire physiological range of concentration ratios, indicating that membrane structures play essentially no role.

The last type of material flow through membranes is *active transport*. Like facilitated diffusion, this mechanism is thought to be based on carrier molecules. Active transport shows saturation kinetics and other enzyme-like properties. The main new feature of this mechanism is its ability to carry material against a concentration gradient, to move substances to places where they are already in excess. Unlike the two gradient driven mechanisms discussed above, this uphill transport requires energy from the cell's reserves.

An example of active transport is the extensively studied mechanism for sodium-potassium exchange. Cells need to concentrate potassium in the cytoplasm for a number of critical biological functions. Extracellular sodium can be used to offset the charge imbalance. Although sodium does not enter the cell easily, a small amount does leak in continuously. An ingenious active transport process simultaneously extrudes sodium and brings in potassium, moving each ion in only one direction, against its concentration gradient. The mechanism needs both intracellular sodium and extracellular potassium to function, which suggests that the carrier is somehow primed by association with both types of ion.

3.1.3 Cytoplasmic Constituents

Within the plasma membrane is the cytoplasm, a fluid with an ionic composition similar to that of sea water, containing thousands to millions of different species of large biological molecules and a variety of specialized structures. The primary activities in the cytoplasm are the derivation of energy from food sources (catabolism) and the use of this energy to synthesize proteins and other complex molecules necessary to the cell's existence (anabolism). Energy derivation by respiration and photosynthesis is accomplished in eucaryotes by two types of large cytoplasmic organelles, mitochondria and chloroplasts respectively.

The mitochondrion has an irregular tubular shape about 1 micrometer (10^{-6} meters) in diameter and up to several micrometers in length. The entire structure is bounded by a membrane which is highly permeable to all small cytoplasmic molecules, facilitating the exchange of material by which mitochondria contribute to the cell's energy reserves. A more selectively permeable and intricately folded inner membrane is thought to house the sites where energy is produced by oxidative phosphorylation (see below). In nonphotosynthetic eucaryotes most of the cell's ATP (adenosine triphosphate), the major energy storing molecule, is produced on the inner membranes of mitochondria. Within the inner membrane, in a region called the matrix, another process generates the precursors of oxidative phosphorylation. Consistent with their probable origin as procaryotic symbiotes, mitochondria contain small amounts of self-replicating genetic material (DNA) and the apparatus (ribosomes) necessary to synthesize proteins from the information in that DNA. Much of the information for construction of mitochondrial components also resides in the nucleus, however.

Although they also contain mitochondria, plant cells produce additional ATP in some stages of photosynthesis, a multifunctional process carried out in organelles called chloroplasts. Although more compact in shape, the chloroplast is comparable in size to the mitochondrion and also contains DNA and ribosomes for specifying and building some of its components. Within the enclosing external membrane, chloroplasts contain internal membranes in a complex arrangement of interlocking stacks. Various aspects of photosynthesis have been tentatively identified with different types of internal regions in the chloroplasts. Similar but simpler membrane arrangements, lacking the enclosing chloroplast structure, are the sites of photosynthesis in those procaryotes capable of using light as an energy source.

Except for mitochondria and chloroplasts, all the membranous eucaryotic organelles are related, both among themselves and with the nuclear and plasma membranes. These membrane systems are ultimately derived from the endoplasmic reticulum, through a kind of membrane recycling system that serves several purposes. At one end of this system is *endocytosis*, a process in which the plasma membrane surrounds an extracellular particle, enfolds it, and separates to form a membrane-bound *vacuole* inside the cell. This loss of plasma membrane to the cell's interior is compensated by the reciprocal process of *exocytosis* in which a particle of waste or other export product is bound in a membrane sac which fuses with the plasma membrane and dumps the particle outside the cell.

Some export products originate in the *Golgi apparatus*, a stack of membranous sacs, within which the products are packaged. A cell producing a hormone, for example, has extensive Golgi apparatuses, from which packages of hormone are continually budding off and undergoing exocytosis. The Golgi apparatus is a specialization of the endoplasmic reticulum, a network of channels which is continuous in many places with both the nuclear envelope and plasma membrane.

The Golgi apparatus is also thought to be the source of *lysosomes*, membrane enclosed packages of digestive enzymes which would attack cellular structures if allowed to float free in the cytoplasm. A lysosome is capable of fusing with a vacuole containing a food particle, breaking down and expelling the useful raw materials, and eventually undergoing exocytosis. An elegant picture of membrane management and recycling is thus emerging, as more is learned about the interior architectural dynamics of eucaryotic cells.

Ribosomes, sites of protein synthesis, are numerous in all types of cells, tens of thousands occurring even in a simple bacterium. In eucaryotes, ribosomes are distributed in the cytoplasm in two different ways. Many are found floating free in the cytoplasm. The protein products of these ribosomes tend to be those that remain in the cell, to serve as enzymes or in structural roles. Other ribosomes are closely packed along edges of some regions of the endoplasmic reticulum. These ribosomes bound to the membrane system are usually involved in the synthesis of protein components of export material, which can be dumped directly into the endoplasmic reticulum. Ribosomes themselves are constructed from many different protein molecules, plus several RNA molecules. When not actively engaged in protein synthesis, a ribosome separates into a larger and a smaller subunit. The size and composition of these subunits differs for eucaryotic, procaryotic, and organelle ribosomes.

3.1.4 The Nucleus

As already mentioned, the envelope that encloses the nucleus is a manifestation of the endoplasmic reticulum membrane system. The nuclear envelope is actually a double wall in which a membrane has been folded back on itself at various points. Because quite large biological molecules must have easy exit from the nucleus in order to carry genetic information to the ribosomes in the cytoplasm, the nuclear envelope has numerous gaps or pores. The nuclear membrane often disappears during the early stages of cell division.

Structurally, the content of the nucleus is distinguished by the absence of cytoplasmic organelles and the presence of *chromatin*, a fibrous substance named for its tendency to bind certain colored dyes. Chromatin fibers are normally widely dispersed in the nucleus. Only during cell division do they coalesce into the familiar shapes of chromosomes. Chromatin is composed of about one third DNA, the primary repository of genetic information, and two thirds protein.

The protein component of chromatin can be divided into two major classes. The *histones* are a group of five different protein molecules which vary little in composition from one organism to another. Histones are thought to be involved in the structure of chromatin fibers, complexing with the DNA to help solve the problem of packing nearly two meters (in man) of linear DNA into a nucleus with a diameter of about 6 micrometers. *Nonhistone chromosomal proteins* are quite variable in composition, even in cells within the same organism, and may contribute from 5 to 25 per cent of the composition of chromatin. These proteins may be involved in activating or repressing various portions of the DNA in cells which restrict their activities to specialized functions.

A prominent feature of most nuclei is one or more *nucleoli*. A nuclear organelle, the nucleolus has been implicated in the manufacture of the RNA components of ribosomes.

3.2 BIOENERGETICS

3.2.1 Energy and Chemical Bonding

Full appreciation of cellular information processing activities requires some knowledge not only of their structural context but also of the sources of energy which make them possible. The prerequisites for assembly and interpretation of informational macromolecules include energy stored in the form of high energy chemical bonds and in the form of reduced coenzymes. The most commonly employed high energy bond is that which attaches the third phosphate group (a phosphorous atom bound to three oxygen atoms) in ATP.

Coenzymes are small molecules, like NAD (nicotinamide adenine dinucleotide), which can be *reduced* (by addition of hydrogen) and later *oxidized* to release both the hydrogen and energy. (Coenzymes resemble enzymes in that they are not consumed in the reactions they facilitate, but differ in that they change form during those reactions. More will be said about the role of enzymes themselves in biochemical reactions during the subsequent discussion of proteins.) The elaborate reaction pathways of both respiration and photosynthesis have among their primary functions the generation of compounds like ATP and the reduction of coenzymes like NAD.

The concept of *free energy* is basic to bioenergetics. Free energy is the fraction of energy in a total system that is available to do work. Reactions in biological systems may be classified according to whether they produce or consume free energy. Catabolic (degradative) reactions are downhill, or exergonic, and liberate free energy, whereas anabolic (synthetic) reactions are uphill, or endergonic, and consume free energy.

Chemical reactions produce or consume energy by making and/or breaking chemical bonds. In biological systems these bonds have a wide range of strengths, from strong covalent linkages to weak attractive forces which cause parts of molecules to associate, but not always strongly enough to generate a single new structure. The *hydrogen bond* is a weak force that is critical in the structures assumed by molecules that carry information.

A hydrogen bond occurs when a positively charged hydrogen atom that is covalently bound to a strongly electronegative atom like nitrogen or oxygen is attracted to another electronegative atom nearby. The hydrogen donor and hydrogen acceptor atoms, as they are called, become associated by mutual attraction to hydrogen. If the second electronegative atom is part of the same molecule as the first, the three-dimensional shape of the molecule can be affected by the hydrogen bond; this shape can be critical to an informational molecule's function. On the other hand, enough individual hydrogen bonds can keep two large structures, like the two strands of the DNA double helix, together in stable association.

The central role of the high energy bond in ATP has been mentioned. It is possible to calculate the free energy involved in making or breaking this bond as 7.3 kilocalories/mole at standard temperature, pressure, and acidity. When ATP loses its third phosphate group, becoming ADP (adenosine diphosphate), it usually donates most of that bond energy to some biosynthetic operation. ATP conversion to ADP also provides energy for a variety of other cellular activities, such as sodium/potassium active transport. Conversely, in various steps of respiration and photosynthesis ADP is recycled back to ATP, using energy derived from the degradation of foodstuffs or from sunlight.

3.2.2 Respiration

When a mole of glucose is burned in the presence of oxygen, complete combustion produces water, carbon dioxide, and about 680 kilocalories of energy in the form of heat. By a much less direct process known as *aerobic respiration*, cells produce water and carbon dioxide from glucose and oxygen, but manage to trap a substantial fraction of the liberated energy in the forms of ATP and reduced coenzymes. Analogous but simpler and less efficient processes carried out in the absence of oxygen are known as *anaerobic respiration*. An important example is the fermentation of glucose by yeast, yielding ethanol and carbon dioxide.

Aerobic respiration occurs in three major stages, each involving about 10 reactions, all of which require their own special enzymes. The intricate biochemical details of all this machinery are not prerequisite to understanding the biological systems considered later in this book. But it is nice to have a general idea of where and in what form energy is produced (and consumed) in each of the stages.

The first stage has been dubbed the EMP pathway, after its codiscoverers Embden, Meyerhof, and Parnas. In this essentially linear sequence of reactions, a molecule of glucose (a 6-carbon sugar) is converted to two molecules of a 3-carbon substance known as pyruvate. Two NAD molecules are reduced in the process. In the first two EMP reactions, ATP is actually used (converted to ADP) to get things started. In later steps, twice the amount is regenerated, yielding a net ATP production of 2 molecules for each molecule of glucose. Note that this represents only about 15 kilocalories of the 680 kilocalorie potential of glucose; most of the energy is still locked up in the pyruvate.

The transformation of pyruvate to acetyl-CoA marks the transition to the second stage of respiration, the Krebs or citric acid cycle. In this circular series of reactions, which occurs in the matrix of the mitochondrion, no ATP is generated at all. Although there is one molecule of the analogous GTP (guanosine triphosphate) generated from GDP during each cycle, the major function of this stage is the reduction of four coenzyme molecules for each acetyl-CoA processed. The carbon dioxide endproduct of the degradation of glucose is also produced in the Krebs cycle.

The last stage of respiration occurs on the internal membranes of mitochondria and is the least well understood aspect of the process. Hydrogen ions and electrons are transferred through a linked succession of carrier molecules, ultimately combining with oxygen to yield water (the other endproduct of glucose degradation). In the course of this electron transport there is *oxidative phosphorylation* (addition of phosphate to ADP in the presence of oxygen). Eighteen ATP molecules are produced by this process for every original molecule of glucose with which respiration began. After various adjustments for physiological conditions, it has been estimated that eucaryotic aerobic respiration recovers as much as 80% of the energy stored in glucose.

3.2.3 Photosynthesis

The glucose used for respiration comes from conversion of a variety of hydrocarbon foodstuffs, which in turn ultimately come from synthetic reactions driven by photosynthesis in plants. Since the source of carbon for these reactions is the carbon dioxide endproduct of respiration, and since oxygen is liberated in the photosynthetic process, an elegant reciprocity is evident in the energetics of ecosystems. The ultimate source of all biological energy, photosynthesis, is more complex and less fully understood than respiration. The basic overall equation of photosynthesis is the reverse of that in respiration, with carbon dioxide and water being combined to yield glucose and oxygen. The energy required to drive this uphill reaction comes from ATP and reduced coenzymes, which are produced in some stages of photosynthesis as well as in respiration.

The first step in photosynthesis is the activation of a light sensitive photopigment molecule, usually chlorophyll, by a quantum of sunlight. Absorption of this light energy causes an excited state in the chlorophyll molecule, leading to a flow of electrons away from it and through other carriers (similar in broad terms to the flow of electrons in oxidative phosphorylation). ATP synthesis is coupled to this electron flow, in the phenomenon termed *photosynthetic phosphorylation.* Simultaneously, water molecules are split to yield oxygen and the hydrogen ions required to reduce coenzymes.

The above aspects of photosynthesis require light, in contrast to the so-called dark reactions, which use some (but not all) of the energy produced in the light reactions and which also take place in chloroplasts. Through a complex series of intermediates, carbon dioxide is transformed into a simple 5-carbon carbohydrate which serves as a precursor to a wide variety of biological molecules, including starches and lipids as well as glucose and other sugars.

3.3 CELLULAR DIVISION

3.3.1 The Cell Cycle

The elaborate bioenergetic machinery just surveyed allows cells to store energy and use it to produce things like enzymes and membranes and copies of DNA. In an environment with enough energy, the consequence is cell growth. But there are limits to cell size, since the environmental interface grows with the square of the cell radius whereas the volume grows with its cube. The way around this seeming impediment to progress is cell division. When a single cell organism divides, it produces two new organisms, each genetically identical to the parent. When a cell in a multicellular organism divides it also produces two new cells with the same characteristics. In a special kind of division in higher organisms, some cells generate *gametes* (egg and sperm cells), which can fuse through fertilization to form a genetically complete *zygote*, the single cell precursor of a new organism.

From all external indications, cell division is a relatively rapid process that is usually complete within an hour. Yet even rapidly dividing cell populations require a half day to a day between divisions. Only a small fraction of this time may be devoted to restocking biochemical pools. It appears that much of an active cell's life is devoted to direct preparations for division, notably the duplication of genetic material.

Current theories divide the cell cycle into four major phases. Of these the last, called M phase for *mitosis* (the actual division process), is the shortest and lasts 20 to 60 minutes. Duplication of genetic material occurs in S (for synthesis) phase, requiring a relatively fixed period of around 8-10 hours in the middle of the cycle. Complete copies of the chromosomes, including their non-DNA components, are manufactured during this phase. A shorter period of 3-4 hours, following S and preceding M, is designated G2 (for the second "gap"). This is a period of extensive protein synthesis.

The most variable phase in the cell cycle is G1, between M and S. Slowly growing cell populations may have individual cells that spend years in this phase. Some cells even reach the point where they never divide again. In this group are the "dead ends" of cell lines, like the erythrocyte which has lost its nucleus. But broadly functional cells with intact genetic material can also cease dividing; one example is the nerve cell in the brains of higher vertebrates.

The cell cycle has received intensive scrutiny in recent years, because many investigators believe that interference with its normal regulatory mechanisms may be a factor in cancerous tissue, where cells seem to have forgotten when to stop dividing. Discovering the signals which control the onset of cell cycle phases in both normal and cancerous cells is a major current challenge in cell biology.

3.3.2 Mitosis

Although it is sometimes loosely applied to the entire cell division process, mitosis actually refers to the series of operations in which eucaryotic chromosomes condense and separate into two identical nuclei within the original cell's plasma membrane. Mitosis is usually followed closely by the physical separation of the binucleate cell into two daughter cells, a process called *cytokinesis*. But sometimes, as in the development of certain insect embryos, many mitotic events precede cytokinesis, creating a temporary situation in which there are thousands of nuclei inside a single large cell. Mitosis is usually divided for purposes of analysis into at least four major stages. These are depicted schematically, for a cell nucleus with four chromosomes, in Figure 3.2.

During *prophase* the chromatin condenses and coils up into distinct chromosomes. Since the genetic material has already been duplicated (during the S phase of the cell cycle), each chromosome is represented twice, as two *chromatids* connected by a *centromere*. In early *metaphase* a structure called the *spindle* forms. The paired chromatids attach to spindle fibers at their centromeres and line up in a regular fashion along the equator of the spindle. In *anaphase* the chromatid pairs separate at their centromeres and move toward opposite poles (centrioles) of the spindle, thus insuring that each new nucleus will have one copy of every chromosome. Finally, in *telophase* the nuclear membrane is reconstituted around the two collections of chromosomes,

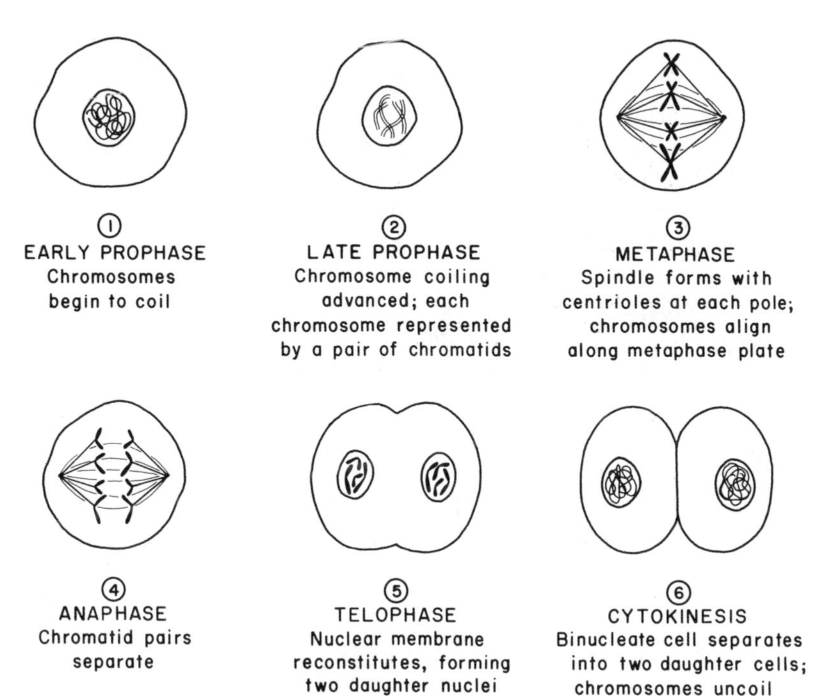

Figure 3.2 The stages of mitosis.

producing two identical nuclei. The chromosomes uncoil and revert to the distributed chromatin state.

3.3.3 Meiosis

In man and other higher organisms the normal complement of (unduplicated) chromosomes exists in *homologous* pairs. Corresponding positions on the two chromosomes affect the same aspect of cell function but may contain slightly different genetic information. This *diploid* configuration contrasts with the *haploid* configuration (each chromosome represented once) found in all procaryotes and during some phases of the eucaryotic life cycle. One such phase is the haploid gamete, an egg or sperm cell which contains just one representative of each homologous pair and unites with a gamete of the other type to regenerate a diploid organism. The process of gamete formation from diploid precursor cells is called *meiosis*. It is more complicated than mitosis.

The dozen or more major stages of meiosis need not be detailed; but a few important aspects of the process may be noted. First of all, the duplicated chromosomes must be sorted into homologous pairs, requiring the homologues somehow to "find" each other. While undergoing assortment and condensation, there is an opportunity for homologous chromatids attached to different centromeres to exchange equivalent chromosomal segments. This process of *crossing over* can produce on a single chromatid combinations of genes which did not previously exist on the same chromosome in the parent. Such recombination of genetic information maintains a constant supply of new options for selection during evolution.

After crossing over occurs, the first meiotic division segregates homologous centromeres, leaving two haploid sets of (duplicated) chromosomes, to which homologues have been independently assigned. There can now be a delay of up to many years (depending on the organism) before the second meiotic division. This second division is much like mitosis except that the duplicated genetic material in each nucleus represents only half the original genetic information. If there is a subsequent cytokinetic separation into four distinct haploid cells (there may have already been a separation into two cells in some organisms), four gametes result. In most female animals, however, all the original cytoplasm becomes associated with just one of the haploid chromosome sets. The others are eventually discarded. Apparently, one fourth of the cytoplasm does not suffice for the extensive biochemical requirements of an egg cell.

3.4 INFORMATIONAL MACROMOLECULES

Cells contain four major types of large biological molecules. Nucleic acids and proteins are intimately involved in the storage and processing of cellular information, and are treated in detail below. Essential information about the other two molecule types, carbohydrates and lipids, is here presented in summary form.

Carbohydrates are composed of carbon, hydrogen, and oxygen, usually in ratios of about 1:2:1. The simplest carbohydrates are small sugars (monosaccharides) like glucose, fructose, and galactose. These 6-carbon sugars are primary sources of readily available energy. The 5-carbon sugar ribose is an important constituent of nucleic acids. Monosaccharides can combine in pairs to form disaccharides, like sucrose and lactose, or in larger numbers to form polysaccharide starches like glycogen. Because they require a few more steps in degradation, starches are a longer term form of energy storage. Some polysaccharides have primarily structural roles, notably cellulose which is widely employed in plant tissue.

Lipids, or fats, are also used to store energy, but in a form which is even less accessible than starches. These "neutral" (uncharged) fats are composed of the 3-carbon sugar glycerol plus several carbohydrate chains known as fatty acids. Another important group of lipids are the phospholipids, which have already been encountered in their role as a component of membranes. Structurally similar to lipids are members of another biologically important group of compounds which includes sterols (e.g., cholesterol) and steroids (e.g., vitamin D).

3.4.1 Nucleic Acids

Named because they were first discovered in the nucleus (during the last half of the 19th century), nucleic acids play fundamental roles in the storage, transmission, and interpretation of genetic information. There are two general types of molecules, *ribonucleic acid* (RNA) and *deoxyribònucleic acid* (DNA). Both are built according to the same general pattern. There is a linear backbone consisting of the regular alternation of a sugar molecule and a phosphate group. In RNA, the sugar is ribose, in DNA, 2-deoxyribose (an oxygen has been removed from the ribose carbon atom numbered 2). Because the phosphate group attaches to the ribose carbon atom numbered 3′ on one side and to the atom numbered 5′ on the other, the backbone can be traversed in two directions, called 5′ to 3′ and 3′ to 5′; it is also common to refer to the 5′ and 3′ ends of a nucleic acid molecule.

Attached to the carbon atom numbered 1 of each sugar in the backbone is one of a group of four *bases*, small single or double ring structures containing several nitrogen atoms. The two double ring *purine* bases are *adenine* and *guanine*. One single ring *pyrimidine* is *cytosine*. The fourth base, another pyrimidine, is *thymine* in DNA, but *uracil* in RNA. The only basic structural distinctions between DNA and RNA are deoxygenated ribose and this difference in one of the bases. Further, since the backbone is entirely regular, a nucleic acid molecule can be completely described by the sequence of bases (usually denoted by their first letters) in a specified direction, such as 5′-A-G-G-C-T-...-C-T-3′. Genetic information is coded in such base sequences. Small fragments of DNA and RNA molecules are diagrammed in Figure 3.3, to show both the connections within and to the backbone as well as the structures of the various bases.

Components of nucleic acids which are frequently found as independent molecules have been given special names. A combination of base and sugar is a *nucleoside*, like adenosine (adenine riboside) or deoxycytidine (cytosine deoxyriboside). A combination of nucleoside and a phosphate group is a *nucleotide*, the complete basic link in the nucleic acid chain; examples are 5′guanylic acid (guanine ribotide) and thymidylic acid (thymine deoxyribotide). Note that a nucleotide is also a nucleoside monophosphate and is frequently so designated. Addition of second and third phosphate groups to riobosides yields di- and triphosphates, such as the familiar ADP and ATP. Thus the energy storing molecules are very similar to some links in the nucleic acid chains they help to build and interpret.

Various weak attractive forces, notably hydrogen bonds, may cause nucleic acid molecules to contort or associate in complex three dimensional arrangements. The most powerful factor is *base pairing*. Because of their structures, properly aligned cytosine and guanine bases can form three intermolecular hydrogen bonds. Similarly, two hydrogen bonds can form between adenine and either thymine or uracil. This base pairing underlies the most common form of

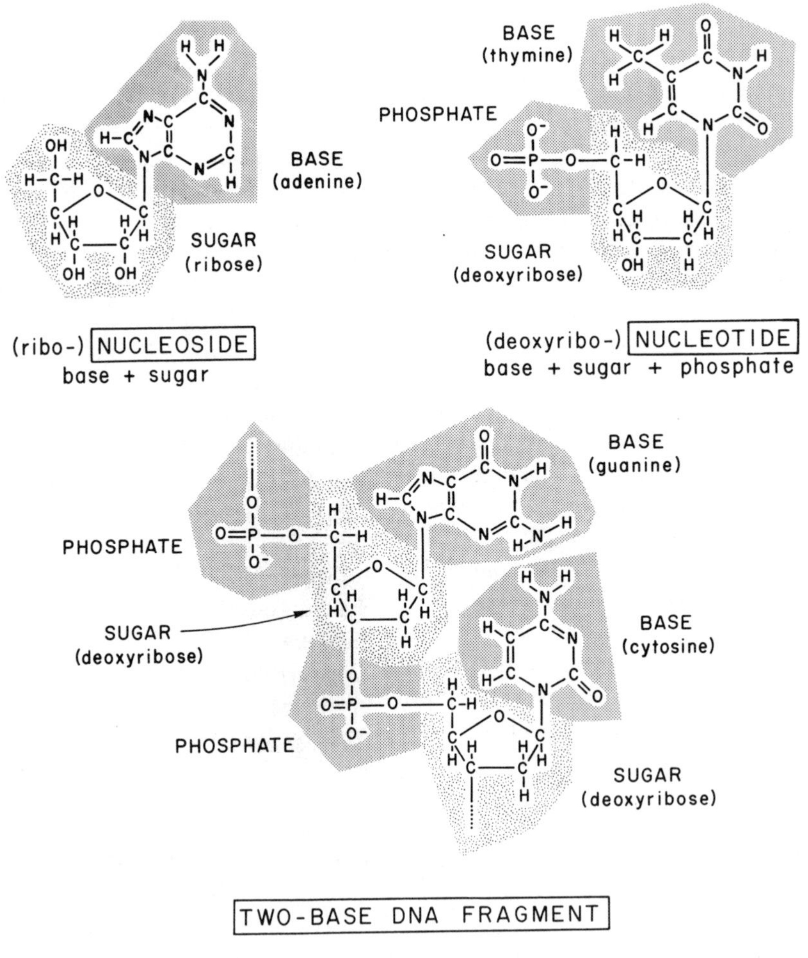

Figure 3.3 Some nucleic acid fragments.

DNA, the double helix. Two chains of DNA coil around each other in a sort of double spiral staircase. One strand exactly complements the other by pairing at every base. The two backbones are *antiparallel*, one running 5′ to 3′ and the other 3′ to 5′. Base pairing also occurs among some sections of individual nucleic acid strands, between RNA molecules of various types, and between RNA and DNA. Several information transfer mechanisms considered in Chapter 4 depend on base pairing for preservation of information.

3.4.2 Proteins

Much of a cell's genetic information is involved in specifying the structure of proteins, whose activities govern most aspects of the cell's existence. Proteins consist of one or more *polypeptide chains*, which may be specified by different genes. These chains are named for the peptide bond which links the elements in the chain together. These elements are *amino acids*. An amino acid always consists of a central carbon atom with bonds to (1) a hydrogen atom, (2) a positively charged amino group (nitrogen and hydrogen atoms), (3) a negatively charged carboxyl group (carbon and oxygen atoms), and (4) a variable side group usually designated R. Although R can potentially take on many more forms, only 20 distinct side groups are normally found in polypeptide chains. The peptide bond between two amino acids is formed by a union of the amino group of one with the carboxyl group of the other (releasing water). The structures of a few simple amino acids and some connecting peptide bonds are shown in Figure 3.4.

Like a nucleic acid, then, a polypeptide chain can be fully described by listing its component amino acids, usually known by three letter abbreviations, in a specified direction (normally starting from the amino terminal). Even more so than with nucleic acids, hydrogen bonds and other attractive forces (especially local charges in side groups) cause polypeptide chains to assume intricate three dimensional configurations, as well as to associate with one another. The sequence of amino acids in a protein's polypeptide chain(s) is known as its *primary structure*.

The *secondary structure* of a protein is the configuration of its polypeptide backbone, such as a helical arrangement. *Tertiary structure* is the often very irregular three dimensional arrangement of secondary structures, which can result in highly specific allosteric regions. Finally there is a protein's *quaternary structure*, which refers to how its component polypeptide chains are associated. It should be emphasized that all these structural complexities are determined entirely by the primary amino acid sequence and the automatic operation of intra- and intermolecular associations. The genetic specification of a protein contains no direct information about its ultimate three dimensional shape.

Proteins play two major types of roles in cells. The first is structural and has been illustrated several times in this chapter. The contributions of proteins to membranes, to the architecture of ribosomes, and to the structure of chromatin have been mentioned. In all of these roles, a protein's three dimensional configuration is critical.

The vast majority of distinct protein types function as *enzymes*. These biological catalysts are required for nearly all cellular reactions to proceed quickly enough to support life. The reaction components, known as *substrates*, are often widely dispersed and usually incapable of spontaneous reaction even upon chance encounters. Enzymes bring substrates together, positioning them

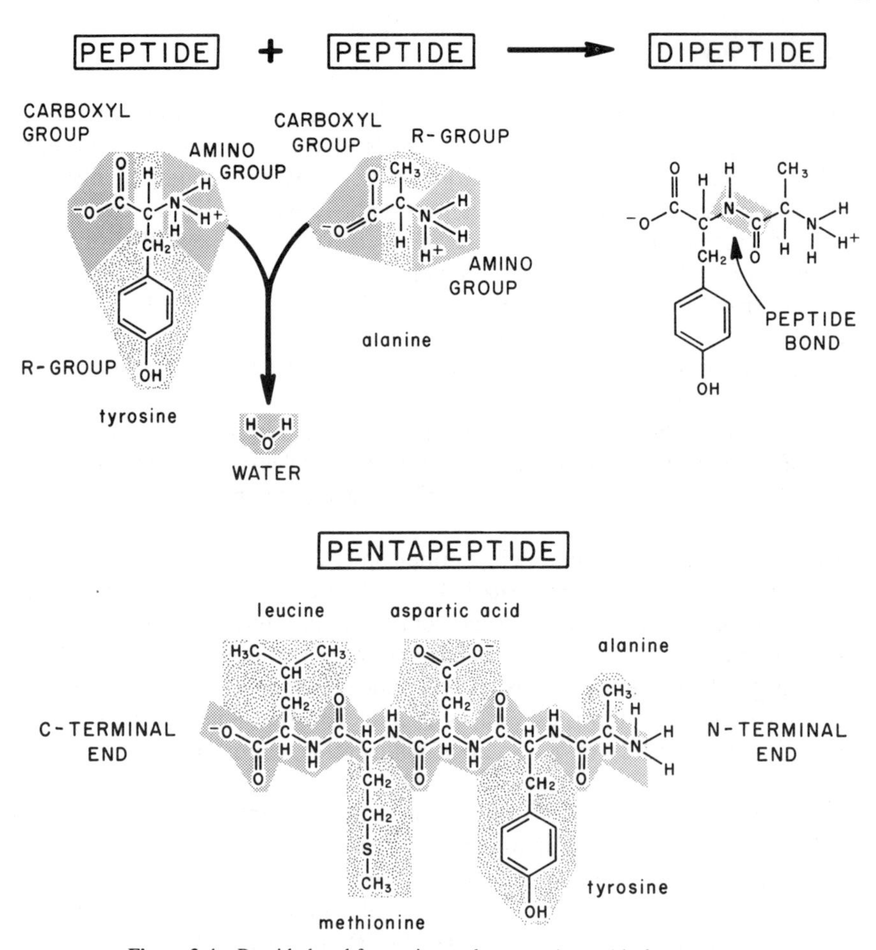

Figure 3.4 Peptide bond formation and some polypeptide fragments.

optimally for making and/or breaking bonds. These highly specific recognition and catalysis functions require precise enzyme shape, particularly in the regions of substrate attachment, known as *active sites*. Frequently, changing just one amino acid near the active site will completely inactivate an enzyme.

Because most reactions depend on the presence of correctly configured enzymes, these protein molecules serve as key points of control in the modification of cell activity in response to changes in environment. In *endproduct inhibition*, for example, a product which has become plentiful may combine with an enzyme necessary for its production, changing the enzyme's shape and reducing output of the product. Control of cellular activities through modification of enzyme availability (as contrasted with enzyme effectiveness) is discussed in Chapter 4.

PART TWO

Information Processing in Cells

Chapter Four

Genetic Algorithms

In the early decades of the twentieth century, many scientists believed that genetic information was encoded in the amino acid sequences of protein molecules. In the period between about 1944 and 1953, several now classic experiments identified DNA as the primary genetic material. After intensive investigation, the double helical, base-paired structure of DNA was discovered in 1953 by Francis Crick and James Watson. This structure was quickly seen to provide several requisite features of primary genetic material, including stable information storage and faithful self-replication.

Also in the early fifties, Crick forwarded a theory which assigns a fundamental role in genetic information processing to RNA as well as DNA. This "central dogma" of molecular biology asserts that information flows along two major pathways: (1) from DNA to itself in *replication*; and (2) from DNA to protein, in two stages by way of an RNA intermediate. These two stages are *transcription*, in which DNA base sequences are recoded in complementary RNA base sequences, and *translation*, in which the RNA transcripts are used to determine amino acid sequences of polypeptide chains. It has since been learned that other types of RNA participate in translation, and that these molecules are also produced by transcription from DNA.

At the time of its promulgation, the central dogma was believed to be a rather complete account of cellular information flow; and there was clear emphasis on the unidirectionality of the path from DNA to protein. Actually, two additional pathways are known, one of which modifies the unidirectional assumption. First, RNA is the primary genetic material in some viruses and is self-replicating in some cases. In other RNA viruses, copies of the chromosome are produced from a DNA intermediate, complementary to the original RNA molecule. It is also possible that information flows from RNA to DNA in some normal cellular processes. So, to the primary informational paths of (DNA) replication, transcription, and translation, have been added *RNA replication* and *reverse transcription.* Since these latter two pathways have been relatively little studied, they will not be treated further here. The primary pathways are covered in the first three sections of this chapter.

4.1 REPLICATION OF DNA

Like virtually all cellular mechanisms, DNA replication is enzyme dependent. There are actually several classes of enzymes involved in the various aspects of DNA processing examined here and in Chapter 6. Of primary importance are the polymerases, enzymes which make polymeric chains from the nucleotide building blocks of nucleic acids. The *DNA polymerases*, used in replication and repair, are discussed below. *RNA polymerases* are considered in the section on transcription (4.2).

In addition to polymerases, DNA processing requires nucleases and ligases. *Nucleases* are complementary to polymerases, breaking nucleic acid chains down into their component nucleotides by removing bonds between sugars and phosphate groups. *Exonucleases* can attack only a free end of a polynucleotide and are usually specific for either 3′ or 5′ decomposition. By contrast, *endonucleases* are specialized for breaking internal bonds, generating two molecules from single stranded nucleic acids. An endonuclease incision in one strand of a DNA double helix, however, leaves the helix intact but with a gap or "nick" in the one strand. Exonucleases can then operate on the two free ends at the nick.

However they arise, nicks can be sealed by *ligases*. These enzymes create the same kinds of bonds as do the polymerases, but only at isolated breaks in otherwise intact doubly stranded molecules. Both polymerases and ligases require energy for bond construction, energy which is derived from ATP.

4.1.1 DNA Polymerases

A cell typically has several different types of DNA polymerase molecules, which make different functional contributions to DNA processing. It is possible that most organisms have distinctive DNA polymerase components. It is certain that there are major differences between procaryote and eucaryote polymerases, the latter including specialized mitochondrial DNA polymerases, for example. Because the multiplicity and variability of polymerases were not discovered until about 1970, and since even isolating an enzyme of this type is a major biochemical challenge, much remains to be learned about DNA polymerases and how they work.

The first thoroughly studied polymerase was isolated in the early sixties by Arthur Kornberg and his colleagues. This molecule is now known as *E. coli* DNA Polymerase I, since it was discovered in the human intestinal bacterium *Escherichia coli*, a favorite experimental subject for molecular geneticists. Polymerase I is a very large molecule, containing only one polypeptide chain. Although it will cause DNA replication under proper conditions in a test tube, Polymerase I is no longer considered the primary agent for such activity, since mutant bacteria lacking the enzyme still replicate their chromosomes. The major functions of Polymerase I seem to be in DNA repair, where its associated exonuclease activities may be important. The primary mechanism for chromosome replication in *E. coli* probably involves the third polymerase isolated, known as Pol III, in conjunction with another protein, Copol III; the complex is called Pol III*.

Although little is known about the structure or specific operation of many DNA polymerases, the conditions required for effective function of these enzymes are well established and generally similar. First of all, a polymerase requires a template, an already connected strand of DNA on which to build the complementary strand. Second, the proper building blocks must be on hand, namely the deoxynucleoside triphosphates, which consist of deoxyribose, a DNA base (A, C, G, T), and three phosphate groups. The second and third phosphate groups have been added to the nucleoside monophosphates (nucleotides) by previous conversion of ATP to ADP, and thus have high energy bonds. This energy is released for use in polymerization when the two phosphate groups are split off as the nucleotide is added to the growing strand. The third requirement of most DNA polymerases is a properly oriented "primer" strand. The polymerase cannot initiate DNA synthesis; it can add new nucleotides only to the 3′ end of an existing strand. This requirement has significant consequences for the replication mechanism, as discussed below.

4.1.2 Initiation of Replication

The replication of a chromosome always begins at one or more designated starting points, known as *origins*. How such points are recognized by DNA polymerase molecules is unknown but probably involves some special sequence of bases. In the small *E. coli* chromosome there is one fixed origin. In much larger eucaryotic chromosomes, replication begins at several origins and proceeds in parallel. Thus, even though replication is a much slower process in eucaryotes, the parallel processing made possible by multiple origins may permit complete replication of the larger chromosome in less time than required for the procaryote. Additional acceleration of eucaryotic replication is conferred by the division of the genetic material into several chromosomes which replicate independently.

At an initiation point, the DNA double helix unwinds locally, allowing each strand to serve as a template for a complementary new strand. This *semiconservative* mode of replication, in which each new double helix contains one intact strand from the original molecule and one new strand, was firmly established in 1958. The alternatives, conservative replication (two new strands paired) and dispersive replication (new and old material mixed within strands), were ruled out by analysis of the distribution of a detectably heavy isotope of nitrogen in the DNA formed from original "heavy" molecules replicating in a supply of "light" building blocks. After one round of replication all molecules were of intermediate weight, eliminating conservative replication; after a second round half the molecules were "light," eliminating dispersive replication.

Initiation of replication poses a problem for most forms of DNA polymerase, since there is no primer strand to which the enzyme can attach nucleotides. In most cases the problem is probably solved by a small amount of transcription; RNA polymerase (see Section 4.2.1) does not require a primer strand to create a base-paired RNA fragment on each DNA strand in the region of the origin. DNA polymerases usually find RNA an acceptable primer and begin to add DNA nucleotides to the 3′ end of it. The RNA primer can later be excised by nuclease action, leaving a gap which can be filled by DNA polymerase.

4.1.3 Mechanics of Replication

The fundamental structural feature of replicating DNA is the *replicating fork*. This is a Y-shaped region in which the two arms contain separated initial strands on which new strands are synthesized. Because replication proceeds toward the vertex, the base of the Y will be the next region in the original molecule to undergo strand separation. Unless the origin is at the end of a linear chromosome, replication usually proceeds bidirectionally, the arms of the two replicating forks joining to form an "eye" shaped region. This and other aspects of the replication process are illustrated in Figure 4.1.

Since DNA polymerase adds nucleotides only in the 5′ to 3′ direction, special arrangements are needed at each replicating fork in order to generate the strand whose overall elongation is in the 3′ to 5′ direction. The solution is discontinuous replication. As shown in Figure 4.1, when a new region of the original helix opens up, small successive pieces (about 1000 nucleotides in length) are added to the strand growing "isodirectionally." Meanwhile, similar small pieces of new DNA are grown "antidirectionally," against the overall movement of the fork, to produce the other strand. Thus replication is always 5′ to 3′. The problem is that there is no DNA primer further down the strand from which to begin replication counter to the direction in which the fork is moving. As with initiation, the problem is solved by RNA primer fragments, a

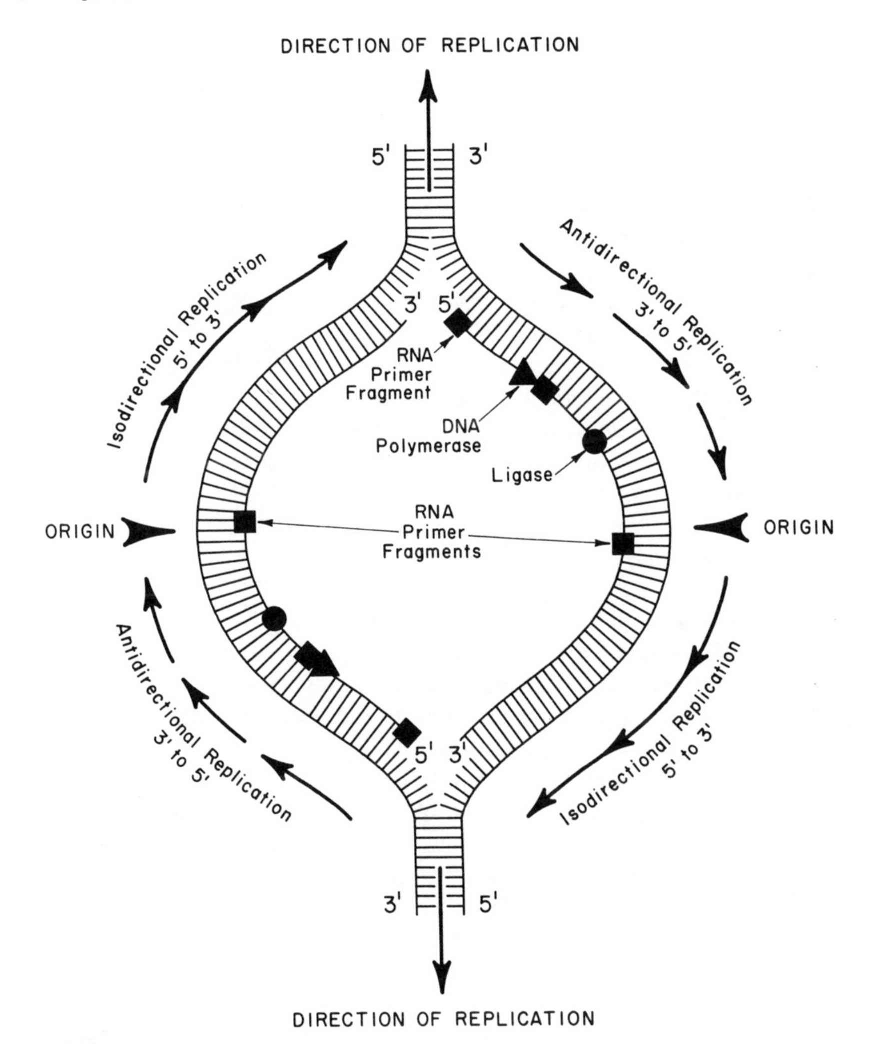

Figure 4.1 Some aspects of DNA replication.

few dozen nucleotides in length, onto which the DNA polymerase can attach DNA nucleotides. The fragments are later removed by nucleases, the gaps filled by DNA polymerase (there is primer available as soon as the next piece has been grown), and the nicks sealed by ligases. The whole process transpires at an overall rate on the order of 200 nucleotides per second (for Pol III) at each fork.

As if all this were not complicated enough, other activities are also involved in replication. Untwisting and unwinding proteins have been identified and appear to be essential to proper strand separation in some kinds of chromosomes. Membrane attachment sites seem to be involved in replication in some procaryotes. And some DNA polymerases appear to perform a "proofreading" function, in which mismatched bases (which are quite unlikely anyway) are excised and replaced with proper bases before replication proceeds any further. Proofreading requires some exonuclease activity, which has been found in a variety of DNA polymerases.

The mechanics of the replicating fork are not the whole story of chromosome replication. In eucaryotes, where the chromosomal proteins must also be multiplied and correctly combined with the new DNA, it is unlikely that there will be a complete and accurate picture of chromosome replication for some time. Even for the simple single chromosomes of viruses and procaryotes, it seems that individual solutions to special replication problems have proliferated. Some of these solutions will be mentioned in connection with the discussion of such chromosomes in Chapter 6.

4.1.4 Repair Synthesis

It has been estimated that the sequence of thousands of bases in a bacterial gene has a 50% probability of remaining unchanged after being duplicated more than 100 million times. The incredible accuracy with which genetic information is transmitted from one generation to the next depends on more than just a very reliable replication mechanism. There are also a variety of ways in which the cellular machinery can repair defects in DNA. Some of these functions are thought to be carried out by DNA polymerases acting in concert with other types of DNA processing enzymes. As with replication, the redundant duplex structure of DNA allows recovery of information about an absent or damaged strand from its complement.

The best understood kinds of repair to damaged DNA are the two mechanisms which counteract the dimerization of pyrimidines induced by ultraviolet light. In this phenomenon, two adjacent thymine bases, for example, can become fused, making replication and transcription inaccurate or impossible. The first repair mechanism, called photoreactivation, occurs in the presence of visible light. The light activates an enzyme which breaks the link between the bases, by a mechanism which is not fully understood.

The second mechanism, called excision repair, does not require light but does need several DNA processing enzymes. Initially, some endonuclease with the ability to recognize the dimer nicks the strand to the 5′ side of it. Then some DNA polymerase attaches to the available 3′ end and uses the complementary strand as a template to synthesize a repaired region of the originally defective strand. The polymerase may have exonuclease capability, so it can also digest the defective portion, or another exonuclease may be involved, so the digestion could precede or parallel the synthesis. In either case, the final step requires a ligase to seal the remaining nick at the 3′ end of the new segment.

4.2 TRANSCRIPTION

Transcription is fundamentally similar to replication in several respects. Both processes use information, acquired by base pairing with DNA templates, to sequence new nucleic acid molecules. Both processes require a polymerase and energy derived from ATP-potentiated nucleoside triphosphates (one of which, in transcription, is ATP itself). And both processes build new chains in the 5′ to 3′ direction only.

The differences between replication and transcription are fundamental too. The latter is and can afford to be a slower and less accurate process, since many copies of a transcript are produced, of which a few can be nonfunctional without serious consequence. The product of transcription is a single stranded nucleic acid with ribose as the sugar and uracil instead of thymine. These RNA transcripts are much less stable than DNA double helices, a factor which can be important in a cell's adaptation to environmental changes (see Section 4.5.1). Carrying the information specified by one or a few genes, transcripts are obviously much smaller than the chromosomes produced by replication. Further, only one strand of the DNA is transcribed in any particular region of the chromosome, since it is generally not possible for the base paired complement of a gene to carry information useful for anything other than replication. Interestingly, the "sense" (as opposed to nonsense) strand can be found on different physical DNA strands in different regions of the chromosome. The mechanism by which only the sense strand is selected for transcription involves special DNA base sequences, which also identify the point at which transcription is to begin. Finally, replication and transcription use different polymerases.

4.2.1 RNA Polymerase

The enzyme responsible for transcription has already been seen to play an essential role in replication, by providing primer fragments. In fact, an important difference between DNA and RNA polymerase is the latter's ability to initiate chains. RNA polymerase can also terminate chains, an essential feature if only small regions of the chromosome are to be used as templates for individual transcripts. Both initiation and termination involve recognition of DNA base sequences, an ability apparently absent in DNA polymerase. On the other hand, RNA polymerase has neither proofreading nor repair functions, which is consistent with the lesser accuracy of transcription.

The structure of *E. coli* RNA polymerase is much more complicated than that of DNA Polymerase I, but is now seen as comparable in complexity to Pol III*. The basic RNA polymerase molecule, known as the *core enzyme*, is comprised of five polypeptide chain subunits. Two of these subunits are identical, two others are similar to each other; the fifth and by far the smallest subunit is not always present and may not be essential to the enzyme's function. Addition of another subunit, called *sigma*, produces the *holoenzyme*, a form of

RNA polymerase which appears transiently during initiation of transcription. Subunit structures of similar complexity, but different specific composition, are found in the RNA polymerases of other bacteria and of those few eucaryotic cells from which the enzyme has been isolated. Eucaryotes actually use three distinct kinds of RNA polymerase molecules, one for each type of RNA transcribed (see Section 4.2.3).

It is not surprising that RNA polymerase has an affinity for DNA. But the binding is of two different types, depending wholly upon the presence of sigma. The core enzyme binds weakly and randomly along the entire DNA molecule; transcription is not initiated from such contacts. The holoenzyme binds much more strongly, and only to those regions of the DNA from which transcription is to be initiated. These sites are called *promoter regions*, or just promoters. The nucleotide sequences of some promoter regions have been determined in recent years, which has led to some understanding of the kinds of sequences for which the holoenzyme has an affinity. These sequences are on the order of 30 bases in length and show certain regularities, such as repetition of subsequences, often in reverse order to yield a kind of palindrome. Many promoters contain several such polymerase binding sites, allowing more efficient use of the template through rapid multiple onsets of transcription. After binding to the promoter, RNA polymerase no longer requires the sigma subunit. Sigma dissociates from the core enzyme as transcription begins, and can be immediately recycled for use at another binding site.

4.2.2 Mechanics of Transcription

After RNA polymerase is firmly attached to the promoter site, transcription can be initiated. The enzyme appears to have two distinct active sites, one specific for the first ribonucleotide in the chain and another for adding nucleotides to a growing chain. The initiating site requires a purine; so all transcripts begin with base A or G. On the DNA, initiation will thus be at a T or C base which is actually several nucleotides down from the point of polymerase attachment and possibly many nucleotides before the onset of the informationally significant sequence of DNA bases (the gene proper). Yet all RNA molecules do not begin with A or G, followed by some number of irrelevant bases. This apparent anomaly is explained by the fact that completed transcripts can be modified by other enzymes; in particular, the meaningless initial segments can be trimmed off.

With the first RNA base in place, the core polymerase enters a chain elongation mode of operation. Available ribonucleoside triphosphates are paired with appropriate DNA bases, the extra diphosphate pairs split off, and the nucleotides bonded to the growing polymer. On the order of 50 nucleotides are added each second. The major steps in transcription are shown diagrammatically in Figure 4.2.

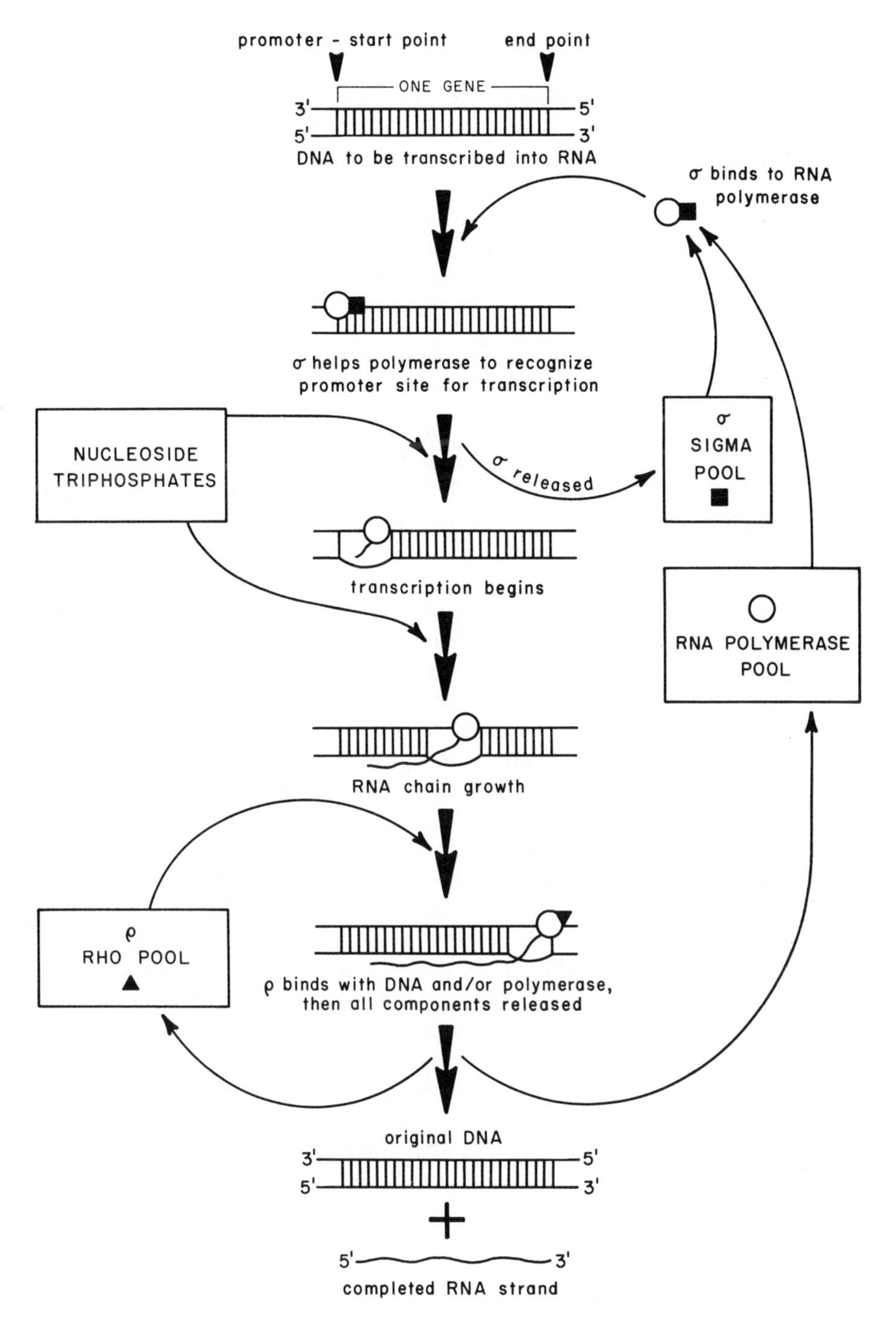

Figure 4.2 Major steps in transcription.

Termination of transcription requires the participation of another polypeptide chain, as well as some stopping signal in the DNA base sequence. As with initiation, it is possible that transcription reads beyond the end of the gene and that extra RNA bases are later removed. The polypeptide chain that assists with termination is called *rho*. Unlike sigma, rho does not become an integral part of the core polymerase. The way in which rho assists with recognition of DNA encoded stop signals is not understood. The stop sequences themselves have not been as fully characterized as promoter initiation sites.

4.2.3 Products of Transcription

All three classes of cellular RNA are produced by transcription. Only *messenger RNA* (abbreviated mRNA) carries information for sequencing amino acids in polypeptide chains. But both *ribosomal RNA* (rRNA) and *transfer RNA* (tRNA) are employed in the translation of mRNA. Transfer RNA is discussed in the next section.

Because they specify chains ranging in length from dozens to thousands of amino acids, mRNA molecules are, in terms of size, the most variable group of nucleic acids. In the eucaryotic nucleus, messenger transcripts are a major component of what is known as "heterogeneous nuclear RNA." Their form can change substantially, however, before they are employed in translation. Nonmessage material may be removed from the beginning, end, and even the middle of a transcript. Frequently a string of about 200 adenine nucleotides may be added to a transcript before it leaves the nucleus. The purpose of these poly-A tails is not known; two speculations are that they (1) facilitate transport of mRNA out of the nucleus or (2) help determine the life span of the mRNA in the cytoplasm. Eucaryotic messages usually last from hours to days.

Procaryotic mRNA is much less stable, remaining effective for a matter of minutes. Although there may be some tailoring of the transcripts before they serve as messages, the modifications are probably not as pronounced as in eucaryotes. There is another important difference between procaryotic and eucaryotic mRNA. The eucaryotic message specifies an amino acid sequence for a single polypeptide chain, whereas the potentially much longer procaryotic transcript can code for as many as 20 distinct polypeptides. These enzymes typically have interrelated functions in cellular metabolism. The placement of their genes together on the DNA, into what is known as an operon, has significant implications for transcriptional control processes, as will be discussed in Section 4.5.

Although there are only three or four major types of ribosomal RNA in any given cell, this form of RNA accounts for about 80% of cellular RNA simply because one molecule of each type occurs in every ribosome. The structure of ribosomes is considered further in connection with translation; the concern here is with the transcription and subsequent processing which produce rRNA molecules. As with mRNA, it is best to discuss rRNA transcription in eucaryotes and procaryotes separately.

The smallest form of eucaryotic rRNA contains about 120 nucleotides. It is transcribed from many identical genes (hundreds to thousands, depending on the species), which may be found in tandem repetition, separated by short "spacer" segments of untranscribed DNA. The transcript undergoes relatively little tailoring before being incorporated into ribosomes. By contrast, the two larger rRNA molecules (and in some cases another small one) are produced by extensive processing of the transcript of a single gene, which again occurs in hundreds of copies. About 10 enzymatically controlled modification, cleavage and degradation steps are required to produce mature rRNA molecules from this large precursor transcript. The transcription occurs in the one or more nucleoli found in eucaryotic nuclei.

In procaryotes, the three distinct rRNA genes are adjacent and transcribed together. Cleavage and a small amount of tailoring produces the final products. The gene group is one of few usually found in multiple copies on the bacterial chromosome (5 to 10 in *E. coli*); the copies are distributed rather widely, unlike the tandem repeat arrangement in some eucaryotes. Redundancy at the gene level thus appears to be a universal solution to the requirement for vast amounts of rRNA.

4.3 TRANSLATION

4.3.1 Transfer RNA

The tens to hundreds of distinct types of tRNA in a cell contain about 80 bases each, making it the smallest kind of nucleic acid routinely used in information processing. An intricate three dimensional configuration confers on these molecules prodigious recognition and binding capabilities, rivaling those found in large enzymes. Like rRNA genes, those coding for distinct tRNA molecules may be present in multiple copies; this seems to be typical in eucaryotes and less common in procaryotes. As is the case for virtually all transcripts, tRNA precursors undergo tailoring and modification before they are functional. About 40 superfluous nucleotides are removed from the ends of the original (*E. coli*) tRNA transcripts. A CCA amino acid coupling sequence (see below) is then added to the 3′ end of every molecule.

The post-transcriptional modifications of greatest significance to tRNA structure involve the modification of bases. A mature tRNA molecule has about 10 "unusual" or "minor" bases, which depart from the structures of A, G, C, and U. These deviations are probably critical to the molecule's intricate three dimensional interactions. All post-transcriptional modifications to tRNA (and, for that matter, to all transcripts) are presumed to be carried out by specific enzymes; identification and operational understanding of these enzymes is only beginning.

The primary structure of a tRNA molecule is thus a sequence of ribonucleotides, some of which are unusual, terminating in CCA at the 3′ end. In many of the individual molecules whose sequence has been determined, there are regions of potential base pairing which can cause a "clover leaf" kind of secondary structure, like that shown in Figure 4.3. Notice that there are three prominent loops, connected by base-paired arms; the size of the extra arm (at the upper right) varies from a few to more than a dozen bases, depending on the particular type of tRNA.

The secondary structure of a typical tRNA molecule appears to expose two of the critical attachment sites, to be discussed further below. Amino acids invariably bind to the 3′ end, whereas a three base "code word" (or codon) will base pair with its counterpart "anticodon" in the middle tRNA loop. Were this the whole story, it would be very hard to imagine how a particular amino acid could be associated with the proper anticodon. Intramolecular attractive forces, however, confer a distinctive *tertiary* structure on each type of tRNA molecule, an example of which is also shown in Figure 4.3. Its tertiary structure allows a tRNA to interact exclusively with enzymes that associate with the proper amino acid, and allows it to fit precisely into the ribosomal matrix so that the three available mRNA bases are in fact the current codon. In a mere 80 nucleotides tRNA thus manifests four recognition sites: (1) amino acid attachment, (2) amino acid attachment enzyme, (3) anticodon, and (4) ribosome attachment.

Attachment of the proper amino acid depends on two of the above sites, and actually occurs in the cytoplasm independently of the subsequent translational events on the ribosome. For every proper combination of three-dimensionally unique tRNA and amino acid (with its own recognizable configuration), there is an enzyme that joins them. These enzymes, the *aminoacyl-tRNA synthetases*, actually have a two step function. First, they use energy derived from the conversion of ATP to AMP (with release of diphosphate) to complex with a combination of amino acid and the AMP. In a reaction with the appropriate tRNA, AMP and enzyme are then released, yielding a combination of tRNA and amino acid, an *aminoacyl-tRNA* complex.

4.3.2 Ribosomes

An abundance of aminoacyl-tRNAs and suitably tailored message transcripts is not sufficient for translation. The pivotal element is the ribosome. This still poorly understood complex of RNA and protein seems to function like a sort of super enzyme, positioning and manipulating the participants in translation. Ribosomes in procaryotes, eucaryotic cytoplasm, and eucaryotic organelles all have the same basic organization. Apparent structural differences are mainly

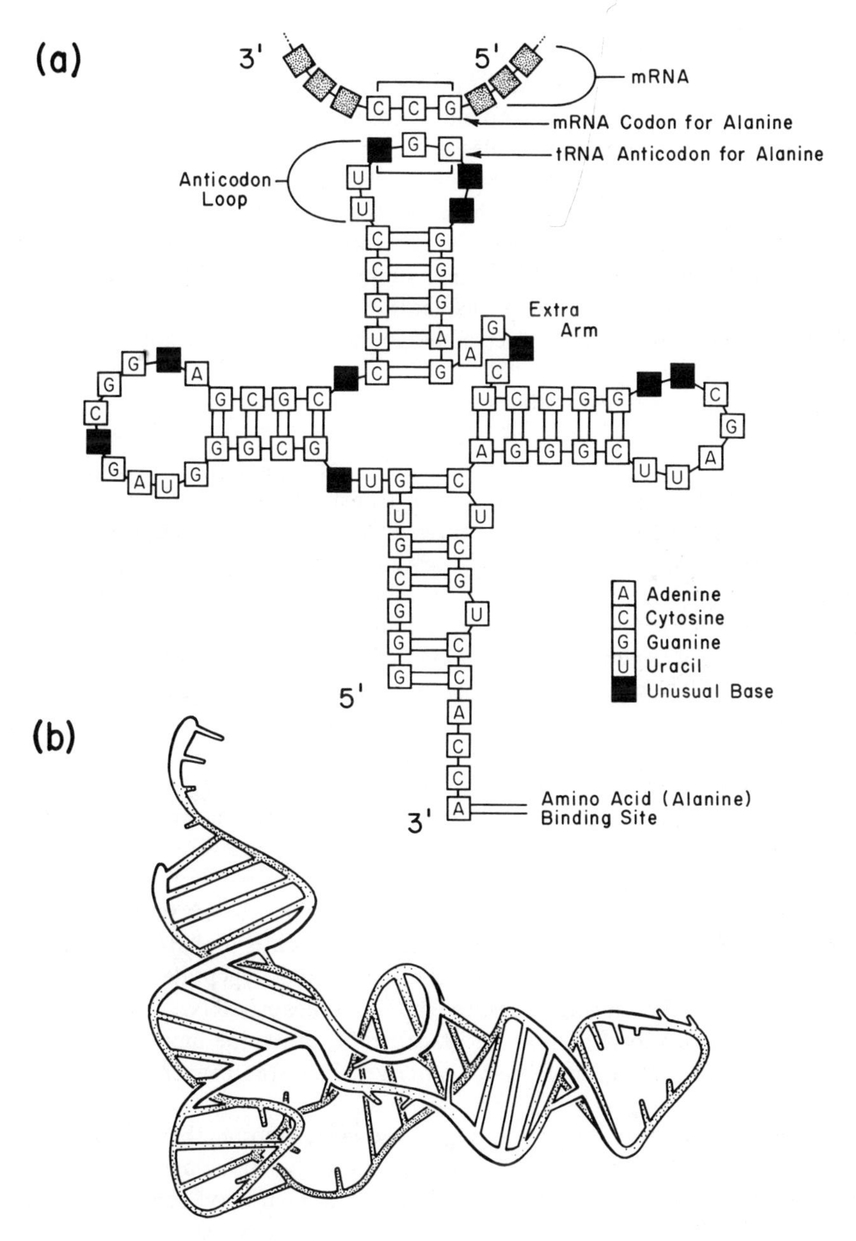

Figure 4.3 (a) Secondary and (b) tertiary structure of transfer RNA.

in size, the eucaryotic cytoplasmic ribosomes having larger components and correspondingly larger rRNA molecules.

When not actively involved in translation, most ribosomes separate in two subunits. The smaller subunit contains the smaller of the two large rRNA molecules and at least 20 distinct protein molecules. The remaining two (in some eucaryotes, three) rRNA molecules are complexed in the large subunit with about 30 other protein molecules. The assembly of ribosomal subunits is a precisely coordinated process, in which the first proteins associate before rRNA tailoring is complete, and yet there is no affinity for tRNA until the ribosome subunit is wholly functional. The small subunit has a recognition site specialized for initiating translation; subsequent combination with the large subunit allows successive amino acids to be added to the polypeptide chain.

4.3.3 Mechanics of Translation

Since it requires that information stored in one modality (nucleotides) be transformed into another (amino acids), translation is a much more complicated process than replication or transcription. Translation mechanics, which are essentially the same in procaryotes and eucaryotes, can be divided into three clear phases: initiation, chain elongation, and termination. These processes are presented diagrammatically in Figure 4.4.

Translation is begun by the formation of an *initiation complex* which consists of the mRNA, the small ribosomal subunit, and a particular aminoacyl-tRNA. The initiator codon in the mRNA is invariably AUG, which codes for the amino acid methionine. How the small ribosomal subunit recognizes initiation sites, and differentiates them from other AUG codons specifying methionine in the middle of a polypeptide chain, is not understood. AUG initiator codons are rarely the first three bases of mRNAs. Three protein "factors" (designated F1, F2, F3) which are necessary for initiation may participate in this recognition. The dual role of AUG means that all polypeptide chains start with methionine; but in many cases it is probably enzymatically removed during post-translational tailoring of the polypeptide. In procaryotic and organelle initiation complexes, the initial aminoacyl-tRNA actually carries a slightly modified form of amino acid, called N-formyl methionine. Energy for formation of the initiation complex is derived from the conversion of GTP to GDP.

Once the first amino acid is in place, the large ribosomal subunit joins the initiation complex and chain elongation begins. Polypeptide chains are built in the direction from the amino terminal toward the carboxyl terminal, at rates approaching 20 amino acids per second. The mRNA is read from the 5′ end toward the 3′ end. Since codon-anticodon pairing is antiparallel (as in the DNA double helix), anticodon triplet nucleotides are given in the 3′ to 5′ direction.

The large ribosomal subunit has two major active sites, the *peptidyl* and *aminoacyl* sites. The aminoacyl-tRNA most recently added to the chain

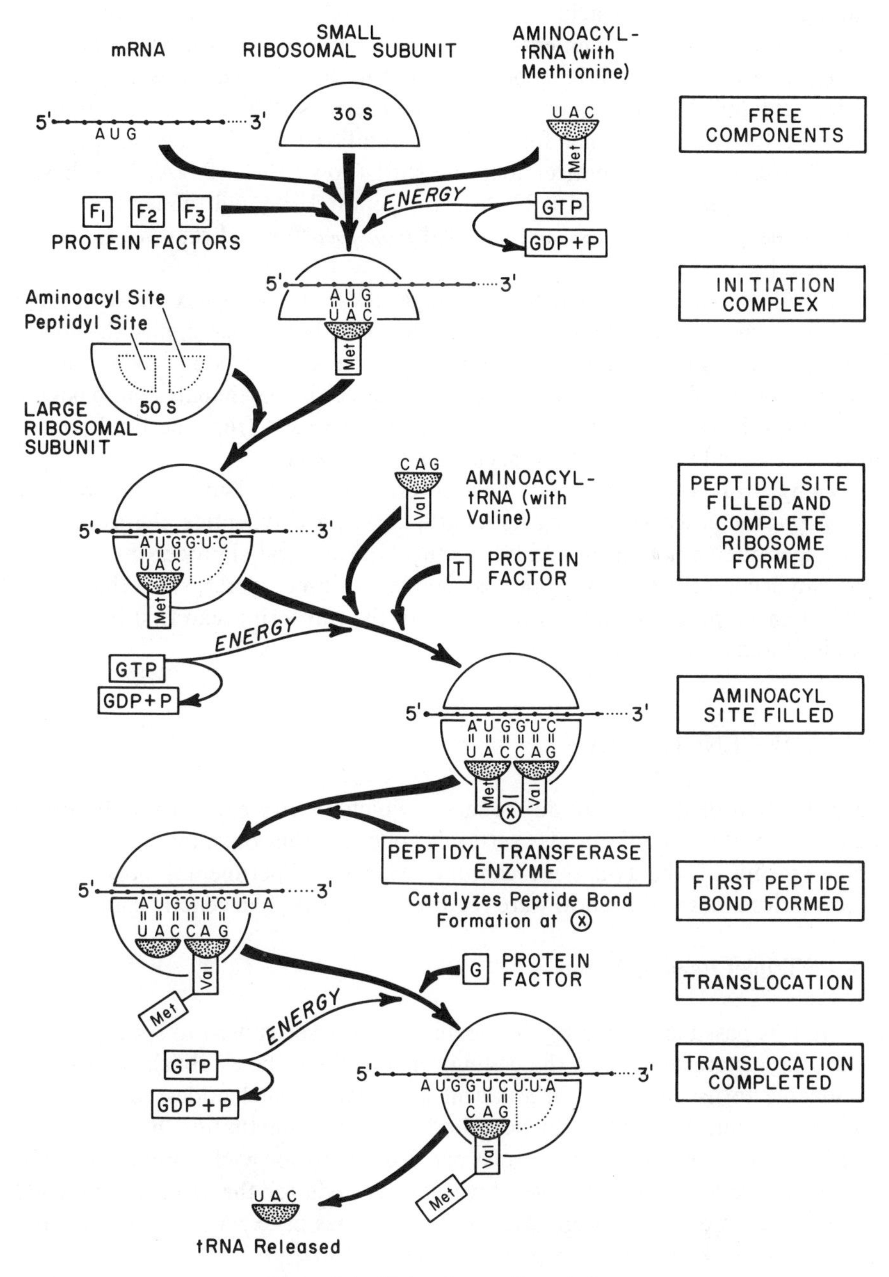

Figure 4.4 Major phases of translation.

occupies the peptidyl site, with the entire polypeptide synthesized to that point trailing from its amino acid. The aminoacyl-tRNA with an anticodon that matches the next codon then enters the aminoacyl site with the assistance of a protein factor designated T. A peptide bond is then formed between the new amino acid and the previous one, freeing the previous tRNA for expulsion from the peptidyl site. The bond formation is catalyzed by the enzyme peptidyl transferase. Finally the entire ribosome shifts down the mRNA three bases, placing the latest aminoacyl t-RNA in the peptidyl site and the next codon in the aminoacyl site. This last step is called *translocation* and requires GTP and a protein factor designated G.

As will be seen in the discussion of the genetic code in the next section, there are three codons which do not specify amino acids but rather signify the end of the message and terminate translation. When one of these "stop" codons enters the aminoacyl site, a protein factor designated R participates in blocking further synthesis, the completed polypeptide is released from the final aminoacyl-tRNA, and the ribosome dissociates into subunits.

A single mRNA molecule can have several ribosomes, known collectively as a *polyribosome*, working along its length, carrying polypeptide chains in various stages of completion. In addition to this type of parallel processing, translation can occur before transcription is complete in procaryotes, since there is no nuclear envelope barrier and since the end of the transcript read first in translation is generated first in transcription.

4.4 THE GENETIC CODE

Identification of the nucleotide sequences which specify each of the 20 amino acids in protein synthesis is certainly one of the outstanding scientific achievements of the twentieth century. Massive experimental programs in laboratories all over the world were required to "break" the genetic code.

4.4.1 Triplet Reading Frame

Before it is possible to investigate the actual code words used to specify amino acids, the general nature of the coding scheme has to be established. It seems reasonable to assume that an adjacent sequence of nucleotides constitutes a code word; but how many nucleotides? Is there punctuation between code words; do they overlap? Is there just one code word for each amino acid; or is the code *degenerate* in that several words may code for the same amino acid? Many of these questions were answered by a series of experiments carried out by Francis Crick and his colleagues, reported in 1961.

The investigators worked with mutants of T4, a virus that attacks *E. coli*. Mutations which alter the normal form (or "wild type") of the T4 *r*-gene region produce noticeably faster destruction (lysis) of bacterial cells upon viral infection; *r* stands for *r*apid lysis. The mutations studied by Crick's group were induced by a group of chemicals called acridines, which act by deleting or inserting one or a few nucleotides. The first such mutant was designated *FC*0.

Bacteria were repeatedly infected with successive generations of viruses originating as *FC*0. Eventually some viruses appeared to revert to the wild type. Careful investigation revealed that the apparent reversion was actually caused by a second mutation that suppressed the effects of the first. *FC*0 was designated a + mutation; the other mutation was isolated and designated —. After much work, some 80 different strains of *r* gene mutants of T4 were isolated and classified as + or —. Nearly any + mutant would suppress the effects of nearly any — mutant, and vice versa. But no mutant could suppress others of the same type. This suggested that one group of mutants (say the —) involved deletions of single nucleotides, and the other (+), insertions. Either a deletion or an insertion alone would cause the reading of the genetic code to be permanently out of phase from the point of the mutation onward. Thus most amino acids would be incorrectly specified, causing the polypeptide product to be nonfunctional and the mutant character to appear. Nearby + and — mutants, however, could compensate for each other. If the proper "reading frame" for the code were restored after only a few incorrect code words, a polypeptide with normal function would often result.

The above results suggested that the genetic code is not punctuated, that proper decoding depends on finding a starting point and reading adjacent, nonoverlapping code words, one after another. The size of these code words can be determined by combining several + (or —) mutants. It was found that exactly three mutants of the same type could cancel each other out, but never two or four. The conclusion was that a code word contained some multiple of three bases, probably exactly three. In any case, the genetic code is read in groups of three.

The experiments establishing the triplet reading frame also suggested that most of the code words specified an amino acid, since "nonsense" codons would have prematurely terminated polypeptide synthesis, usually producing an inviable product. With 64 available 3-base sequences and only 20 amino acids, the code is clearly degenerate. Triplet codons suggest an appealing economy in nature. The 16 possible 2-base code words are not enough, whereas use of the 256 4-base sequences would be excessively wasteful. The same consideration weighs heavily against a higher multiple of three, although the above experiments did not conclusively rule out such a possibility.

4.4.2 Codon Assignment

With the triplet reading frame established, attempts to assign the 64 putative codons to individual amino acids could be undertaken. This work was done in several laboratories during the mid-1960s. Four successively more powerful experimental techniques were developed. The first three depended on methods for cell free protein synthesis in the test tube. Proper combinations of messenger, aminoacyl-tRNAs, ATP, GTP, and various protein factors were discovered which resulted in such synthesis, although not in anything like the amounts found in natural cellular systems. Equally important was the ability to prepare synthetic mRNA for these artificial protein synthesis systems. The enzyme polynucleotide phosphorylase can be induced to combine ribonucleoside triphosphates without the presence of a DNA template. This situation makes it possible to synthesize mRNA from restricted subsets of bases, limiting the possible number of different codons in the "transcript."

The ground breaking experiments were done with mRNAs containing reiterations of a single base. The first result was that protein synthesis using polyuradilic acid (UUUUUUU...) resulted in a polypeptide chain containing only the amino acid phenylalanine. Hence the first codon assignment: UUU codes for phenylalanine. Similar experiments established that CCC codes for proline and AAA for lysine. Since there is no transcription from GGGGGG... (the bases hydrogen bond to each other), the potential of the single-base mRNA method was exhausted after the establishment of only three codons.

The second method used mixed copolymers as artificial transcripts. Random mRNA sequences employing just two bases contain eight potential code words, whose expected frequencies can be calculated from the proportions of the bases and correlated with the proportions of amino acids incorporated into polypeptide. Although only highly probable codon assignments can be determined from such results, the method did contribute substantially to the breaking of the genetic code.

The third method involving synthetic mRNA made use of newly discovered methods for making regular repeating sequences of ribonucleotides. The polypeptide synthesized on CUCUCU..., for example contains alternating leucine and serine, whereas cysteine and valine are produced on UGUGUG.... Combined with previous knowledge, results of such experiments considerably advanced knowledge of the genetic code.

The fourth method for discovering codon assignments was the last to be developed and also the most powerful. A means was found to bind a known sequence of exactly three bases (i.e., a codon) to a ribosome; the combination attracts and binds only the tRNA specific for the amino acid specified by the codon, since only that tRNA has the proper anticodon. Final points of ambiguity in codon assignments were resolved in this fashion, although most of the code could have been established using this method, had it been available earlier.

Since all of the above work was carried out in the test tube, the possibility remained that the genetic code employed inside cells could differ in some way. There are now extensive data, based on analyses of natural polypeptides and the ways their amino acid sequences change when nucleotides change, which establish that the genetic code in a variety of organisms is identical to that discovered in the test tube. One now-classic study of this type provided key evidence for the fact that the sequence of codons in a gene is found in the same order as the sequence of amino acids in the corresponding polypeptide. This *colinearity* of gene and product, although always considered likely, was not an inevitable consequence of the codon assignment experiments.

Colinearity was established by work with the gene coding for one of the two polypeptide chains in the *E. coli* enzyme tryptophan synthetase, which is involved in the biosynthetic pathway for the amino acid tryptophan. Techniques for sequencing polypeptides allowed precise placement of the effects (amino acid substitutions) of various mutations in the gene. Sophisticated gene mapping techniques (of the sort discussed in Chapter 6) showed exact positional correspondence of these effects with base changes in the DNA.

It now appears that the genetic code is universal, that a given codon specifies the same amino acid throughout the biological kingdom. Experiments in which mRNA from one species, when inserted into cells of another, directs synthesis of foreign proteins lend credence to the idea of code universality. Since the genetic code presumably evolved from simpler schemes and underwent variation during this evolution, code universality raises the question of what is so special about the current code. The possible value of the genetic code now in use is one topic addressed in the next section.

4.4.3 Code Patterns and Wobble

The genetic code is shown in Figure 4.5. Inspection of the code when displayed in this fashion reveals a number of interesting patterns. First of all, there are at least two, and as many as six, codons for each amino acid except methionine, which must make do with the single AUG codon. This versatility in the specification of any given amino acid sequence may be important; it certainly figures in suppression, as discussed later. Since the methionine codon has to do double duty as the "start" signal, it is rather surprising that only one codon is used; perhaps this is related to some aspect of initiation of translation. The three "stop" codons also provide some flexibility in chain termination.

Other patterns in the code include the fact that structurally similar amino acids have similar codons. Aspartic acid and glutamic acid, for example, are close relatives; their codons are GAU-GAC and GAA-GAG. A change in the third base, say from U to A or G, can cause the wrong amino acid to appear at that point in the polypeptide, but perhaps without serious consequence to protein function.

Among codons specifying a particular amino acid, the greatest variability is usually in the third base. In half of the 16 sequences of two initial bases, the amino acid is already determined; choice of the third base is irrelevant. This flexibility in the last of the three bases to match up probably helps protect against errors when the pairing of the first two bases is strong enough to tolerate a mismatch at the third position. This notion has been formalized by Crick as the *wobble hypothesis*. The "wobble table" is a list of the permitted

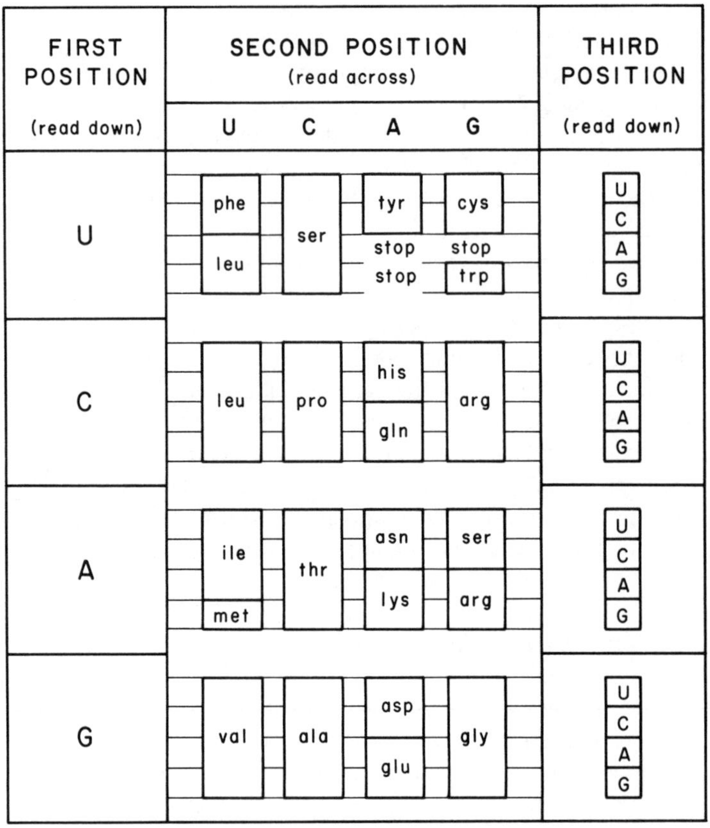

FIRST POSITION (read down)	SECOND POSITION (read across)				THIRD POSITION (read down)
	U	C	A	G	
U	phe leu	ser	tyr stop stop	cys stop trp	U C A G
C	leu	pro	his gln	arg	U C A G
A	ile met	thr	asn lys	ser arg	U C A G
G	val	ala	asp glu	gly	U C A G

ala	– alanine	gly	– glycine	pro	– proline
arg	– arginine	his	– histadine	ser	– serine
asn	– asparagine	ile	– isoleucine	thr	– threonine
asp	– aspartic acid	leu	– leucine	trp	– tryptophan
cys	– cysteine	lys	– lysine	tyr	– tyrosine
gln	– glutamine	met	– methionine	val	– valine
glu	– glutamic acid	phe	– phenylalanine		

Figure 4.5 The genetic code.

mismatches at the third base: an anticodon base of G will pair with U or C, U with A or G, and I (for inosine, an unusual purine found in many anticodons) with A, U, or C. Only the A-U and C-G third base combinations between anticodon and codon are exclusive.

The wobble concept admits the possibility that the set of anticodons, and hence distinct tRNAs, for a given amino acid in a given cell may not be in one-to-one correspondence with the codons. In some cases, fewer anticodons could match all the codons; in other cases, there might be extra anticodons. In fact the number of distinct tRNA species varies from one organism to another, in the range of 40 to 80 in procaryotes but running into the hundreds in eucaryotes.

4.5 REGULATION OF TRANSCRIPTION AND TRANSLATION

A cell's environment may change dramatically with time. Certain types of available nutrients may call for specific enzymes. Or the availability of certain raw materials may make their manufacture unnecessary. It is thus useful to be able to regulate the supply of various groups of enzymes, which can be done by controlling the rates of their transcription and/or translation. Modifications to these processes may also help overcome the effects of deleterious mutations. There is currently only fragmentary data and scant theorizing about regulation of genetic information processing in eucaryotes. This section considers some well studied phenomena in procaryotes.

4.5.1 The Operon Model

In the late 1950s the French biochemists Jacob and Monod sought a genetic explanation for a familiar adaptive phenomenon in *E. coli*. The bacterium feeds preferentially on glucose, even in the presence of other sugars. But when only lactose is available, three enzymes which process this sugar suddenly appear, allowing the cell to use lactose as a primary food source. This phenomenon is known as *enzyme induction*. If glucose is reintroduced in the growth medium, the lactose processing machinery shuts down.

Lactose could induce the production of its processing enzymes by initiating translation of continuously available transcripts. But it can be shown that induction affects transcription. It is, after all, more economical to avoid making both the transcripts and the enzymes unless they are needed. The speed with which such transcriptional control mechanisms respond depends in part on the short lifetime of procaryotic transcripts; other mechanisms may predominate in eucaryotes.

Lactose induces transcription by combining with and inactivating *repressor* molecules, which are always present in small number and normally block transcription. Jacob and Monod proposed that repressors combine with a site on the chromosome between the promoter and the three lactose-enzyme genes, thereby preventing RNA polymerase from moving down and transcribing the DNA. The experimenters called the site of repressor attachment an *operator* and the group of genes under control of the promoter-operator combination an *operon*.

The lactose or *lac* operon and its control points are diagrammed in Figure 4.6. The three genes code for beta-galactosidase (which splits lactose), galactoside permease (which facilitates transport of lactose into the cell), and galactoside acetylase (which has an unknown function). Notice that the repressor is itself a protein, translated from the transcript of the *regulatory gene*. Although this gene happens to be adjacent to the lac promoter, other regulatory genes can be quite remote from their operons.

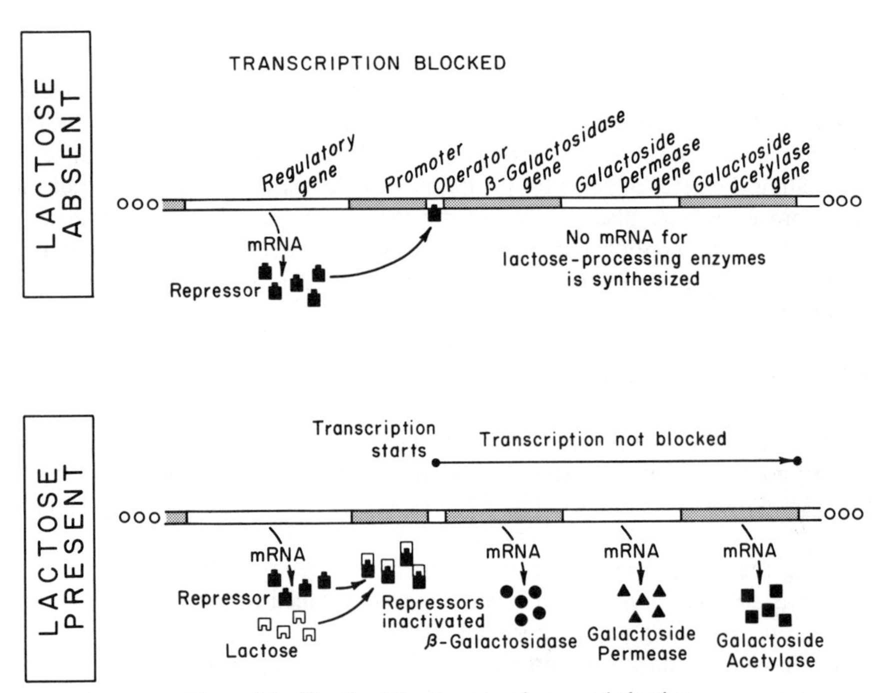

Figure 4.6 The *E. coli* lac operon and enzyme induction.

Enzyme induction is not a binary phenomenon. Uncombined repressor molecules are continually associating with and dissociating from the operator. Increasing amounts of lactose inactivate more repressor; the operator is then occupied less often, allowing a greater proportion of RNA polymerase molecules to pass through and transcribe the operon. A picture thus emerges of a finely tuned feedback control system.

Mutations in the regulatory gene which interfere with repressor production or function, or in the operator region which destroy the binding site, lead to unregulated, or *constitutive*, synthesis of the lactose processing enzymes. Some procaryotic genes are normally constitutive, although transcriptional control seems to be the rule rather than the exception. The lac regulatory gene is constitutive, for example, as are the genes that code for basic metabolic enzymes like those serving the EMP pathway. Constitutive protein synthesis is not influenced by environmental conditions; but its overall rate may be governed by a variety of factors, including the number of RNA polymerase attachment sites in the promoter, the lifetime of the transcript, and the accessibility of the start codon for initiation of translation.

4.5.2 Varieties of Transcriptional Control

The lac operon has now been found to exemplify only one of several related control processes affecting transcription in procaryotes. A first distinction is between *inducible* and *repressible* systems. Food molecules like lactose tend to induce the production of normally absent enzymes which are essential for utilization of the food. By contrast, the environmental appearance of an amino acid, like tryptophan, represses the normally active transcription of trp operon enzymes, which are required in the biosynthetic pathway that produces tryptophan. This situation resembles end-product inhibition, as described at the end of Chapter 3, except that repression is more economical in that it controls enzyme (actually, transcript) availability instead of enzyme function.

The basic difference between the mechanisms of induction and repression (in systems which employ a repressor molecule) is the state of the uncombined repressor. In inducible systems, repressors are normally active, combining with operators to prevent transcription unless they are inactivated by combination with an *inducer* (e.g., lactose and other sugars). In repressible systems, repressors are normally inactive and do not act at operators to prevent transcription unless they are activated by combination with a *co-repressor* (e.g., tryptophan and other amino acids).

The *E. coli* trp operon is shown in Figure 4.7. Transcription of the five genes is repressed by attachment to the operator of a repressor which has been activated by combination with tryptophan. Scarcity of tryptophan derepresses the system and transcription begins. But there is an unusual feature in this operon because most transcripts are aborted even before the start of the first gene is reached, unless the cell is virtually starved for tryptophan. This second level of control involves a sequence of about 160 bases between the promoter-operator and the start of the first gene; this sequence is called the *attenuator region* because it normally attenuates transcription, allowing only occasional readthrough by RNA polymerase. The way in which almost total absence of tryptophan causes inactivation of the attenuator, and thus nearly 100% transcription of the complete operon, is not fully understood. Presumably, attachment of a control molecule is involved.

As has just been seen, not all transcriptional control systems are based on molecules which attach to DNA to block mRNA synthesis. Such *negative control* systems contrast with those (like trp attenuator inactivation) exhibiting *positive control*. An operator may normally block transcription until an *activator* combines with it. If the activator must be complexed with an environmental substance before it can promote transcription, the system is of

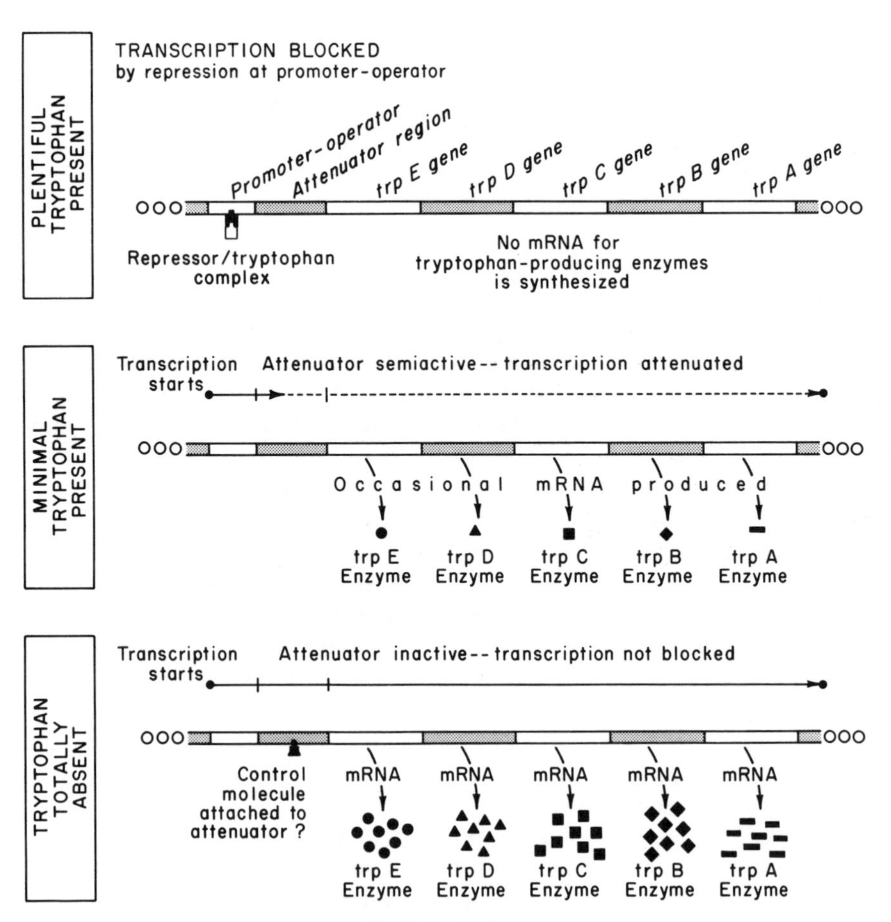

Figure 4.7 The *E. coli* trp operon.

the positive inducible variety; induction of some sugar operons in *E. coli* works in this fashion. In a positive repressible system, the activator normally combines with the operator but is inactivated (turning transcription off) by a co-repressor.

To complicate the picture further, some operons have genes in two or more different regions of the chromosome, possibly with multiple operators affected by the same repressor or activator. In other operons, extra promoters are found between genes, which allows constitutive or independently controlled transcription when the "front" of the operon is shut down.

Not all transcriptional control involves operators. Positive control sites are frequently found in promoters. One such site is now seen to account for a major gap in the original Jacob-Monod model of the operon, namely that the cell does not digest lactose if there is also any of the preferred glucose present. Even though the operator is presumably unblocked, something else is needed for lac transcription. The additional requirement is a Catabolite Activator Protein (CAP) which has positive inducible control action at a site within the lac promoter.

Thus inactivation of the repressor and presence of CAP are both prerequisite to lac transcription; the first of these is guaranteed by the presence of lactose, whereas the second requires the absence of glucose. The latter mechanism is indirect. CAP must actually be complexed with a nucleotide called cyclic AMP (cAMP) before it can activate the promoter. Synthesis of cAMP (from ATP via an enzyme called adenylcyclase) is somehow inhibited by one of the products of glucose catabolism. So more glucose means less cAMP and hence less activation of the lac promoter by CAP. The same phenomenon occurs in the promoters of other sugar operons, insuring that the bacterium will consume glucose first whenever it is available.

The CAP-cAMP activator thus differs from those activators and repressors which function at one or two operators. Although the mechanisms underlying both specific and more general attachment of activators and repressors are not fully understood, nucleotide sequencing of operators and promoters has revealed regular patterns at such sites, not unlike those at sites for RNA polymerase attachment (see Section 4.2.1).

4.5.3 Suppression

Suppression of the effects of deleterious mutations is carried out by a wide variety of mechanisms in procaryotes. Those which involve control of translation are emphasized in this discussion. It is first necessary to distinguish among three ways in which the effect of a mutation can disappear or seem to disappear. The first is *reversion*, in which the original nucleotide sequence is restored by a second mutation at the same point. A second mutation which restores a different codon for the proper amino acid may also be considered reversion. *Pseudoreversion* occurs when the second mutation occurs in the same codon as the first but, instead of restoring the original amino acid, specifies one sufficiently similar to confer near normal activity on the polypeptide.

In contrast to both forms of reversion, suppression involves mutations in two different codons. If these are in the same gene, the situation is called *intragenic suppression*. One example of this kind of suppression is the compensating frame shift mutation, of the sort used by Crick to determine the triplet reading frame. Another common form of intragenic suppression occurs when a second mutation restores protein function by compensating for a shape change introduced by the first mutation. A final example is the case where an AUG initiator codon has been destroyed by mutation, making it impossible to start translating the transcript; suppression can here take the form of a second change which creates a new initiator codon near the site of the original.

In *intergenic suppression* the second mutation occurs in a different gene. This may be an indirect compensating mutation, in which a change in the product of some other gene may allow it to substitute for the product of the first gene, or may open up some alternate metabolic pathway supplanting the one destroyed by the first mutation, or may alter cytoplasmic conditions in such a way as to allow the defective product of the first gene to function. In any case, indirect intergenic suppression does not change the product of the gene whose mutational consequences are being suppressed.

Direct intergenic suppression affects the translational machinery so that the mutated transcript is misread in such a way as to restore functional status to the polypeptide product. This phenomenon has been most thoroughly studied in the suppression of "nonsense" mutations in the *E. coli* gene which codes for the enzyme alkaline phosphatase. A "nonsense" mutation occurs when a codon specifying an amino acid undergoes a base change which turns it into a stop codon. Unless such a change occurs quite near the end of the polypeptide, the resulting truncated amino acid sequence is not likely to be functional. The particular "nonsense" mutations in question are called "amber" after an early term for the UAG codon.

If another mutation, called $su3^+$, is present in a bacterium with an amber mutation in the alkaline phosphatase gene, some (subnormal) enzyme activity is restored. Similarly a T4 virus with an amber mutation in the gene that codes for a protein in its shell will successfully infect *E. coli* cells only if they contain the $su3^+$ mutation. Such suppression is accounted for by abnormal tRNA.

Normal *E. coli* chromosomes contain genes for two forms of tyrosine tRNA, both having the same anticodon, AUG, which (by wobble) recognizes both tyrosine codons, UAU and UAC. These tRNA molecules are called the major and minor species because they are produced in substantially different amounts. The $su3^+$ mutation affects one of the two or three gene copies for the minor tRNA, changing the anticodon to AUC, which will base pair with the UAG stop codon. In this cell then, there are three species of tyrosine tRNA, one of which inserts tyrosine at one kind of stop codon. This relatively rare event affects all translation equally, causing the cell to grow rather slowly because some proportion of normal polypeptides are not terminated properly. (Some procaryotic genes have been discovered which end with two different stop codons, possibly to safeguard against the undesirable side effects of nonsense suppression by mutant tRNAs.) Some alkaline phosphatase activity is restored, however, since some of the mutant transcripts will be fully translated and be normal except for the appearance of tyrosine instead of the amino acid originally specified at the mutation site.

Although it is tempting to think of this rather amazing compensatory process as an intentional solution to a problem, it must be kept in mind that this and all suppression phenomena are the result of a chance co-occurrence of two mutations, whose high survival value, relative to either mutation separately, tends to propagate the suppressor strain. Mutant tRNA suppression is not limited to nonsense mutations; other types of nucleotide changes have been found to be subject to the same sort of suppression. Other kinds of direct intergenic suppression which work through modifications to the translation machinery may involve genes that help determine the structure of ribosomes by coding for their RNA or protein components. Mutant ribosomes may permit more flexibility in the base pairing of codons and anticodons.

Chapter Five

Simulation of Cellular Activities

In this chapter, the first to deal with specific programs for simulating models of biological information processing, attention is directed to relatively elementary biological activities. The first section considers three simulation systems for studying enzymatic control of biochemical reactions. Section 5.2 deals with studies which use simulation to explore models of the cell cycle. In this and subsequent chapters on simulation work the framework and terminology of Zeigler (1976), as previously presented in Section 2.3, are extensively employed.

5.1 SIMULATION OF BIOCHEMICAL REACTIONS

In all of the modelling and simulation activities discussed in this section, the *real system* under consideration is the complex network of biochemical reactions that make up a cell's metabolic pathways. Chemical reactions in general alter molecular structures by making and/or breaking chemical bonds. In biological systems, bond alteration is almost always assisted by one or more specific enzymes (as discussed in Chapter 3), whose availability is in turn subject to a variety of genetic control mechanisms (as discussed in Chapter 4). The work considered in this section deals primarily with enzymatically controlled biochemical reaction networks; it does not usually extend to the genetically controlled production of the enzymes themselves.

The temporal behavior, or *kinetics*, of the simplest types of chemical reactions can be derived analytically from the *flux equations*, which specify the nature of the interactions among the reactants, and from associated numerical parameters, which specify the initial concentrations of the reactants and the rate at which the reaction proceeds in a particular direction. Such *rate constants* usually vary with environmental conditions like temperature and pressure. A reaction which may proceed in both of the two possible directions, under appropriate conditions, is termed *reversible*.

As an example, consider the unidirectional reaction in which one molecule of A combines with one of B to produce one of C. The flux equation is

$$A + B \xrightarrow{k} C \tag{5.1}$$

where k is the rate constant. The associated rate equation is

$$d[C]/dt = k[A][B] = -d[A]/dt = -d[B]/dt \tag{5.2}$$

where the concentration of a reactant is denoted by its name in brackets. If the reaction is reversible, generating A and B from C with reverse rate constant k', the rate equation becomes

$$d[C]/dt = k[A][B] - k'[C] = -d[A]/dt = -d[B]/dt \tag{5.3}$$

Notice that the rate of change in the concentration of C is now a function of its own concentration as well as of the product of the concentrations of A and B.

Equations (5.2) and (5.3) are susceptible to analytical solution. But even a slight increase in complexity makes continuous simulation (numerical integration) the only feasible method. Addition of an enzyme to the above system, for example, may lead to reactions in which the enzyme (E) reversibly associates with A, the AE complex reversibly associates with B, and AEB reversibly yields E and C. The kinetics of this system are governed by a system of six differential equations.

All three of the simulation projects discussed in this section are designed to facilitate the description of reaction systems and their subsequent analysis by simulation. These projects are thus simulation *systems* (as the term was used in Section 2.2.3) rather than programs written to study one particular model. In selecting for study one group of reactions among the thousands occurring in a cell the investigator establishes an *experimental frame*, exploiting the simulation system as a tool for exploring a particular model. The assumption that it is meaningful to study any group of reactions in isolation must of course be questioned critically in each case.

The *base model* for all of these investigations not only consists of the entire cellular metabolic network, but also exists at the level of individual molecular interactions. To derive a *lumped model*, the critical simplification procedure is typically to group the discrete state changes (actually appearances and disappearances) of molecules of any one type. This information is represented in a single continuous variable, the reactant's concentration. The use of a digital *computer*, both to create data structures which describe models and to carry out the actual simulation, varies among the three systems. The following individual discussions highlight relative strengths and weaknesses in these simulation systems, particularly with respect to their interactive capabilities.

5.1.1 The Garfinkel System

For nearly two decades David Garfinkel, with his colleagues and students, has been using simulation to study biochemical reaction systems (Garfinkel 1968, 1977). The 1968 paper describes a FORTRAN program which implements a machine-independent language, subsequently named BIOSSIM (Garfinkel et al. 1978) for simulation of chemical kinetics.

The versatility of BIOSSIM, and the way in which a modeller communicates with the system, can best be understood by working through a simple example (very similar to that given in the 1968 paper). Each of the indented lines below is a statement, in the implementation of a model and specification of a simulation run, for a system involving two-equation enzymatic modification of a substrate (S) to produce a product (P). Each statement is discussed as it is presented. (It should be noted that positioning of data is usually critical; the "T" in the first statement below is in "card column 1." Also note that "/" is usually a delimiter rather than a division operator.)

```
T        A SIMPLE REVERSIBLE ENZYME SYSTEM
```

A title for the model is designated by a T in the first column. Other comments may be interspersed at will, using the FORTRAN notation of a C in the first column.

```
         E + S = ES
```

In the first equation, enzyme and substrate combine to yield the ES complex. The "=" designates a reversible reaction; otherwise a "—" is used. There may be up to 12 reactants on each side of an equation. If there are no reactants on the left, those on the right are understood to appear spontaneously; absence of right side reactants is associated with disappearance of material from the system.

```
         1/1.0E8 2/1.0E3
```

The (forward and reverse) rate constants for an equation may be given as any valid FORTRAN constants; 1 and 2 are half-reaction identification numbers. Another way of specifying rate constants is shown later.

```
         ES = E + P
```

In the second reaction, also reversible, the ES complex dissociates to yield enzyme and product.

```
S        P                                2.0
```

Two molecules of P are actually produced in the second reaction. This is indicated by assigning a stoichiometric coefficient (S in column 1) of 2.0 to P.

```
E
```

An E in the first column terminates the equations portion of the model specification. In an option not shown, any reactant may have its concentration held constant by giving its name in a statement beginning with an H.

The remainder of the example provides additional numerical data and governs aspects of the output from a simulation run.

```
03      /                      1.0E5
04      /                      1.0E8
```

The above two statements establish the rate constants for the second equation, numerically identified as the third and fourth half-reactions. Initial concentrations of reactants are specified in the next three statements.

```
        E                      1.0E-1
        S                      1.0
        ES                     0.0
```

Initial concentrations may also be specified as coefficients of reactants in the equations themselves; but this usage is confusing since the numbers then look like stoichiometric coefficients. Dimensions of numerical values (e.g., moles, liters, seconds) are not known to the system. The user is responsible for consistent quantification. If a parameter has been specified more than once, the last value is used. The program detects missing parameters and, at user option, either aborts or uses values of zero.

```
LABEL RUN 2 OF 5
TIME    5.0E-3
```

The above two lines provide an arbitrary identification of the simulation experiment and a duration for computation time. A format similar to the TIME statement is used for optional specification of the initial step size (default 10E-9) and the allowable truncation error (default 10E-6) for the numerical integration routines (see below).

```
G       E                      5.0E-4E
G       ES                     5.0E-4C
```

Statements beginning with G request a graph of concentrations of the named reactants at the specified time intervals, using the indicated symbols. All curves are superimposed on a single plot. There is also an option for dumping tabular information about desired reactants at specified intervals. Finally, a Z in column 1 terminates model and simulation specifications. Subsequent runs can be stacked, with specification only of model characteristics or simulation parameters that are to be changed.

The actual simulation run of a model specified in BIOSSIM is accomplished in three phases. First, the BIOSSIM statements are passed to a "generator" program, which produces FORTRAN data structures and expressions for the differential equations underlying the model reactions. (The equations can be modified by user intervention.) The resulting FORTRAN code is then processed by the local FORTRAN compiler, producing machine language routines suitable for linking to the run-time monitor or "operating" program. In the third phase, this monitor makes a variety of validity checks and, if appropriate, initiates and governs the simulation run.

A first-order Euler method is used to solve the differential (now difference) equations. The step size is initialized to the user specified value (or its default), but is adjusted as necessary to remain within the user specified error tolerance (or its default). Time is incremented until the end of the run is reached. The operating program also monitors reactant concentrations and will abort upon encountering a negative value, unless the user has asked that such situations be ignored. Tabular (dump) and graphical output data are accumulated and processed. Finally, there is provision to interface the operating program with user supplied special purpose FORTRAN subprograms.

Garfinkel and his fellow investigators have used BIOSSIM to study a variety of biochemical models. The implementation of the language has evolved in the process of its application; a recent version is available through the SHARE Program Library Agency (Garfinkel et al. 1978). Attempts to study relatively complex models have led to the development of some ancillary computer-based modelling techniques, discussed in the 1977 and 1978 references. The latter paper also provides a complete set of references for, and a brief description of, a major recent project, modelling normal and abnormal metabolism in rat and dog hearts.

Critical evaluation of the Garfinkel system is here based on the last full description in 1968. Reprogramming (mentioned in the 1978 reference) could have partially or fully eliminated some of the drawbacks discussed below. First, however, several quite attractive features of BIOSSIM should be emphasized. Since it is written entirely in FORTRAN, the system is highly portable. Further, the simple numerical integration routines minimize computation cost yet provide sufficient accuracy (maximum error of 0.1%) for most biological modelling applications. Model specification, in terms of flux equations and numerical parameters, is generally simple and straightforward; but stoichiometric coefficients rather than initial concentrations should appear in the equations.

The major deficiencies of BIOSSIM, particularly for the inexperienced user, are its rigid format and batch oriented approach to simulation. It appears that failure to position critical characters (e.g., "/") in particular columns can cause misinterpretation of the statement and/or termination of the run. The description of the system suggests the presence of some diagnostics; but few details are given.

Because each individual simulation run must be coded on a deck of cards (or equivalent) and then handed to a remote process involving three job steps, BIOSSIM inevitably lacks nearly every desideratum of interactive simulation outlined in Section 2.2.3 above. There is no dynamic modification of model specification, no on-line editing, no simulation monitoring, and no storage and retrieval of models other than in their BIOSSIM language forms. The work described in the next section was primarily oriented toward adding such interactive facilities to the already desirable features of BIOSSIM.

5.1.2 The Huneycutt System

For his Master's thesis research at the University of Alberta, Huneycutt (1976) employed the Garfinkel design as a point of departure. In addition to an interactive superstructure, Huneycutt's Interactive Biochemical Simulation System (IBSS) added some higher level control structures to the basic reaction kinetics. Since this project has not previously been described in the literature (except very briefly by Sampson and Dubreuil 1979) IBSS is given a rather detailed treatment here.

Huneycutt actually designed two interactive interfaces with the simulation routines of IBSS. One of these, involving selection of items from displayed menus of options, was suited to sophisticated CRT type terminals. The other is usable from a standard teletype variety of terminal and intended for the more portable version of the system. The present discussion deals only with the system as implemented with the "teletype monitor."

The IBSS syntax for defining a reaction system and its simulation parameters is in part a slightly streamlined and rationalized version of Garfinkel's BIOSSIM language. The following 7 IBSS statements, for example, accomplish essentially the same tasks as the 15 BIOSSIM statements in the example of the last section.

```
INIT
INPT        E + S <—> ES /1E8, 1E3
INPT        ES <—> E + 2P /1E5, 1E8
INPT        E = .1
INPT        S = 1
RUN         20
GRPH        0, 20, E, ES
```

Note the use of "<—>" for reversible reactions ("—>" indicates unidirectionality) and "=" for initial concentrations. Duration of the run is specified in terms of number of time steps (in this case 20) instead of computer execution time. Further, note that the stoichiometric coefficient of the product P is shown directly in the second equation, that the forward and reverse rates follow each equation, and that a single GRPH command specifies the parallel plotting of two (or more) reactants, in this case for the full 20 time steps of the run. Finally, although the above command sequence has been aligned for readability, IBSS generally tolerates a much freer format than BIOSSIM.

The two systems offer comparable features for the maintenance of constant concentration (in IBSS such reactants are designated BUFFERED), for user specification of initial step size and maximal error, and for input of reactant material from the external environment. The on-line character of IBSS entails its additional run-time commands, HALT and CONT(inue). Modification of a model, via the editing facilities described below, may occur between halting and continuing.

"Reaction monitor statements" and "macro commands" provide important extensions to the range of models that can be studied by a biochemical simulation system. In IBSS both these facilities are used independently of the reaction definition and simulation parameter statements illustrated above. The system automatically links all three kinds of user specifications during execution.

Reaction monitor statements consist of a label, a relational expression comparing the current value of a model variable with that of another variable or a constant, and a macro command reference. The comparison is checked at each time step; whenever it is true the appropriate macro command (or sequence of commands) is executed. For example:

```
C3      P .GT. S      M17
```

has label C3 and causes execution of macro commands, beginning at M17, when the concentration of P first exceeds that of S;

```
CMD9     d(E) .EQ. 0      M3
```

executes macro command M3 when the first difference of the concentration of E becomes zero; and

```
A      TIME .GE. 100      M22
```

branches to M22 when the model time step reaches 100.

The macro commands themselves provide a simple programming language capability. (The term *macro* refers to this higher-level aspect of the commands, and should not be confused with the macro features typically found in assembler languages.) The values of any model variables can be tested, adjusted, or reset. Reaction monitor statements can be deleted or inserted. The following macro command sequence, excerpted from the specification of the *lac* operon model to be discussed later, illustrates a variety of the capabilities.

```
M10: SEXR       TRANSLAG, 1.0
M11: DMS        C1
M12: IMS        C4 LAC .LT. 1.0 M40
M13: CONT
M20: INPUT      TRANSLAG = 0
M21: COMP       CAMP .LT. 4.0 M23
M22: INC        EXT, BGLD, 1.0
M23: CONT
```

Macro commands M10 through M13 are executed whenever the concentration of lactose, monitored in monitor statement C1 (not shown), exceeds a specified threshold. M10 *S*ets the *EX*ternal-input *R*ate for a substance whose concentration is to be used as a timer; having thus been started, the timer's accumulated concentration is compared against the desired final value by a monitor statement. TRANSLAG refers to the transport lag before production of *lac* mRNA is initiated by the appearance of lactose. M11 and M12 respectively *D*elete the existing *M*onitor *S*tatement C1 and *I*nsert a new *M*onitor *S*tatement C4 (which looks for a new lactose level). M13 *CONT*inues the simulation from the point of interruption by the now deleted C1.

Macro commands M20 through M23 are executed when the monitored TRANSLAG timer expires. M20 resets the timer. M21 *COMP*ares the concentration of cyclic AMP (which will be high if glucose is absent) to a threshold; if CAMP is sufficiently large, M22 *INC*rements the production rate of the *lac* enzyme betagalactosidase, denoted BGLD.

When not actually simulating an implemented model, IBSS offers the user a variety of simple file management commands. Programs specifying individual models can be stored, retrieved, destroyed, and combined. The last option, "merging" of models, provides a facility for separately testing relatively independent portions of a complex metabolic network, then combining them automatically.

Another set of IBSS commands lets the user scan and edit the currently specified model at any time prior to, or during interruption of, a simulation run. Data concerning individual reactants, equations, or macros can be displayed. Reactants and their coefficients can be changed, or whole terms added to or deleted from equations. Macro commands and equations can also be added or deleted. Altogether, including execution and file management functions, the IBSS monitor interprets about 30 distinct types of commands. Since some of these have a rather complex syntax, the user must acquire experience with the system in order to exploit its capabilities fully.

The IBSS software is organized around a half dozen modules which communicate via a common data base composed of tables that describe a model and drive the system. The I/O-Monitor module defines an implementation dependent format for user interaction. The simplest form of this "front end" module is the teletype monitor referred to above. The System Monitor is a module which accepts suitably translated user commands from the I/O-Monitor. The System Monitor controls definition, modification, storage, and retrieval of model specifications, as well as initiation and supervision of simulation runs. All execution sequences in IBSS, except for those within the I/O-Monitor, begin and terminate in the System Monitor, which thus serves as the control center of the simulation system.

Remaining modules include the Equation Compiler, which translates a model description into table entries, an Integrator, which numerically solves the differential equations and calls monitor statements when appropriate, and a Run Record, which inserts simulation run data into the system's mass storage area at regular, user-specified intervals. Finally, there is the Mass Data Handler, which manages storage and retrieval of programs specifying models, and retrieval of data saved by the Run Record module.

The IBSS data structures are a set of cross referenced tables, designed to allow both rapid access during simulation and dynamic model modification during user interaction. Extensive use of pointers makes it possible to access any information in the system by a path consisting of at most two pointers and an index. Flags are used to identify table entries associated with deleted model characteristics, reducing the need to restructure tables when the model changes during simulation.

There are four tables, for Formulas, Equations (half reactions), Reactants, and Data. A formula name provides access to the proper entry in the highest-level table. A Formula record consists of a string of pointers for each associated half reaction, forward and/or reverse. The pointers refer to an Equation table entry and a series of paired entries in the Reactant and Data tables, one pair for each term in the half reaction. The Data table records the stoichiometric coefficient of the term. In the Equation table record there is a rate for the half reaction and a string of pointers to the Reactant table, one for each reactant. The Reactant table is ordered by reactant name. Each record contains this name, the current concentration of the reactant, any rate(s) of external supply or disappearance of the substance, an indicator of whether the reactant is buffered, and a set of three pointers which access entries in the Equation and Data tables that facilitate computation of the first derivative of the reactant concentration.

IBSS was subjected to a number of test runs, some of which paralleled those reported by Garfinkel (for comparison of numerical accuracy). Huneycutt also used IBSS to design an elaborate model of transcriptional control at the *lac* operon in *E. Coli* (see Sections 4.5.1 and 4.5.2). Such an undertaking was possible only because the IBSS reaction monitor and macro command facilities introduce a level of feedback control over enzyme availability comparable to that in natural genetic systems. As specified in IBSS, the model consisted of 7 flux equations, 6 external inputs, 4 reaction monitors, and 30 macro commands (some of which were shown above). The model dynamics included both the induction of the three *lac* enzymes and CAP repression in the presence of glucose.

Unfortunately, no results of experimentation with Huneycutt's *lac* operon model can be reported. A combination of time constraints and biological inexpertise made it impossible to discover realistic parameter values and associated predictions of model behavior. The computer scientist needs to work together with the biological specialist in undertakings of this scope.

Evaluation of IBSS must begin with the acknowledged debt to Garfinkel, whose BIOSSIM provided the basis for Huneycutt's work. The two systems use the same simple numerical integration routines and achieve the same, limited accuracy. The format for description of flux equations and associated numerical parameters is similar, and would quickly become routine for a user of either system, even one with no programming experience. Both systems are written in FORTRAN and therefore are relatively portable. In their most portable versions, both have limited graphical output capabilities (line-printer plots). The two major new contributions by IBSS have been the focus of much of this section. The modelling contribution is the superstructure of adaptive control. The simulation contribution is the interactive mode of operation.

5.1.3 The Cassano System

Presumably unaware of Huneycutt's work, Cassano (1977) also undertook to extend the model description and interactive simulation facilities of a system like Garfinkel's. The Cassano system is organized around four main programs, running on two different computers. A model is implemented via a SNOBOL program called TRANSLATE, which runs on a large central computer (IBM 370/165). The output of TRANSLATE is transferred on magnetic tape to a minicomputer (PDP 11/45) for subsequent simulation, dynamic model modification, and graphical output.

Models are described to TRANSLATE in a straightforward syntax that has the same reaction definition, external input, and buffering capabilities of the two previous systems. Reversible reactions are described as two separate half-reactions. Thus the first equation in the model previously used as an example would be defined by the following two statements in the Cassano language:

```
E + S → ES     (1.E8*'E','S')
ES → E + S     (1.E3*'ES')
```

Compensating for this somewhat extended notation are provisions for arbitrary FORTRAN expressions as rate constants and for compact definition of systems governed by "standard" enzyme kinetics.

The second main program, called DIFEQ, interacts with the user in order to establish such parameters as step size and initial concentration values. DIFEQ then carries out the numerical integration of the implemented model, using a more sophisticated algorithm than those of the previously considered systems. DIFEQ actually runs under control of the MONITOR program, which can effect model modification during pauses in a simulation run. Finally there is PLOTFILE, which graphs user selected variables on a CRT terminal. Plots obtained during a run can influence adjustment of model parameters.

Cassano describes the implementation and simulation results for a model of purine metabolism involving 12 (half) reactions. Parameter values were obtained from the literature and "reasonable guesses." Although the results showed a "qualitative consistency with the current understanding of the dynamics of purine metabolism," the variety of sources employed for parameter value assignment prohibited any detailed interpretation. A model generating useful predictions would be based on the laboratory work needed to establish a coherent set of parameter values in a single biological system. Thus, as is frequently the case, simulation and laboratory experimentation are mutually dependent factors in advancing the understanding of a biological information processing system.

The Cassano system has desirable features in common with BIOSSIM and IBSS, including a natural syntax for model description and interactive simulation monitoring (as in IBSS but not BIOSSIM). Facilities for easy description of complex kinetic relations enhance the modelling capabilities of this system. On the other hand, it lacks the adaptive superstructure of IBSS. The biggest shortcoming of Cassano's system (at least from the standpoint of the community of potential users) is its "home grown" character. Direct portability would presumably require a comparable pair of machines using comparable operating systems. Some reprogramming would probably make the entire system available on the PDP 11 class of machines (as is the interactive neural network simulation system discussed in Section 12.1.2). Clearly this lack of portability does not detract from the utility of the system at its home installation. But even there, the requisite tape transport between model specification and simulation must sometimes become a nuisance.

5.2 SIMULATION OF THE CELL CYCLE

As discussed in Section 3.3.1 above, the cell cycle theory proposes that cells go through a series of at least four phases prior to and including actual mitotic division. There have been a number of models of the cell cycle, some of which have been studied by simulation. Two relatively recent simulation studies are considered in this section.

There have also been some simulation systems, along the lines of those treated in the previous section, developed for the purpose of specifying and testing models of the cell cycle. Two of these are the batch-oriented system of Donaghey and Drewinko (1975), whose cell cycle model is the first discussed below, and the interactive simulation system of Evert (1975). The model description languages and other features of these systems are treated only briefly in the referenced papers. In the context of the biochemical simulation systems just discussed, the cell cycle systems have no interesting new features. For these reasons the remainder of this chapter focuses on individual models of cycles in specific cell types.

5.2.1 A Model of the Lymphoid Cell Cycle

Donaghey and Drewinko (1975) developed their simulation system primarily to study a model of the cell cycle in a cultured line of cells derived from tumerous tissue in a human lymph node. Studies of such cell lines are important in at least two ways. First, there is usually extensive laboratory data on tissue cultured cells, allowing unusually thorough validation of models of cell behavior. Second, any increase in understanding of the abnormal cell division behavior of tumor cells is of evident value to medical science.

The *real system* for this modelling enterprise is a particular group of cells, growing and dividing in laboratory culture. The *experimental frame* for Donaghey and Drewinko limits attention to various measures of the cell cycle, including the following: mean and distribution of full cycle duration; length of individual phases of the cycle; and proportions of the cells in the various phases, including a "resting" (G0) state. Such data may be also be collected for a culture of cells that has been artificially synchronized by chemical additives which block DNA synthesis. When the effect of the additive disappears, most of the cells will enter S (DNA synthesis) phase and proceed approximately in step for a few cycles. Although it is easier to measure cell cycle parameters in synchronized cultures, the synchronization process can itself distort the cycle. Donaghey and Drewinko tested their model against data for both asynchronous and synchronous cell populations.

The *base model* for this work includes all the metabolic characteristics of the individual cells. Extensive simplification generates a *lumped model* in which a given cell is characterized only by its current state in the cell cycle. Further, each model cell initially represents a group of about 10 real system cells, all in the same state. As cells divide, the simulation system periodically revises the representation scheme, keeping computer storage requirements approximately constant. Since the model is discrete in both space and time (another simplification), the discrete time simulation paradigm is employed.

The model cycle has nine states and a stochastic state transition function. After mitosis, approximately 10% of the daughter cells go into an absorbing state, G0, and do not divide again. The remaining 90% of the cells proceed along a linear path through the basic cycle: early and late G1; early, middle, and late S; early and late G2; and back to M(itosis). A given cell's time in any one of these eight states is governed by a probability distribution. Each of the three S subphases, for example, has a duration that is normally distributed with a mean of 4.5 hours. The other states have skewed triangular distributions with means ranging from 3 hours (late G1) to .567 hours (M). An average of 1.6 cells emerges from M for each cell that enters that state, since cells do not always divide successfully.

For experiments simulating asynchronous growth, 100 cells were given initial states distributed uniformly across the cycle (i.e., proportional to state duration). A typical run lasted for 250 hours of simulated time, with data reported once per hour. To simulate synchronization, cells were prevented from moving into S phase from late G1; those in early S were halted there, while all other progressions through the cycle proceeded normally. Comparisons of simulation and laboratory results (e.g., asynchronous cycle duration of 27.94 vs. 27

hours) gave the authors sufficient confidence in their model to plan experiments simulating effects of chemotherapeutic agents on the tumor tissue.

5.2.2 A Model of the Fibroblast Cell Cycle

Alberghina (1978) has developed a model of cellular dynamics during growth of fibroblasts (connective tissue cells) in the mouse. This is not the only *real system* for which Alberghina's model has demonstrated replicative validity; with adjustment of parameters, it can also simulate bacterial growth and division. The *experimental frame* for this work is similar to that of the Donaghey and Drewinko project, the major variables being times spent in various phases of the cycle. The *lumped model* is less simplified, however, since a cell's state is characterized by amounts of DNA, ribosomes, and polypeptides in the Alberghina model. This model is actually at a level of resolution intermediate between that of Donaghey and Drewinko and those of the more detailed cellular models treated in Section 7.1. As in those projects, the time base is here continuous.

Alberghina worked first with a simplified version of her model, suited to study of growth parameters in a mass of cells. She then refined the model for study of the division cycle in a single cell. The first version of the model is specified by three differential equations, respectively governing the number of ribosomes, the number of amino acids polymerized into proteins, and the amount of DNA (expressed as a fraction of the normal diploid total).

$$d\text{RIB}/dt = \text{R1}(\text{A1}(\text{DNA})\text{-RIB}) - \text{RIB}/\text{T1} \tag{5.4}$$

$$d\text{PAA}/dt = \text{R2}(\text{RIB}) - \text{PAA}/\text{T2} \tag{5.5}$$

$$d\text{DNA}/dt = \text{PAA}/\text{A2} - \text{DNA} \tag{5.6}$$

In the above equations, T1 and T2 are time constants for degradation of RIB and PAA respectively. R1 and R2 are the rates of ribosomal production and activity respectively. A1 is the amount of RIB required for self-sustaining production. And A2 is the amount of PAA required per cell.

Tests of this preliminary model with an initial mass of 1000 cells showed good agreement with mouse fibroblast cell dynamics in both active growth and "resting" modes. In the growth situation, the cell cycle length was about 15 hours.

To study the behavior of a single cell, the above model was modified to reflect a "two subsystem" view of the cell cycle. The A1(DNA) term in Equation 5.4 is replaced by PAA, making the ribosome-protein subsystem self contained. Equation 5.6 is replaced by a set of rules governing activity in the DNA replication subsystem. When PAA reaches a specified threshold, there is stochastic initiation of DNA replication at rate R3; replication stops 1/R3 minutes later. When replication terminates, a specified delay period occurs before cell division.

Simulation experiments with this model also produced data in good agreement with observed cell cycle dynamics. Division occurred at about 900 minutes, the same point in simulated time as the population average in the preliminary version of the model. Alberghina discusses two tasks for the future: a return to the population level via a model which keeps track of individual cell ages and dynamics, and a further development of the single cell model in order to incorporate known details of ribosome synthesis.

On the evidence of this section, modelling and simulation of the cell cycle is still in its infancy. The Donaghey-Drewinko and Alberghina models make only modest explanatory contributions; they are both largely epiphenomenal in character. It is not difficult to mimic average phase durations for some cell type whose cycle has been studied in the laboratory. Putting a model cell through such simulated paces neither explains the mechanisms underlying the cycle nor predicts the behavior of cell types not yet studied. Useful models must embody testable hypotheses about reality. Donaghey and Drewinko hypothesize that the cell cycle phases can be subdivided; the next step is to discover productive hypotheses about the mechanics underlying the probability distributions of the subphases. Alberghina takes some of the mystery out of the duration of G1 by incorporating a macroscopic treatment of the biochemistry; and her "independent subsystem" hypothesis is explanatory in character. But the expected durations of S and G2 remain simple parameters of the model. Thus, although both of these modelling enterprises are moving in the right direction, there is considerable distance to go.

Chapter Six

Storage and Transmission of Genetic Information

Much of the biological theory covered thus far has been of a sufficiently general character to apply to most organisms, although care has been taken to distinguish between procaryotes and eucaryotes (and Chapter 4 treated operons particular to *E. coli*). This chapter initiates study of more specialized information processing mechanisms in various classes of organisms. First, there is an introduction to the ways in which genetic information can be localized, or "mapped," on a chromosome.

6.1 GENETIC MAPS

Since chromosomal DNA is an unbranched chain of nucleotide pairs, it is not surprising that the genetic information in a chromosome can be depicted as a linear sequence of genes. It *is* surprising that such "genetic maps" were developed long before the structure, or even the identity, of the genetic material was known. Mapping techniques need not involve sophisticated biochemical measures. Early twentieth century investigators constructed genetic maps from simple observations. Such observations depend on the presence of a *marker*, a physical characteristic attributable to a particular mutation of the gene in question. Eye color and wing shape markers were commonly used in the early genetic mapping work with the fruit fly *Drosophila melanogaster*. Progress in biochemistry now permits detection of much less obvious markers, involving as little as a single amino acid change in a polypeptide.

Discussion of genetic mapping requires some more precise terminology. The word *gene* is variously and loosely used, denoting sometimes a particular physical trait, sometimes a specific nucleotide sequence, and sometimes a position on a chromosome. For the last of these meanings a better term is gene locus, or just *locus*. The locus of the genetic information determining some physical characteristic can differ among organisms of the same species. Sometimes a region of the chromosome gets reversed by a process known as *inversion*. Or part of one chromosome may be joined to another, by a mutation called *translocation*. There are even *deletion* mutations, in which part or all of a locus physically disappears. (Such large-scale structural mutations are distinguished from the point mutations that involve a single nucleotide pair.)

It is thus useful to differentiate between the locus and the "content" of a gene. The different nucleotide sequences that may occur at a given locus are called *alleles*. Different alleles may not result in detectable physical differences. Some nucleotide changes do not even alter the amino acid sequence. In other cases, the information encoded in an allele may never be physically expressed. When geneticists can breed strains of an organism that have detectably different alleles at a given locus, these markers can be used in determining the relative position of that locus, by means to be discussed shortly.

Knowledge of genetic loci can be useful in several ways. First, a detailed map of a "normal" chromosome can assist in uncovering the nature and extent of various mutations. Another result of genetic mapping is increased insight into the organization of genetic information, such as the clustering in bacterial operons of genes with related functions. The location of genetic information can affect the timing of its expression (see Section 6.2.2). Finally there is basic scientific curiosity; an organism cannot be thoroughly understood without knowledge of the identity and location of all its genes. A complete genetic map has so far been achieved only for a few simple viruses.

6.1.1 Mapping by Recombination Analysis

Most genetic mapping has been accomplished through experiments in which recombination of genetic material brings together in a single chromosome two or more markers that were previously found on distinct chromosomes; alternatively, a recombination experiment may separate markers previously found on the same chromosome. Recombination does not exist for the convenience of genetic-map makers; it is the major source of new patterns of genetic information, patterns which represent competing alternatives in the evolution of species.

The physical recombination mechanism varies with the level of organism being studied. In diploid eucaryotes, crossing over occurs during meiosis (Section 3.3.3). Bacteria can exchange chromosomal material in several ways

(Section 6.3.2). Probably the simplest form of recombination occurs during simultaneous infection of a cell by two viruses of the same type. Since the principles of genetic mapping by recombination analysis are the same in all organisms, the "viral cross" will be used to illustrate the rest of the discussion.

Viral markers include structural variations, which can be seen with the electron microscope, and behavioral differences, like those associated with the T4 *r* region discussed in Section 4.4.1. As a hypothetical example, consider two viral chromosomes, one of which has a marker *d* at some locus and the other a marker *k* at another locus. A geneticist would describe these two genetic compositions as *d*+ and +*k*; the "+" denotes the "normal" allele at the other locus in each case. If chromatids from these two chromosomes come together in a host cell, it is not unlikely they will exchange some comparable genetic material, by breaking and rejoining at corresponding points, as shown at the left in Figure 6.1a.

In a large number of such potential exchanges, some proportion of the new chromatids will have both markers (*dk*) and some will have neither (++); both of these types are *recombinants*, containing alleles which did not previously occur together. Recombination analysis assumes that crossover is a random event that occurs with equal probability along the whole length of the chromosome; this assumption is only a reasonable approximation of the truth. Under the assumption, however, the proportion of recombinants after simultaneous infection should be directly related to the distance between the loci. If the markers are closer together (like *h* and *k* at the right in Figure 6.1a) random crossover will produce fewer recombinants. At the other extreme, every crossover would do so if the markers were at opposite ends of the chromosomes. Genetic maps are actually calibrated in units of recombinant proportions, so that finding 80% *dk* and ++ chromosomes would place *d* and *k* 80 map units apart. The same result should obtain in a cross of *dk* and ++ parental types, where *d*+ and +*k* would be the recombinants.

The above analysis is oversimplified and incomplete. First, it was assumed there would be at most one crossover point. If the parental chromatids actually break and rejoin more than once, as shown in Figure 6.1b, some crosses between the markers could have their effects reversed by a second cross. Such multiple crossovers are usually sufficiently rare that recombinant frequencies provide adequate data. Map refinement is also possible, since data can be gathered about a third marker *h* that lies in the interval between *d* and *k*, as shown in Figure 6.1a. In theory, the distance in map units between *d* and *k* should be the sum of the distances between *d* and *h* and between *h* and *k*. A sufficient number of such comparisons can produce a reasonably accurate genetic map.

The above type of experiment is called a *two factor cross*. Three such experiments are needed to determine the relative positions of *d*, *h*, and *k*. And it is still unknown how these three loci are arrayed relative to any other marker. If *r* stands for a position at the right end of the chromosome, two more crosses may

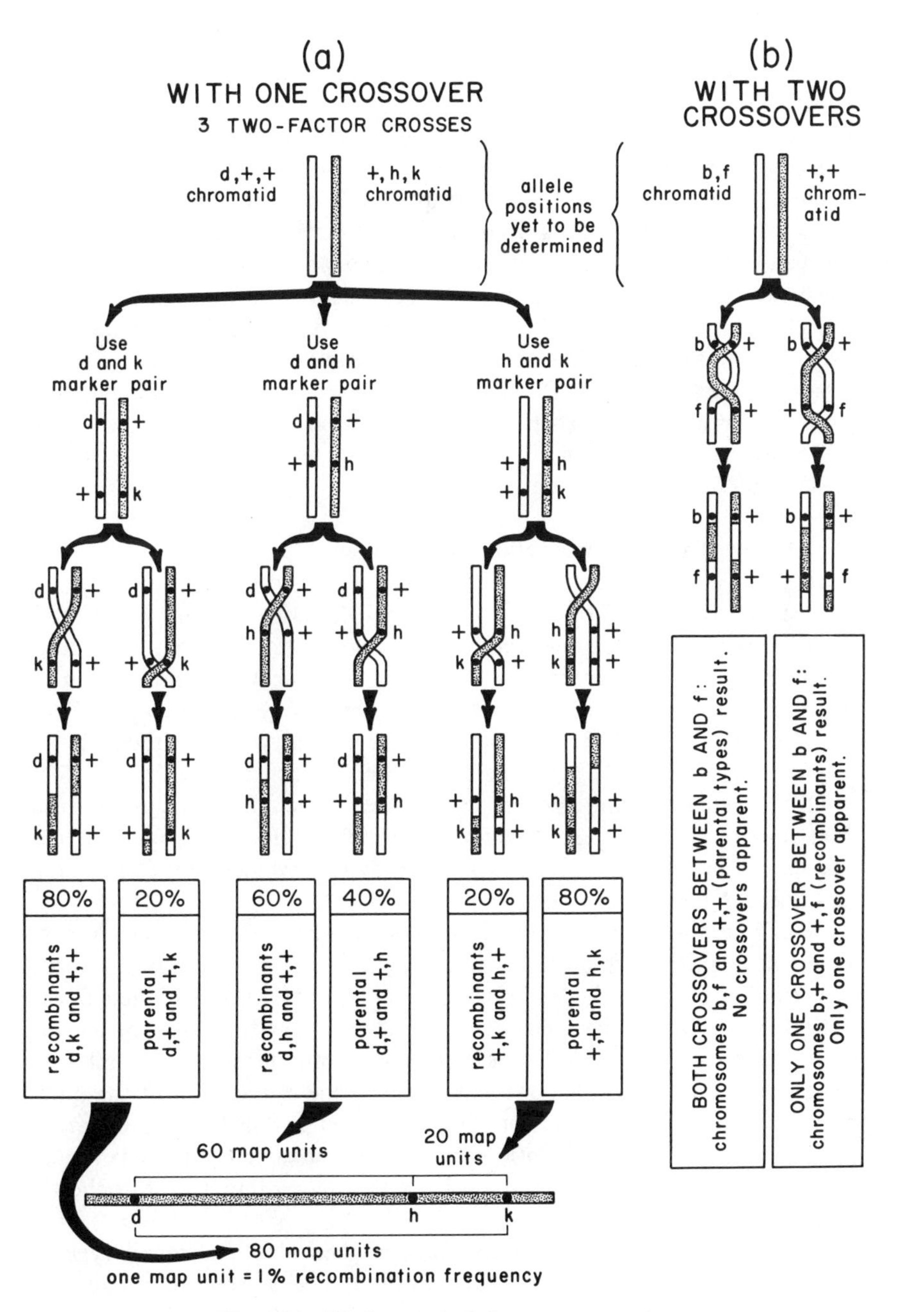

Figure 6.1 Viral crosses and chromosome mapping.

be required to determine if the order of the four loci is *dhkr* or *khdr*. Further intricacies arise when the map is circular, as is true for most bacteria. The radically inconsistent map distances, which place one "end" of the chromosome adjacent to the other, eventually provide sufficient evidence for a circular map. (It will be seen later that a circular map does not always imply a circular chromosome!)

The number of experiments can be reduced, and map accuracy increased, by the use of *three factor crosses*. For the loci used above, one such cross would be between *dhk* and $+++$. Some of the rare double recombinants now show up as $d+k$ and $+h+$; and the relative order of the three loci can be immediately established from the proportions of all six recombinant classes.

Since recombination is not limited to gene boundaries, crossing over can occur between two markers in the same gene. Although such events are rare because the two mutation sites are quite close together, intragenic recombination can be exploited to map the *fine structure* within individual genes. Extensive experimentation with the *r*-gene region of the T4 virus led to identification and sequencing of more than 1000 mutation sites in the two adjacent genes designated *r*IIA and *r*IIB.

6.1.2 Mapping by Complementation and Deletion

The existence of two distinct *r*II genes in T4, with separate polypeptide products, was actually not determined by recombination analysis. A method called the *complementation test* shows whether two nearby mutations are in the same or different genes, and thus allows mapping of gene boundaries. Some *mixed infections* by pairs of *r*II mutants appear normal, which (without recombination) could not happen if a single gene were affected by both mutations, since all of its copies would have one or the other defect. The conclusion is that each of the two required gene products (viral polypeptides) is being produced by the one normal gene on each type of mutant viral chromosome. The two mutations thus *complement* one another.

Although complementation and recombination have some effects in common, the mechanisms are quite different. Complementation generates apparently normal behavior through a transitory interaction of gene products during mixed infection. Unless there has also been a recombination event generating some wholly normal offspring, the next generation of viruses (after complementation allows normal reproduction) will all carry one or the other of the mutations and not show normal behavior. The experimenter thus knows the markers are in different genes when *both* normal infection and exclusively mutant offspring are observed, after simultaneous introduction into a host of the two mutant strains.

Complementation can be used to discover gene identities in all types of organisms. But the procedure requires the simultaneous presence of two chromosomes, or at least two copies of the genetic region under study. This situation is not easy to arrange in normally haploid organisms, like bacteria, but does sometimes arise during some types of recombination events. Complementation is relatively easy to study in diploids since chromosome homology is the normal situation. One parent might have a mutant allele (denoted m), at both homologous loci in one place and normal alleles at a nearby location; in genetic notation, this situation would be written $m+/m+$. If the other parent has the complementary composition, $+m'/+m'$ (where m' is another mutant allele), the offspring, which always get one chromosome from each parent, will be $m+/+m'$ and can appear normal because both normal gene products are present. This picture can be complicated by dominance and other forms of diploid gene interaction.

The last mapping method to be discussed uses mutants which are missing whole regions of the normal chromosome. Once the boundaries of such *deletions* are established, recombination work can quickly find the approximate location of some new mutation. The method depends on the fact that no normal recombinants can emerge from a cross of a point mutation with the deletion mutation if the point mutation falls in the area of the deletion. A group of over 30 deletions in the *r*II region was used in this way to accelerate the mapping of the region's fine structure. Crosses with the seven largest deletions quickly placed a marker in a small portion of the region. Crosses with smaller deletions narrowed the search even further, allowing assignment of a new mutation to one of 47 small pieces of the *r*II region, after perhaps as few as half a dozen experiments.

6.2 VIRUSES

Viruses are by far the simplest repositories of genetic information. Add to that their usually short life cycles (under an hour) and the ease of laboratory study, and it is not surprising that virtually complete understanding of some viruses is now in sight. Nearly all viruses follow the same basic structural plan: a single nucleic acid molecule, encoding from three to a hundred genes, surrounded by a protective shell or "coat" usually comprised mostly of protein. Within this general framework there exists a size range spanning an order of magnitude as well as a variety of specializations in coat shape. The tiny R17 virus has a spherical coat with a diameter of 20 nanometers, comprised of about 180 interwoven copies of a single polypeptide. The comparatively huge smallpox virus has a diameter of 250 nanometers, a quarter that of *E. coli*. Yet not even relatively massive size endows a virus with any independence of function. Only by entering a host cell and subverting its molecular machinery can viruses reproduce and exchange genetic material.

Viruses are usually classified by the nature of their chromosomes (to be considered shortly) and, independently, by the nature of their host cell. The largest viruses attack plant and animal cells, but so do some small ones. The best studied class of viruses, ranging from very small to modest size, are those that attack bacteria, notably the well known *E. coli*. These dozen or so *bacteriophages* (or just "phages") have life cycles that are well understood and genetic maps that are nearly complete. The rest of this section deals exclusively with phages.

The life cycle of a phage is usually one of two general types. In virulent or *lytic* phages, normal infection is always followed by lysis (breakage) of the host cell wall and release of hundreds to thousands of new virus particles. An example of the stages in such a life cycle is diagrammed in Figure 6.2a, for infection of *E. coli* by a phage known as T7. In temperate, or *lysogenic* phages, the lytic infection pattern can occur but does not always do so. The alternative pathway, shown in Figure 6.2b for the lambda phage infecting *E. coli*, involves insertion of the dormant viral chromosome, or *prophage*, into the host chromosome where it replicates with the bacterial genetic material, perhaps for many generations. Later, a prophage can undergo *induction* and take up the lytic cycle where it left off. The induction mechanism is not fully understood; it can be triggered in the laboratory by ultraviolet light and certain chemicals.

6.2.1 Viral Chromosomes

Arrangements for storage of primary genetic information in viruses are surprisingly varied. First of all, the nucleic acid is in some cases not DNA but RNA, and is a single strand instead of a double helix. There is a group of very small phages, typified by R17, whose chromosome is a single short strand of RNA. The sequence of about 3500 nucleotides is known. Just three genes are represented: one for the RNA replicase enzyme which copies the chromosome, one for the coat protein, and one for an "attachment" protein which assists viral entry into the host cell.

Replication of single stranded RNA chromosomes transpires rather differently from the semiconservative replication of the DNA double helix (Section 4.1). But complementary base pairing is still the means for information preservation. After combining with some host polypeptides, the replicase generates a piece of RNA called the "minus" strand, which is complementary to the original chromosome (the "plus" or "sense" strand). Many copies of each strand are made from their respective complements. Only the plus strand serves as a template for translation of viral polypeptides (transcription is not involved when the chromosome is the message) and is eventually packed inside the coats of the new virus particles.

The majority of DNA phages have chromosomal structures similar to those in procaryotes. But even among these there are some unexpected features. The "sticky ends" of lambda, the "concatemers" of T7, and the "circular permutations" of the T-even (T2, T4, T6) phages, serve to complete this inventory of viral chromosome varieties.

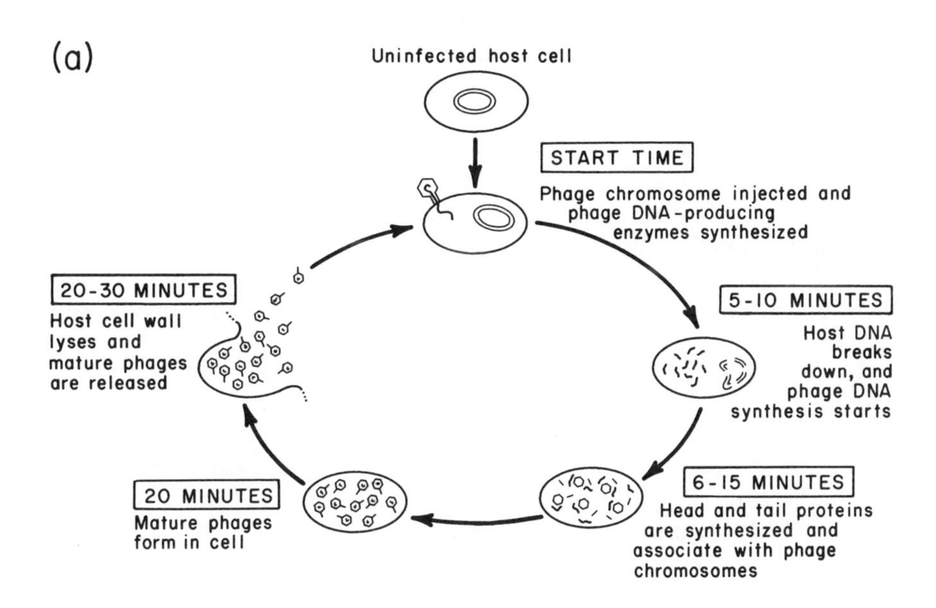

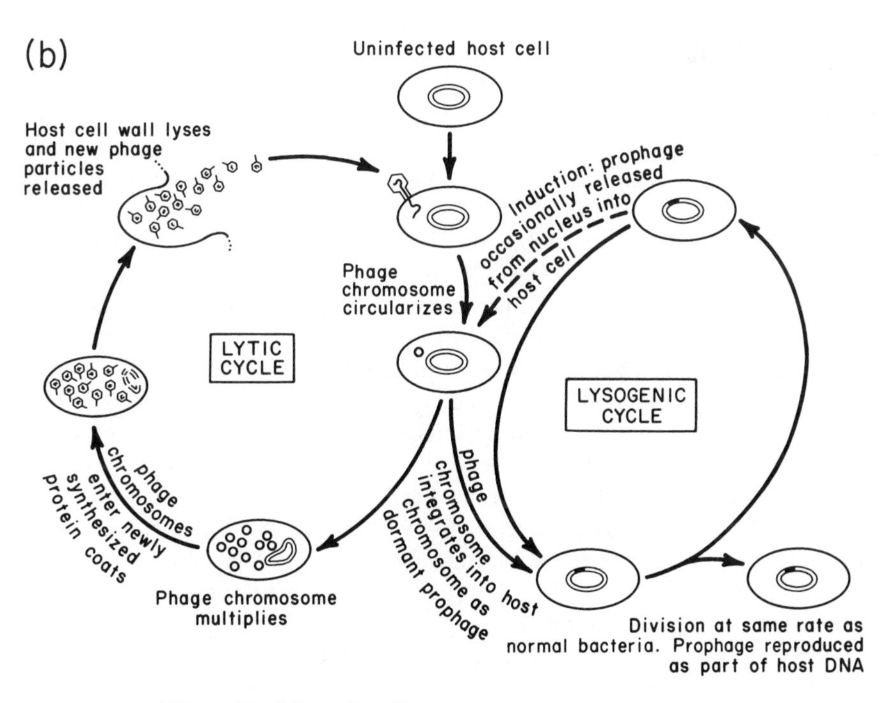

Figure 6.2 Life cycles of bacteriophages: (a) T7, (b) lambda.

The lambda chromosome is stored in the viral shell as a linear sequence of some 46,500 base pairs. It appears that the linear structure is an adaptation which facilitates injection of the chromosome into the host cell. Once inside, the chromosome becomes circular for purposes of replication and transcription. Circularization is a natural consequence of the linear phage's "sticky ends." At the 5′ end of each strand is a single stranded projection of 12 unpaired bases. The two projections are exactly complementary, so that base pairing occurs naturally once the chromosome has room to form a circle. Ligases seal the two resulting gaps. Precise reconversion to the linear form is assured by an endonuclease specified by a lambda gene. This enzyme inserts nicks in the 5′ and 3′ strands, always at same place in the circular map, and always 12 bases apart. The linear piece now has the proper sticky ends and can be packaged in the viral shell.

Replication of the (always) linear T7 chromosome employs an intermediate stage in which a very long double helical molecule contains several copies of the entire chromosome, concatenated one after another (hence "concatemer"). Concatemerization is a natural consequence of DNA replication. Because the first few bases on each 3′ end of the template must be paired to an RNA primer (see Section 4.1.2), there will be a single stranded 3′ tail on each end of each new chromosome. Because the tail sequences are complementary, they will undergo base pairing and ligation to form a linear dimer. Replication of the dimer leads to formation of a tetramer by the same process, and so on. But the final single copies of the chromosome, as found inside the head of the T7 shell, have no single stranded ends. A specialized endonuclease is again employed; but the gaps left by the slightly displaced nicks are filled in through local synthesis of complementary DNA.

The T-even chromosomes are the largest among the phages, about 160,000 base pairs. Like the T7 chromosome they remain linear and form concatemers. Unlike either lambda or T7, T-even chromosome processing does not employ a specific endonuclease. It appears that as much of the concatemer is pulled into the viral head as will fit; then it is chopped off. This operation has been dubbed the "headful mechanism." While electron microscopists were discovering all of this, geneticists were finding an undeniably circular map for the T-even chromosome. The apparent conflict has been resolved as follows. Each "headful" is larger, by several thousand nucleotides, than the actual length of the chromosome map. If a concatemer is denoted ABCDE... ABCDE... ABCDE..., then the first head might get a chromosome with ABCDE...AB, the second a chromosome with CDE...ABCD, and so on. This means that all T-even chromosomes have a few genes repeated at the ends, a so-called *terminal repetition.* Further, in a large sample the repeated genes will be distributed evenly over the entire genetic map. Mapping studies will thus suggest a circular chromosome, although the T-even chromosome population is actually a collection of *circular permutations.*

6.2.2 Regulation of Viral Gene Expression

At all levels of biological organization there are processes that determine which gene products, among all those encoded in a chromosomal repertoire, are available at particular times. Different processes are critical at different biological levels and in different contexts. Thus bacteria spend much of their lives adapting to availability of environmental substances through control of operon transcription (Section 4.5). Bacteria need to deal relatively briefly with special mechanisms for cell division since normal growth supplies most requisites for the daughter cells, other than the new chromosome.

Viruses, however, evince little or no environmental adaptation. The particle is wholly passive until it encounters a suitable target, after which the sole objective is reproduction. A host of new viruses must be created from primitive raw materials. Where bacteria stress maintenance, viruses must be specialists in development. In this respect, the regulation of gene expression in viral chromosomes closely resembles that in the development of multicellular eucaryotes. In both situations, emphasis is on coordination of otherwise autonomous events, rather than on response to changing environmental circumstances.

The appearance of viral gene products in a sequence proper for assembly of new phages is governed by two major types of control mechanisms. One is of the feedback variety, where newly produced substances enhance or decrease the production of others. A simpler form of control depends on the relative positions of genes. The furthest loci on a long operon, for example, will be expressed some minutes after the first ones. These control strategies are illustrated here by some of the better understood cases of viral gene expression.

Although most genetic regulation in viruses appears to be at the transcriptional level, this control point is unavailable to RNA viruses like R17, which rely on control of translation. About 90% of the R17 chromosome codes for amino acids; but there are short untranslated sequences at the ends of the molecule and between the genes. From 5′ to 3′ (the direction of translation) the gene order is: A, the encoding of the attachment (also called attenuator) protein; C, the coat protein; and S (for synthetase), the chromosome replicator protein. R17's basic control problem is the requirement for a nearly 200-fold excess of C protein. Mere repetitive translation of the whole chromosome would waste a lot of time and energy in making unnecessary copies of the other two proteins.

As might be expected, one primary control mechanism limits access by ribosomes to the attachment points at the beginning of each gene. Initially only the C attachment site is "open." The other two are hidden by the intricate secondary structure of the chromosome, which has many "hairpin loops" resulting from local base pairing. A ribosome which completes C synthesis also opens the S attachment site, so there is initially a one-to-one ratio of C to S. This dependence of S production on C synthesis correlates with findings of unidirectional (or "polar") mutational interactions. Nonsense mutations which abort C translation also prevent S production; but nonsense mutations in S have no effect on C.

Equal production of C and S does not last for long. The C protein itself is a translational repressor, binding at the S attachment site and interfering with ribosome attachment. After the first 10 minutes following infection, which is roughly a quarter of the way through the whole cycle, S production has been shut down altogether. Sufficient synthetase exists for all remaining chromosome replication.

The A attachment site is also open only part of the time. This occurs when a new plus strand is being produced from a minus strand template. Translation can begin when the site is first formed and for a while thereafter. Shortly, however, enough of the chromosome has been generated for it to begin assuming secondary structure, shutting down production of A. Translation is thus continuous for C, repressed for S, and intermittent for A.

Given the proper proportion of products, how are the new viruses put together? As in most such cases the process is largely automatic, driven by the conformational affinities of the component molecules. Exclusive incorporation of plus strand chromosomes into phage shells may be executed with the assistance of the A protein, which is thought to bind only to plus strands. The resulting complex may be particularly suitable for incorporation. Interlocking of 180 C molecules to form a spherical shell is a much simpler process than, say, the self-assembly of a ribosome. The one tricky point in R17 assembly is the insertion of the shell's contents *before* its completion. The whole process, from a single invading chromosome to 10,000 assembled phages, takes as little as 30 minutes.

Among the simpler regulation mechanisms in the DNA phages are those of T7. Its linear chromosome contains about 30 genes, most of which have been identified. The genes have been grouped into three classes, which correspond to three consecutive regions of the chromosome. The four Class 1 genes comprise the leftmost 20% of the chromosome. They are transcribed as a single operon by *E. coli* RNA polymerase. One of these "early" genes codes for T7's own RNA polymerase, which is essential for transcription of the remainder of the chromosome. Another early gene product shuts down operation of host RNA polymerase. Thus after about eight minutes (roughly a third of the infection cycle), a whole new subsystem (the "late" genes) comes into play.

The T7 RNA polymerase first transcribes the seven or so Class 2 genes, whose products are mostly concerned with replication of the viral chromosome. In addition to T7 DNA polymerase and nucleases, this phase of operation leads to production of lysozyme, which ultimately breaks down the bacterial cell wall. The third class of T7 genes, comprising the final 60% of the chromosome, are responsible for the dozen or so proteins used to build the viral shell.

T4 has seven times as many genes as T7 and employs more sophisticated procedures for subverting host information processing. There is still use of host RNA polymerase for transcription of a group of early genes. But the products

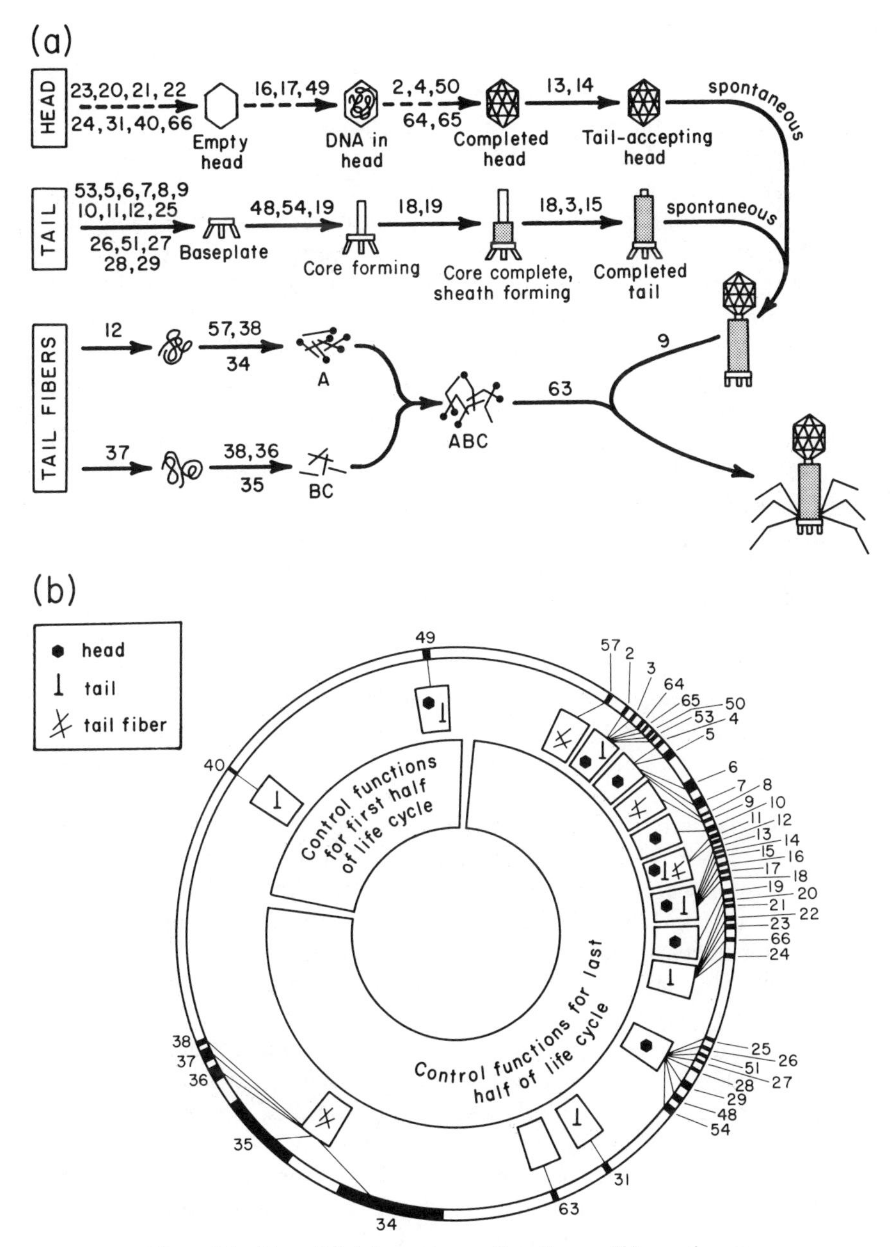

Figure 6.3 Phage T4: (a) morphogenetic pathways, (b) genetic map.

of these genes include a new sigma factor instead of a new polymerase. Other products inactivate host sigma and begin to break down its chromosome. The viral chromosome is protected from these latter enzymes by special modifications to its bases.

Having thus shut down *E. coli* genetic activity, T4 begins transcription of the "middle" genes, grouped in operons that have promoters responsive to the core polymerase of the host combined with phage sigma. Around this time there may also be modifications to host ribosomes and t-RNA molecules, so they work more efficiently in the translation of T4 transcripts. Finally, at least in close relatives of T4, there is a shift to a third sigma factor, which enables host polymerase to recognize the late gene promoters. As in T7, the last genes transcribed specify the enzymes and structural proteins used to build the shell.

Although the details of gene regulation in T4 have yet to be thoroughly understood, the construction of mature phages can be followed more closely here than in any other virus. A tree of converging *morphogenetic pathways* displays assembly line precision. Figure 6.3a is a sketch of the main pathways; numbers on the arrows indicate genes thought to be involved in each step. The approximate positions of these and other genes are shown in the partial T4 genetic map of Figure 6.3b.

Although it has only about 50 genes (compared with T4's 200), genetic regulation in lambda is very complex. This is because, unlike its closer neighbor in size, T7, lambda can undergo either lytic infection or enter a lysogenic state, with the possibility of subsequent induction. The intricate controls governing these alternative modes of existence have been the subject of intensive study, leading to a good understanding of how gene expression is regulated in lambda. The following discussion makes frequent reference to loci in the lambda genetic map, shown in Figure 6.4. (Some of the gene positions are only approximately known.)

The lambda chromosome is organized around four operons, two on each strand of the DNA. The "early right" and "late" operons are transcribed to the right, off the inner strand; the "repressor" and "early left" operons are transcribed to the left, off the outer strand. Promoter regions have been identified for the early right (promoter *Pr*), early left (*Pl*), and repressor (*Pre*) gene groups. The "late" genes may begin with a *Plt* promoter (as shown in the figure); or they may not be a true operon but rather a group transcribed by polymerase which "reads through" the termination sequences of the early right group. The "Q" gene product somehow facilitates this late transcription.

The "N" gene product similarly lengthens transcripts, in this case on both of the early operons, through a mechanism which probably disables the interaction between host RNA polymerase and rho factor required for ribosome release. Until the "N" polypeptide is produced, most transcripts initiated at *Pr* and *Pl* are only one gene long. Thus there is a further distinction between "immediate early" ("N" in the left group and "Cro" in the right group) and the remaining "delayed-early" genes.

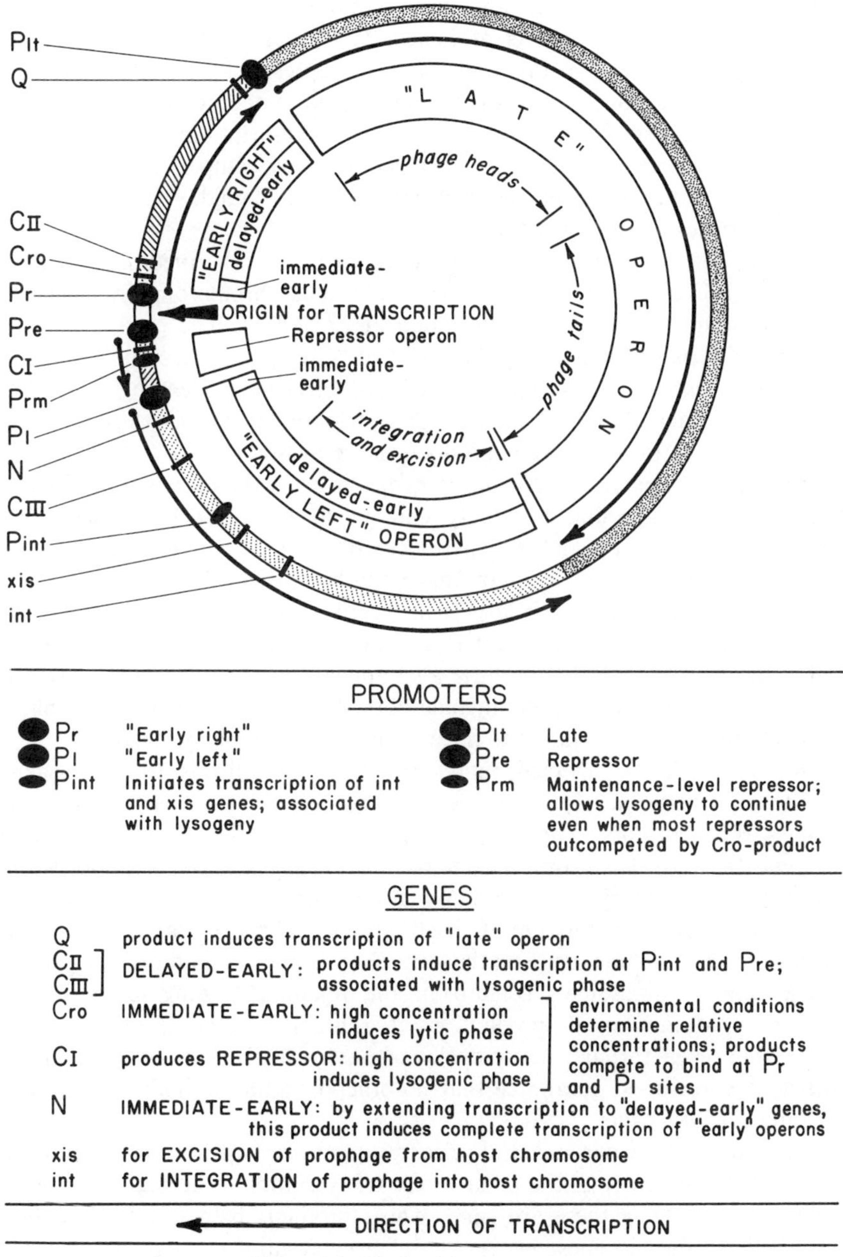

Figure 6.4 The lambda chromosome.

Two other promoters have been identified and found to be associated with lysogeny: *Prm*, just to the left of *Pre*, supports a lower, *m*aintenance rate of *r*epressor transcription; *Pint*, within the early left operon, initiates transcription of several genes, including "int" and "xis," required for integration and excision of the prophage. The products of genes "CII" and "CIII" are needed for initiation of transcription at *Pint* and *Pre* . The repressor itself is the product of "CI." Dimers or tetramers of this polypeptide bind to operators adjacent to the *Pr* and *Pl* promoters, shutting down most other transcription. When available, the "Cro" gene product undermines the repressor by competing for the same operator sites but allowing partial transcription when bound. ("Cro" stands for *C*ontrol of *r*epressor and *o*ther things.)

The interaction of all the above relationships can be observed in the following scenario. Following the insertion of the lambda chromosome into the host cell, host polymerase begins transcription at *Pl* and *Pr*. The products at this first stage will be primarily from the "N" and "Cro" genes. Accumulating "N" product soon extends transcription to delayed-early genes, leading to production of "CII" and "CIII" products. Transcription at *Pre* is thus initiated, leading to production of the "CI" product, the repressor.

The decision to enter the lytic or lysogenic pathway thus depends on whether "Cro" product is produced quickly enough to beat out the repressor in competition for the *Pr* and *Pl* operators. If local conditions (temperature, metabolism, genetic composition of phage and host) dictate a relative excess of "Cro" product, the repressor is largely ineffective. The infection then continues into a lytic phase. The delayed-early gene products assist in replication of the viral chromosome. The "Q" product also appears, leading to transcription of the late operon, whose products include both the building blocks of the phage shell and the lysozyme required to release the assembled phage particles.

On the other hand, sufficient repressor can prevent all this. In the lysogenic sequence of events, nearly all phage transcription ceases when the repressor occupies the operators. The rapid disappearance of the "CII" and "CIII" products also shuts down transcription from *Pre*. However, the repressor has an unusual autoregulatory character. When present only in small amounts it induces low level production of itself through *Prm* initiated transcripts. Meanwhile an attachment site on the viral chromosome and an insertion site on the bacterial chromosome have broken and rejoined, leading to an inserted prophage. "Maintenance" production of repressor perpetuates the stable relationship, even through many cycles of host cell division. There is always sufficient repressor to guarantee that no other invading lambda chromosome can enter the lytic cycle.

But eventually environmental circumstances or agents may dilute or inactivate the repressor to the point where transcription of delayed-early genes begins again. Among these are genes whose products help free the prophage from the bacterial chromosome. As "Cro" product gains an edge on the scarce or ineffective repressor, the lytic cycle resumes.

6.3 BACTERIA

Genetic information processing in phages can be studied only in the context of their bacterial hosts. This consideration adds to the already substantial reasons why *E. coli* and (to a lesser extent) a few other bacteria have become the molecular biologist's primary experimental subject. Eucaryotic cells are much more difficult to maintain in the laboratory and to study in isolation. And even the comparatively docile yeast and fungi have generation cycles at least four times as long as *E. coli*'s fastest. Because bacteria are so basic a source of data about biological information processing, it is appropriate to recap briefly the *E. coli* life cycle and then consider key aspects of the storage and recombination of genetic information in bacteria generally.

The human intestinal bacterium *E. coli* is a rod shaped cell about 2 micrometers long and 1 micrometer in diameter. Growth extends the cell's length, until a cleavage produces two shorter daughter cells. The generation time ranges from 20 minutes to 2 hours, depending primarily on temperature (37 degrees is optimal) and available nutrients (water, glucose, and half a dozen inorganic ions are the minimum requirements). At slower growth rates, the phases of the cell cycle are distinct. During rapid growth, however, the phases overlap, with nearly continuous DNA synthesis and as many as four copies of the chromosome present at once. Chromosome replication appears to require an "attachment site," in the form of an internal membranous structure known as a *mesosome*. Also required is sufficient amount of an "initiator" substance, the accumulation of which may function as a timing mechanism.

6.3.1 The Bacterial Genome

The term *genome* is used here to refer to a collection of storage locations that contains exactly one instance of every piece of genetic information normally found in the given species. The viral genome is typically a single chromosome. The eucaryotic genome is one haploid set of homologous chromosomes plus one representative of each piece of organelle genetic material; it is also possible to refer separately to the nuclear and organelle genomes. The bacterial genome is usually a single, circular chromosome plus an assortment of much smaller pieces of DNA (also circular), known as *plasmids*.

The *E. coli* chromosome is roughly 1000 times the length of the cell itself, which presents an interesting packaging problem. Part of the solution involves tertiary kinks in the double helix, called *superhelical twists*. Of the perhaps 3000 genes on the chromosome, some 500 have been mapped. Figure 6.5 shows some of these loci. The map is calibrated in units to be explained later. Note the lac and trp operons near 10 and 25 units, respectively, and the lambda prophage insertion site at 17 units.

Although the identified loci are distributed rather uniformly along the *E. coli* map, there are some regions where relatively few genes have been

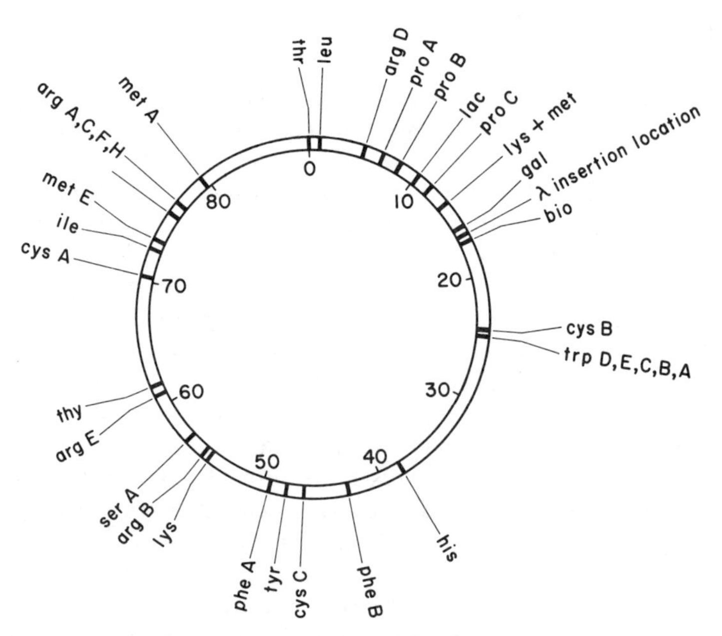

Figure 6.5 Genetic map of *E. coli*.

identified. Perhaps markers for genes in these regions have yet to be discovered. Or they may play roles in chromosomal organization that do not involve transcription, such as a site for attachment to the mesosome. There are certainly short stretches of untranscribed DNA with control functions, such as operators and promoters. Unlike the more enigmatic situation in eucaryotes, however, it would appear that the vast majority of procaryotic DNA consists of nucleotide triplets that specify amino acids.

A bacterial plasmid is a chromosome in miniature, containing from a few to a few hundred genes and capable of autonomous replication. Sometimes a plasmid will undergo several cycles of replication for each cell division, leaving dozens of copies of itself in a single bacterium. Plasmids may be independently classified in at least three ways. First, some types of plasmids can forsake their autonomy and integrate into the main chromosome. This two-state kind of genetic element, of which the prophage is another instance, has been termed an *episome*. Second, plasmids may or may not be *transmissible*, capable of moving (actually sending copies of themselves) to another bacterial cell during the conjugation process to be discussed shortly.

The third and perhaps most significant classification of plasmids relates to the functions of the genes they carry. One group of plasmids has gene products that confer resistance to a variety of environmental hazards, such as ultraviolet light, heavy metal ions, and a broad range of antibiotic drugs. Another group of plasmids is responsible for synthesis of proteins that are lethal to bacteria of the same species; one gene in such a plasmid provides immunity for the cell which carries it.

Probably the best understood plasmid is the *E. coli* F (for Fertility) element. Traits conferred on a cell by this plasmid include thread-like appendages called pili, sensitivity to infection by a group of RNA phages (including R17), resistance to another group of phages (including T7), and immunity to acquisition of additional F elements. The fact that F is a transmissible episome has made it invaluable in recombination mapping of *E. coli*, as discussed in the next section.

6.3.2 Recombination Mechanisms in Bacteria

To make exchange of genetic information possible, part of one bacterial genome must somehow enter another cell. A transient state of partial diploidy, or *merozygosity*, then occurs, during which it is possible for homologous regions of DNA to be exchanged by crossing over. Since any "left over" pieces of DNA are soon discarded, bacterial recombination changes just one functional genome and thus lacks the reciprocity of phage crosses or eucaryotic recombination. Because almost all bacterial reproduction is by mitotic fission, merozygotes are extremely rare. They arise from three known natural processes: conjugation, transformation, and transduction. As discussed below, all three of these processes have been exploited for recombination mapping in bacteria. The scarcity of merozygotes is countered in the laboratory by use of large populations, sensitive screening methods for detecting recombinants, and techniques which enhance merozygote frequency.

Conjugation offers the best opportunity for transferring large amounts of genetic material. This process occurs when an *E. coli* cell that possesses the F element, designated F^+, encounters one that does not, an F^- cell. One of the pili on the F^+ exterior makes contact with the F^- cell and forms a cytoplasmic bridge, or *conjugation tube*, between the cells. A copy of the F element travels through the tube, changing the F^- cell to F^+. In this relatively common process, no merozygote is formed, so no recombination can occur.

Recall, however, that F is an episome. In a population of F^+ cells, about 1 in 10,000 will have the plasmid integrated into the chromosome, at a genetically determined site that varies from one strain to another. These strains are designated Hfr, for "high frequency recombination." The interaction of an Hfr and F^- pair begins in the same way as that involving an ordinary F^+ donor. But the F element that crosses the tube now has trailing behind it a copy of the entire Hfr chromosome. Actually the augmented chromosome is opened for linear transferral in the middle of the F element; so some F genes are transferred first, and the rest last. Since the process is usually interrupted by environmental agitation, only rarely does the recipient acquire the whole episome and become Hfr.

Conjugation is of great help in genetic mapping. Because it is possible to stop transfer (through artificial agitation in a blender) at any desired point, and because the transfer always begins at the same place for any given Hfr strain, recombination can be explored in arbitrarily delimited chromosome regions. Some 20 different Hfr strains have been isolated, with starting points well distributed around the *E. coli* chromosome. A complete transfer requires about 90 minutes. The calibration units on the *E. coli* map in Figure 6.5 above are minutes of conjugation. The origin is based on an Hfr strain in which the plasmid inserts just before the *thr* operon.

Transformation was discovered in the 1920s, when it was found that a few bacteria from a nontoxic strain became lethal after exposure to heat-killed cells of a closely related toxic strain. Since the lethality was inherited by offspring, the genome itself was somehow transformed by the presence of the dead-cell DNA. Transformation is now seen as similar in principle to conjugation, except that the donor DNA is taken up by the recipient directly from the extracellular medium instead of through a conjugation tube. The donor DNA need not come from dead cells. Some bacteria routinely extrude small pieces of DNA.

There are some stringent prerequisites to successful transformation, making it as rare as other sources of bacterial merozygotes. First, the free DNA must be of the proper type and size. Only double stranded pieces containing on the order of 10,000 nucleotide pairs (a few genes) are incorporated. Second, the recipient cell must possess a genetically determined characteristic called *competence*. Competent cells produce enzymes required to bind the donor DNA and draw it into the cell. It appears that just one strand is drawn in, through the use of energy derived from enzymatic destruction of the other (hence the double strand requirement).

Transduction actually refers to a group of three similar recombination mechanisms in bacteria. *Generalized transduction* is mediated by just four known *transducing phages*, only one of which (P1) infects *E. coli*. These phages sometimes make a packaging error, putting a fragment of bacterial chromosome, instead of the phage chromosome, into a shell. The released *transducing particle* can infect a new host. But the inserted chromosome will create a merozygote rather than a viral infection. Generalized transduction is so rare an event that recombinants resulting from it may safely be assumed to have resulted from a single transducing particle. Thus whenever a pair of markers is transduced, they may be assumed to be within a few genes of one another; and the frequency of such events will be directly related to the distance between the markers. (Mapping with transformation is less straightforward, mostly because a competent cell can easily acquire more than one DNA fragment.)

Specialized transduction requires the participation of a temperate phage. The best studied case is the lambda prophage in *E. coli*. Lambda normally inserts between the *gal* and *bio* operons (see Figure 6.5) and, on induction, is excised cleanly from this region. Occasionally a cutting error frees a lambda-chromosome sized piece of DNA that contains either *gal* or *bio* genes and lacks a corresponding amount of the lambda genome. Phages built around this chromosome are termed *defective* since they can insert the DNA into new hosts but not then exhibit normal lambda functions. Other specialized transducing phages can acquire other portions of the bacterial chromosome.

Some differences between generalized and specialized transduction are apparent. First, generalized transduction incorporates a random piece of host chromosome rather than one adjacent to a prophage insertion site. Second, following a specialized transduction error, all new phages are defective, not just the odd one. A less obvious difference occurs in the new host. Defective lambda chromosomes do not exchange genetic material with *E. coli*, but insert themselves as defective prophages, often next to a normal prophage. Thus the augmented bacterial genome may carry two copies of some of its genes and some of lambda's. This form of partial diploidy is heritable.

Genetic mapping via specialized transduction is confined to the regions of the bacterial chromosome bordering on prophage insertion sites. But techniques have been developed which "move" selected bacterial genes into these regions. Moreover, specialized transduction facilitates mapping of the phage chromosome, since the defective phages have deletion mutations.

The third and final type of transduction resembles the specialized variety in many ways, except that the episome is not a prophage but the F plasmid. In F-mediated *sexduction* the reversion of an Hfr cell to F^+ is accompanied by an excision error. The new plasmid, denoted F', contains some genes normally found only on the chromosome. Unlike the result of faulty prophage excision, however, the new plasmid retains all of its own genetic material and is actually increased in size. Later, when the F' cell sends a copy of the augmented plasmid to an F^- cell, heritable partial diploidy arises, with the opportunity for recombination between homologous regions on the chromosome and the F' element. Because of the widely distributed insertion sites for the various Hfr strains, sexduction mapping can be applied to most of the *E. coli* chromosome.

6.4 EUCARYOTES

The vast leap in complexity associated with the transition from procaryotes to eucaryotes has been touched on several times. This section considers further the implications of multiple diploid chromosomes which contain more protein than DNA and a large fraction of nucleotide sequences that are never transcribed. But first, what Crick has called a "mini-revolution in molecular genetics" must be acknowledged. It appears that eucaryotes employ a different organizational scheme even within the nucleotide sequence that contains information for a single polypeptide (the term "gene" can be used only loosely).

The individual polypeptide-coding segments of procaryote transcripts may begin or end with short nonmessage sequences, containing primarily sites recognized by ribosomes. In eucaryotes (most of the early evidence is for mammals), the initial transcript contains the information to specify a single polypeptide, but frequently in several disjoint segments, with long intervening nonmessage sequences. These *introns* are removed during the post-transcriptional processing of RNA; the remaining *exons* are put back together to create a continuous message for ribosomes to translate. This "RNA splicing" of "split gene" transcripts is presumably directed by enzymes.

6.4.1 The Eucaryotic Chromosome

The genetic material of higher organisms was described in Section 3.1.4 as a complex of DNA and proteins, usually found in widely dispersed chromatin fibers. During cell division these fibers condense into individual rod shaped chromosomes, each containing a single long DNA molecule. Replicated chromosomes are temporarily joined at a centromere. Differences in chromosome length and centromere placement frequently permit identification of particular chromosomes. A systematic ordering of an organism's distinct chromosomes constitutes its *karyotype.* The number of chromosomes in the (photographic) display of a karyotype will normally be an integral multiple of the number of distinct chromosomes, according to the organism's degree of *ploidy.*

In the haploid eucaryotes, which include some marine invertebrates, there is a single instance of each chromosome. Some plants have triploid or tetraploid karyotypes. But most higher animals are diploid, with homologous pairs of chromosomes (except in the gamete phase). The human karyotype consists of 23 pairs of chromosomes. Numbers 1 through 22 denote the *autosomes* common to both sexes; the composition of the last pair (identified by letters) determines, among other things, the sex of the organism. The human female has two X chromosomes where the male has one X and one (much smaller) Y. Nearly all animals have such "sex chromosomes," although the unmatched pair is found in the female of some species.

The total number of chromosomes in the eucaryote genome, which ranges from 4 to more than 50, is not a reliable predictor of phylogenetic level. Man has 23 chromosome pairs, the tobacco plant 24, and the goldfish 50. The sheer volume of genetic information in a eucaryote seems to require multiple "packages." What determines the number and size of the packages is not clear. It might be supposed that total genome size would predict organism complexity. But even this correlation is far from perfect. Man carries roughly 2.8 billion nucleotide pairs in the haploid chromosome complement, less than a tenth the number found in some plants and amphibians. Among the amphibians themselves there are 100-fold differences in genome size. Although there are a few tentative theories, no satisfactory explanation for these disparities has been found.

When not condensed for cell division, chromatin still manifests, among the dispersed fibers, some compacted regions. This *heterochromatin* apparently contains the genetic information that is not expressed in a particular eucaryotic cell's specialized functions. Heterochromatic material is also distinguished during replication; it is the very last DNA duplicated, despite its presence in many different chromosomes. The term *euchromatin* is applied to the extended fibers, which are presumably in that state to facilitate transcription of genes whose products are among the cell's requirements.

In female mammals an entire X chromosome is heterochromatic. Expression of both homologues is presumably detrimental, or at least unnecessary since the male gets by with a single X. Interestingly, X inactivation occurs by random choice, after several cell divisions in early development. Thus female mammals express paternal X alleles in some cells, and maternal ones in the others.

Another characteristic of the eucaryotic chromosome is the presence of distinctive classes of DNA nucleotide sequences, some of which are seldom seen in procaryotic or viral genomes. The proportions of these classes vary widely among organisms for no evident reason. The most common class, usually on the order of 70% of the DNA, is the *single copy* or "unique" sequence. Most genes that code for polypeptide chains are presumably present in just one copy, which is not to say that all unique sequence DNA has such a coding role. Although procaryotic genes also occur mainly in single copies, a significant difference in organization within the gene has already been noted.

Middle repetitive nucleotide sequences occupy anywhere up to one third of the eucaryotic genome. This is the most heterogeneous class, consisting of sequences on the order of a few hundred nucleotides in length, with repetition factors ranging from a few dozen to many thousands of occurrences. Although much of this DNA is believed to have a control function, some of it is transcribed. Examples of transcribed middle repetitive sequences are the tRNA and rRNA genes.

Highly repetitive DNA sequences occur on the order of a million times per genome. The well studied cases show tandem repetition of sequences ranging in length from 2 to 10 nucleotides. This material is frequently found near the point of centromere formation, which suggests it may have a more structural than informational role. This idea is bolstered by the fact that centromeric regions are normally heterochromatic.

6.4.2 Segregation, Independent Assortment, and Linkage

The transmission of genetic information from diploid eucaryotic parents to their offspring occurs in several stages, beginning with meiosis (Section 3.3.3). The major principles underlying chromosomal inheritance were first discovered around 1860 by the monk Gregor Mendel. Partly because the very existence of chromosomes was unknown, Mendel's "laws" of inheritance were ignored until they were rediscovered at the turn of the century. Today it is possible to identify Mendel's laws directly with meiotic events.

The *principle of segregation* (Mendel's first law) asserts that when a parent carries distinct alleles at homologous loci, only one of the two alleles will appear in any given gamete. The alleles must segregate because of the reduction from diploidy to haploidy accompanying meiosis. The *principle of independent assortment* (Mendel's second law) asserts that all combinations of alleles from nonhomologous parental loci will occur with equal frequency in the gametes. The loci are independent because every haploid combination of chromosome homologues is equally likely.

As an example, consider an individual with alleles A and A′ at a locus on chromosome 1, and alleles B and B′ at a locus on chromosome 2. The law of segregation says that any gamete will have either A or A′, never both (likewise for B and B′). The law of independent assortment says that gametes will have the four possible combinations (AB, AB′, A′B, A′B′) with equal frequency. Independent assortment is an important source of genetic variance.

Note that the example made explicit an assumption in the second law, that independently assorting loci are on different chromosomes. Such loci are said to be "unlinked." As geneticists discovered exceptions to the second law, they were able to place genes that tended to assort together more than half the time into *linkage groups*. Later it became possible to identify each such group with a particular chromosome. Linkage is not an all or nothing affair because recombination can occur during meiotic crossing over. Two loci on a given chromosome will remain together somewhere between 50 and 100% of the time.

The closer the loci, the tighter the linkage. So the basic principle of mapping by recombination analysis thus applies to individual eucaryotic chromosomes. Actually, chromosomes are so long and recombination so frequent that experimental data do not permit distinction between loci on separate chromosomes and those near opposite ends of a single chromosome. Two remote loci on the same chromosome can be linked to each other, however, through a series of intermediate loci.

The presence of a nonhomologous pair of sex chromosomes in many eucaryotes creates an unusual linkage situation that is responsible for certain unexpected inheritance patterns. A *sex-linked* gene is one which follows these patterns because its locus is on a sex chromosome; the traits specified by the gene may have nothing to do with the organism's sex. For example, a gene with an allele that affects red-green color vision occurs on the human X chromosome. For reasons discussed in the next section, the presence of the mutation does not affect vision when there is a normal allele at the homologous locus. But there is no such locus on the Y chromosome. The mutation is thus "carried" but seldom expressed in women because of the very low frequency of mutant alleles on both X chromosomes in one individual. The mutation is always expressed in men. More than 60 loci on the human X chromosome have been identified, partly by means of their inheritance patterns.

6.4.3 Dominance and Gene Expression

Expression of genetic information in eucaryotes occurs at three major levels. First, there is the long term and essentially irreversible *differentiation* of cells into specialized types during development of multicellular organisms. Large portions of the genome, not required by the cell type, are never again used. (Information processing in development is the subject of Part III of this book.)

At a second level, there is short term regulation of gene expression. This type of control is used in primitive eucaryotes, which respond to environmental circumstances by mechanisms similar to those employed in the induction and repression of bacterial operons. In animals, circulatory systems provide cells with a relatively constant environment; the uniform composition of blood is in turn maintained through hormonal feedback systems which affect various organs that process it (e.g., liver, kidneys). Other hormonal systems govern more radical adaptations, as in pregnancy. Some types of hormones work indirectly by affecting concentrations of cellular substances, like cyclic AMP, which can affect gene expression. Other hormones enter the nucleus and somehow directly influence transcription.

Several mechanisms have been suggested for short term regulation of higher eucaryote gene expression. Candidates for control points include not only transcription but also m-RNA processing and transport, translation mechanisms, and polypeptide conformation. Both the nonhistone chromosomal proteins and some of the middle repetitive DNA sequences are favored participants in transcriptional control.

The third level of gene expression in eucaryotes has to do with the interactions among the particular set of alleles present (and transcribed) in a given organism. Such interactions can occur among alleles at distinct loci, as in intergenic suppression (Section 4.5.3). Cooperating loci can be widely dispersed; mammalian hemoglobin has two subchain components whose genes are on different chromosomes. Some apparently continuous (or "quantitative") traits, like height and weight, are thought to be determined by the combined effects of alleles at numerous loci.

Another kind of allelic interaction occurs between different versions of a gene simultaneously present at homologous loci in diploid organisms. In the simplest case there are just two alleles, denoted A and a, that can occur in three combinations. The combinations AA and aa are called *homozygous*; Aa is *heterozygous*. When the presence of A completely masks that of a, so that it is impossible to distinguish between AA and Aa by any physical or biochemical means, there is complete *dominance*. Allele a is *recessive*, and A dominant. If the heterozygote has distinguishable characteristics, A may partially dominate a, or both alleles may be expressed equally. More complex dominance relationships also occur, especially when there are more than two alleles available.

The distinction between genetic composition and its physical expression, as in the case of dominance, is captured in the terms *genotype* and *phenotype*. An organism's genotype is the full set of alleles present at all its genetic loci, the total genetic information it carries. Alternatively, it is possible to refer to the genotype at a particular locus (e.g., AA, Aa, and aa above) or set of loci. Note the distinction between genotype and genome. The latter term defines a set of positions, possibly multiply occurring, which can contain genetic information; the genotype specifies actual entries at all the occurrences of each position.

The phenotype of an organism is its complete set of detectable properties, both structural and functional. The phenotype associated with a locus is the physical manifestation of the genotype at the locus. In the example of complete dominance given above, there are two phenotypes, one associated with genotype aa and the other with genotypes AA and Aa. Ordinarily a phenotype is determined by interaction between a particular genotype and the current environment. It is possible, for example, that different alleles may be dominant in different environmental contexts. Thus a given genotype may be associated with more than one phenotype, as well as vice versa.

Although the biochemical mechanisms that underlie dominance relations among alleles are not known in very many cases, one well understood situation occurs at the locus on human chromosome 9 which controls ABO blood type. The three common alleles can be designated A, B, and O. A and B are *codominant* with respect to O, so that the six genotypes produce four phenotypes: A (from genotypes AA and AO), B (from BB and BO), AB (from AB), and O (from OO). The relative proportions of these four blood types vary with geographical location and race.

The enzymes produced by the A and B alleles recognize a particular polysaccharide molecule and add different chemical groups to it, producing what are called the A and B antigens. These molecules are subsequently incorporated in larger chemical structures found on the surface of the red blood cell. The term antigen is used because there can exist in the blood stream specific antibodies, generated by the immunological system, which recognize and combine with several antigen molecules of the proper type. The result of such a reaction is "clumping" (agglutination) of the red blood cells and consequent interference with their function.

Since people do not produce antibodies against their own chemical substances, the A phenotype has A antigen and anti-B antibodies, whereas the B phenotype has B antigen and anti-A antibodies. Blood type AB has both antigens and neither antibody; type O has neither antigen and both antibodies. When a person receives a transfusion of whole blood, new antibodies are quickly diluted and not especially harmful. But if the donor red cells contain an antigen that matches a host antibody, they will agglutinate with possibly fatal consequences. Thus blood type O, with no antigens, is the "universal donor," whereas blood type AB, with no antibodies, can accept transfusions of any blood type.

Chapter Seven

Simulation of Genetic Information Processing

Two quite different groups of modelling and simulation projects are considered in this chapter. The first is concerned with modelling the life of bacteria at the molecular level. The second set of investigations applies the basic mechanisms of genetic information processing to problems of adaptive search.

7.1 MODELS OF THE BACTERIAL CELL

In the late 1960s, the life cycle of *E. coli* was the subject of independent modelling and simulation efforts at the University of Michigan and the State University of New York at Buffalo. The latter project is discussed first because of its more specific focus, on events associated with DNA replication. The Michigan work, involving at least five major participants, led to collateral studies of populations and evolution, which are treated only briefly here.

The *real system* for all these studies is the whole array of reactions and controls underlying the growth and division of an *E. coli* cell. Supplementary data from related bacteria are sometimes incorporated. Fewer than half of the reactions involving the roughly 3000 types of *E. coli* molecules are presently understood. The *experimental frame* thus has its maximal boundaries imposed by ignorance. Further restrictions are established by the interests of the investigators. The simplification procedures leading to particular *lumped models* are treated individually below.

7.1.1 A Model of Cell Growth and Division

Margolis and Cooper (1970, 1971) have implemented a model of events leading up to and including chromosome replication and cell division in *E. coli*. Although it has similarities to the cell cycle models of Section 5.2, this approach is distinguished by its meticulous attention to details of DNA synthesis. This focus facilitates comparison of the model's quantitative behavior with some important laboratory experiments.

The model deals with conditions of exponential bacterial growth, where the number of cells doubles every one to three hours. The major factor in the growth rate is the type of nutrients supplied by the environment. This nutritional level determines the rate of accumulation of a "replication initiator." The parameter I specifies the time needed to accumulate a threshold level of initiator. There is then a fixed time interval C for chromosome replication, followed by a fixed delay D before cell division. D is known to be about half of C. I is assumed to be no greater than $C + D$ during exponential growth. A cell starting with no initiator divides after an interval of $I + C + D$. Since accumulation of initiator never stops, the length of time between cell divisions will be I during continuous growth of a culture.

In particular, when $I = C + D$ the amount of initiator will have reached twice threshold just after cell division. Each daughter cell thus receives a threshold amount and can immediately commence chromosome replication. If I is less than $C + D$, replication of the daughter chromosomes begins before parental cell division. In the cases so far considered, a given chromosome has only a single growing point ("fork"). When I drops below $(C + D)/2$ initiator accumulates so rapidly that incomplete daughter chromosomes begin replicating, yielding a three-forked structure that resembles a balanced binary tree (see Figure 7.1a below). Each fork in the model must actually represent the two symmetric growing points at either end of a replicating "eye" in the circular bacterial chromosome, although Margolis and Cooper never make this simplification explicit.

Another simplification, which is described explicitly, lumps all real system cells at a given stage of growth and chromosome replication into a single, "representative" model cell. Frequently overlooked, this laudable technique serves to focus attention on relevant variables and to reduce computer resource requirements during simulation, often without sacrifice of any relevant information. Each lumped cell in the Margolis and Cooper model is characterized by the state of its chromosome, the total mass of all real cells it represents (mass increases exponentially at a rate consistent with I), two "clocks," and two "counters."

The "initiator clock" (actually a measure of initiator quantity) increases by $1/I$ for each unit of model time. When it reaches 1, a new round of replication begins at all available chromosome origins. Termination of replication causes doubling of the "chromosome counter." The "division clock" marks off the D time units between the end of replication and cell division. The "cell counter" doubles after each division. Since this is a discrete event model, the "clocks" do not run continuously but rather participate in the scheduling of future events. The next event in a given model cell is normally one of the following: completion of a replicating chromosomal segment (see below), initiation of a round of replication, or cell division.

When lumped model cells do not necessarily represent equal numbers of real cells, a proportional weighting of model cell characteristics is necessary to produce summary data about the modelled population. Margolis and Cooper observe that an exponentially growing cell population follows a simple distribution with respect to cell age. Weights determined from intervals of this distribution are used as multipliers of quantitative cell characteristics.

At the outset of a simulation run, an initial group of model cells, with evenly distributed "ages" (times to division), is generated as follows. A single cell is first initialized so that it has unit mass, a single unreplicating chromosome, and no initiator. This cell is allowed to grow and divide normally, producing two cells which will be typical of the division products in the growing population (but have taken longer to appear). One of these offspring is then grown incrementally, in steps of I/n, where n is the desired number of model cells. At each increment a copy of the cell, with all of its current characteristics, is made. The set of n copies provides the starting population for the actual simulation run. In the implementation, n can range from 1 (a synchronous culture) to 60. A useful working interval, in the experiments carried out, was 15 to 30.

The heart of the Margolis and Cooper model is the representation of the chromosome. Manipulation of this representation is also the most intricate part of the simulation algorithm. Before the representation and its manipulation can be appreciated, it is necessary to describe the kind of laboratory experiment these investigators wished to simulate. In an exponentially growing bacterial cell culture, several measures can be recorded both before and after modifications to the nutrient medium. The measures include cell mass, numbers of cells, DNA mass, and number of replication forks per cell. The simplest change to the medium either enriches it (a "shift up") or removes some nutrients (a "shift down"), changing the doubling time because cells have to manufacture either fewer or more of their requirements. In the Margolis and Cooper model, shift down and shift up experiments correspond to changing the parameter I.

A more complex set of modifications to the medium will alter the composition of replicating DNA strands in measurable ways. A density label can be obtained through substitution of a heavy ion (e.g., the heavy hydrogen in deuterium) for a normal one. Such a shift in the medium is from *light* (normal) to *heavy*. The immediate result will be hybrid DNA, having an old light strand, a new heavy strand, and a detectably intermediate weight. A radioactive label can be obtained through substitution of a radioactive ion for a normal one. Such a medium shift is from *cold* to *hot*. Measurements of radioactivity can determine the proportions of cold and hot strands. A complex experiment might initiate growth in a cold, light medium, then introduce a timed succession of shifts, first up, then to hot, then to heavy, then back to cold, and so on.

Margolis and Cooper represent the replicating chromosome as a string of seven numbered *segments*. Associated with each segment is its (possibly zero) length, expressed as a percentage of the overall length, and the type of DNA strands in the segment. Five of the six possible strand types can be represented: light (normal), hybrid, heavy, hot (and light), and hybrid hot. Figure 7.1 shows the two examples of chromosome representation given by Margolis and Cooper. The chromosome in Figure 7.1a is presumably the result of a temporary shift to a heavy medium, whereas the one in Figure 7.1b has been replicating successively in light, hot, and (cold) heavy media. The replication algorithm, which adds new material, shifts material from segment to segment, and inserts new forks, is straightforward but rather tedious. The algorithm is fully described in the technical report (Margolis and Cooper 1970), which also includes a complete listing of the FORTRAN program.

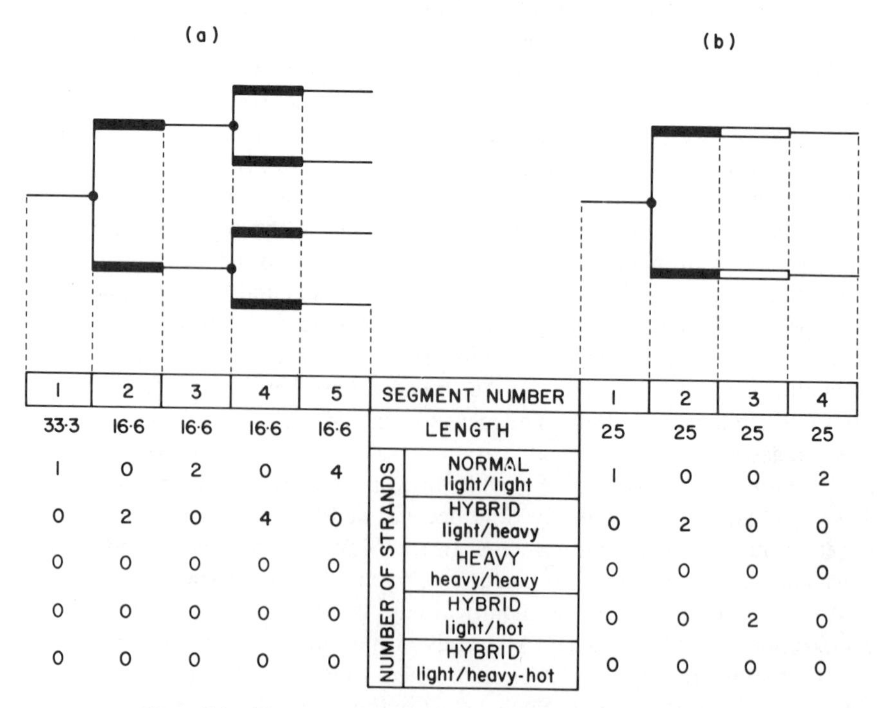

Figure 7.1 Chromosome replication in the Margolis-Cooper model.

Margolis and Cooper describe five groups of simulation experiments. Model behavior was first examined under steady-state conditions, where all of the model characteristics were found to replicate normal laboratory growth. A series of shift up experiments also produced good agreement with laboratory data, particularly with respect to the sudden increase in the cell division rate $C + D$ minutes after the shift. The third group of experiments simulated inhibition of initiator synthesis. Under such conditions there was residual DNA synthesis, from forks already growing at the time of inhibition. Simulation results replicated laboratory data, showing more residual synthesis in richer growth media, where there were more forks.

The last two types of experiments involved more than a single shift. In the "Meselson-Stahl" experiment, there was a simultaneous shift up and to a heavy medium. Plots, automatically produced by an auxiliary feature of the Margolis-Cooper program, showed the expected sharp decrease in light DNA, followed by a more gradual accumulation of heavy material, with the transient peaking of hybrid strands. The "Lark-Repko-Hoffman" experiment employed successive shifts to hot and then (cold) heavy media (see Figure 7.1b). In some variations partial or complete inhibition of initiator synthesis was also introduced. These experiments offered the greatest challenge to the Margolis-Cooper model. Only incorporation of three carefully reasoned "mitigating circumstances" in the simulated model allowed replication of the Lark-Repko-Hoffman results. This work is thoroughly discussed in both references.

7.1.2 A Comprehensive Cell Model

The Michigan *E. coli* model was begun in the 1968 doctoral research of Weinberg (reported in Weinberg and Berkus 1971), continued by Weinberg and others (Weinberg, Flanigan, and Laing 1970; Weinberg, Zeigler, and Laing 1971; Goodman, Weinberg, and Laing 1971), and culminated in the doctoral research of Goodman (1972). This discussion divides the work somewhat arbitrarily into three phases: (1) the original Weinberg model and some additional experiments with it, (2) extensions of the model to colonies of cells and to evolution, and (3) Goodman's thesis research.

The Weinberg model is primarily continuous in character, with a few discrete control structures superimposed. The major simplifying assumption is the aggregation of *E. coli* molecule types into some 30 chemical pools. The primary pools and the pathways for material flow among them are shown in Figure 7.2. Additional pools result from the subdivision of both MRNA and PROTEIN into 10 smaller pools. This refinement allows individual treatment of the transcripts and enzymes responsible for flow of material among the primary pools (e.g., the production of TRNA from NUCLEOTIDEs). Three criteria contribute to the choice of pools. First, a chemical can be incorporated in a pool only if it is no more than one reaction step distant from at least one other member of the pool. Second, pools are to be representative of functional groups of molecules. Finally, it is desirable that quantitative laboratory data exist for the group of substances in a pool.

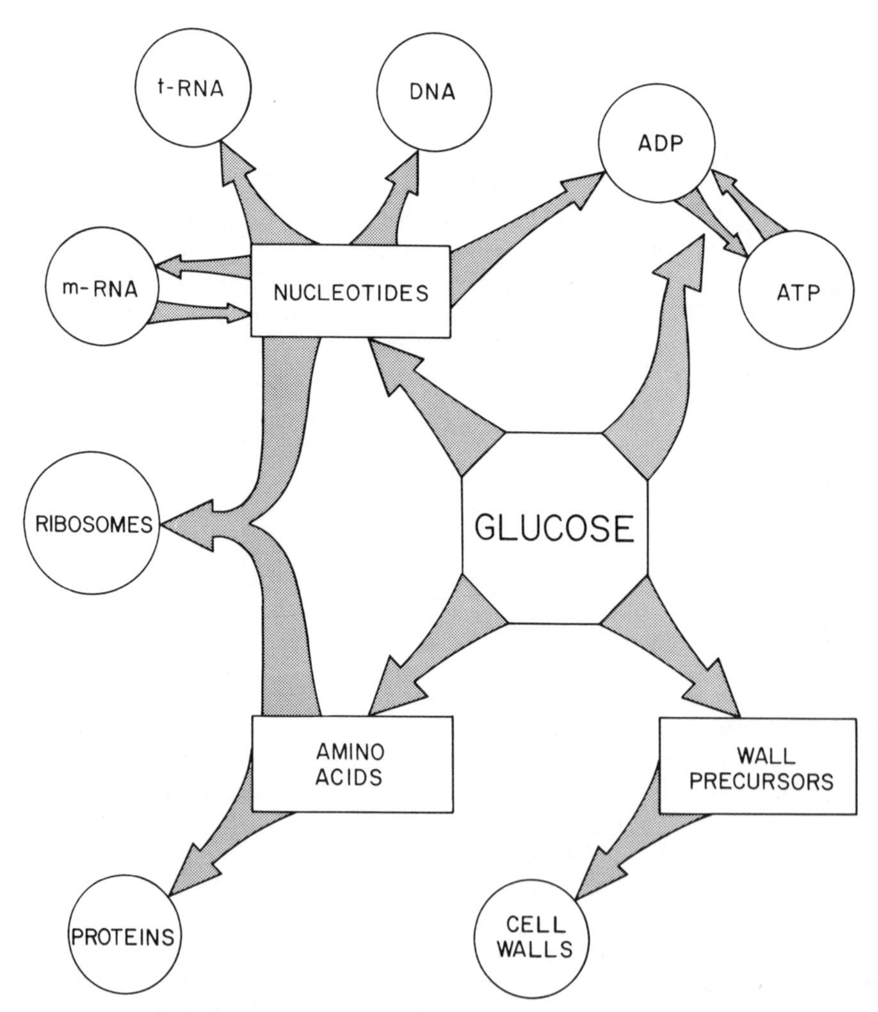

Figure 7.2 Primary chemical pools in the Michigan *E. coli* model.

In addition to the pool concentrations, other state variables of the model include cell volume, amount of replication initiator substance, availability of replication sites, and extent of replication for up to three chromosomes. The model of DNA replication is like that used by Margolis and Cooper. Because of its greater overall scope, the Weinberg model treatment of chromosomal activity is necessarily less detailed than that seen in the previous section.

The heart of the state transition mechanism is a system of differential equations relating changes in pool sizes to the concentrations of other pools and to model parameters. In the FORTRAN program, the differential equations are converted to difference equations and solved iteratively, with a discrete time step corresponding to about one second of real time. Because this form of numerical integration is quite crude, accuracy and stability in the Weinberg model depend critically on the size of the time step. The unnatural oscillation of ADP and ATP concentrations observed in some experiments led to suggestions for a smaller time step and more refined numerical methods.

In the following examples of the difference equations, dT is the time step, the EKs are enzyme concentrations, and variables beginning with K are rate and proportionality constants.

$$\text{dAA} = \text{K}_2\text{*GLUC*ATP*EK}_2\text{*dT - KK1*dRIB - KK2*dPRTN} \quad (7.1)$$

Amino acids (AA) accumulate at a rate proportional to the concentrations of glucose, ATP, and the relevant enzyme. Amino acids simultaneously disappear as they are used for construction of ribosomes and proteins.

$$\text{dEK}_2 = \text{K}_7\text{*(RNK}_2\text{/MRNA0)*AA*ATP*EK}_7\text{*dT} \quad (7.2)$$

Enzyme EK_2 is produced from its corresponding RNA transcript and amino acids, in a reaction governed by ATP and another enzyme. MRNA0 is the initial concentration of all messenger species.

$$\text{dRNK}_2 = \text{(K8K}_2\text{*NUC*DNA*ATP*EK}_8\text{ - KdRNK*RNK}_2\text{)*dT} \quad (7.3)$$

Finally, RNK_2 appears at a rate governed by available nucleotides, DNA, ATP, and another enzyme (presumably RNA polymerase). The second term on the right hand side of Equation 7.3 represents spontaneous decay of RNK_2, at a rate proportional to its own concentration.

In addition to growth in volume, chromosome replication, cell division, and the changes in pool concentrations just considered, the model cell transition function can employ two adaptive control mechanisms. In *allosteric* (end-product) *inhibition* (see Section 3.4.2) an enzyme's efficiency is altered by revision of its rate constant according to the relative concentrations of its three forms: unbound, bound with a single end-product molecule, and bound with two molecules of the product. Thus where E catalyzes the conversion of S to P, there will be four secondary rate constants governing the system

$$\text{E} \leftrightarrows \text{EP} \leftrightarrows \text{PEP} \quad (7.4)$$

Normally the conversion of SPEP to P and PEP is slower than that of SEP to P and EP, which in turn is slower than that of "free" SE to P and E. The bound forms of the enzyme may not function at all. Alternatively, allosteric inhibition can be defeated, either by giving the bound forms normal efficiency or (more economically) by setting the forward rate constants in Equation 7.4 to zero.

The other control mechanism is *repression* of transcription (see Section 4.5.2). As various cellular products become relatively abundant, a somewhat complicated set of calculations can be used to decrease the rate constants associated with production of transcript molecules (e.g., $K8K_2$ in Equation 7.3).

A variety of simulation experiments were conducted with Weinberg's original program. Initially the model cell was "grown" in three simulated cell-culture media, all of which were considered to have optimal temperature and abundant oxygen. The "minimal" medium contained glucose and a few essential salts and minerals. The "amino acid" medium added to this a supply of polypeptide precursors that the cell would otherwise have to manufacture. The "broth" medium included both amino acids and the nucleoside building blocks for nucleic acid molecules. Early simulation results agreed well, for all three media, with laboratory data on growth in cell volume, time to cell division, and number of chromosomes present at division.

Further experiments simulated the shifting of a cell from one medium to another. A shift down to low glucose concentration provided the first test of the cell's behavior in an environment that had not been used in parameter tuning. Again there was agreement with real cell decreases in growth. Shifts up and down between the minimal and broth media showed the importance of allosteric inhibition for realistic adaptation to richer (though not poorer) environments.

In the second phase of work with the Michigan *E. coli* model, two separate projects extended the single-cell model to populations of bacteria. One of these studies (Goodman, Weinberg, and Laing 1971) deals with a colony of cells growing on a flat circular nutrient base (as in a Petri dish). Two useful aggregation techniques yield a lumped model in which each simulated cell represents many real ones.

The first technique resembles that of Margolis and Cooper, who kept track of just one representative model cell at any given stage of the cell cycle. In the colony growth model, space, not time, is the critical variable. Each model cell therefore represents many real cells in a given region. The number and total volume of the represented cells are part of the model cell's description. The other major simplification involves an assumption of symmetry. The colony can potentially grow into a three dimensional conical shape. The model treats only a vertical, triangular cross section, which would be rotated through a circle to produce a model of the entire colony. Combination of these two aggregation procedures results in a lumped model of 31 cells representing several hundred actual bacteria.

Cells in the colony can sense aspects of their neighbor's states. Direct influences are limited to the four model cells horizontally and vertically adjacent to a given cell. Potential interactions include limitations on access to nutrients, responses to crowding, and intercellular transport of chemical messengers. Such interactions can influence the three dimensional shape of the colony by locally altering cell division rates.

In one preliminary set of experiments with the colony model, an initial, centrally located cell was allowed to divide under minimal medium conditions until a stable colony shape appeared. In the "uncontrolled" case, a rapidly growing, multilayer, "cancer-like," colony developed. Control was introduced in the form of an inhibitor which suppressed accumulation of the replication initiator substance. The amount of inhibitor in a cell was proportional to its number of immediate neighbors. The result was a slowly growing, monolayer colony. This type of study has bearing on the theory that one difference between normal and cancerous cells is the inability of the latter to respond properly to crowding.

The other population embedding of the single cell model (Weinberg, Flanigan, and Laing 1970) was done to study evolution of the simulated cells' genetic parameters. The allosteric modification of enzyme activity described above was extended to include a fourth rate, when the enzyme is bound to three product molecules. A cell is represented as a string of 40 rate constants, one group of 4 associated with each of the 10 enzymes responsible for the major chemical pools. The evolutionary algorithm treats this string as 40 genetic loci, where the range of potential values for the rate constant is the set of alleles at that locus. A cell's performance or "fitness" can be evaluated by using the alleles in the single cell simulation program, presented with an environment characterized by fluctuating nutrient supplies. Cells which maintain greater internal stability in the face of external fluctuation are considered more fit and are ranked higher for purposes of survival.

The evolving population is maintained at a constant 40 individuals, divided into 4 subpopulations of 10 each. The cells in a subpopulation are ranked by fitness. In a single "generation," the 4 lowest ranked cells are discarded and replaced by "offspring" generated using the top 4 cells as parents. The recombinants of a crossover between the top 2 replace the bottom 2; and recombinants of the third and fourth parents replace the third and fourth worst individuals. In addition to crossover, mutation and inversion serve to maintain genetic variance. Mutation consists of a small random modification to an allele in an offspring. Inversion reverses a substring of loci, creating a new linkage map. For simplicity, inversion is applied uniformly to subpopulations, so that all 10 individuals always have the same map. The last step in the production of a new generation involves ranking the new subpopulations according to their most fit individuals. Inversion is applied to copies of the two better subpopulations, which then replace the other two.

In the first of two evolution experiments, various initial populations were allowed to evolve for 9 generations, both with and without crossover in effect. The best individual to evolve was substantially superior when crossover was employed, confirming the importance of this mechanism to natural selection. In the other experiment, a single population with very low initial fitness was allowed to evolve for 120 generations. An impressive 10^{10} fold increase in the best fitness was obtained, strong evidence for the power of evolutionary mechanisms.

Goodman (1972) undertook the third phase in the development of the bacterial cell model. He returned to the level of the individual cell and introduced a variety of refinements, extensions, and improvements. Goodman's contributions can be grouped in three areas: (1) changes in the model, (2) changes in the simulation methodology, and (3) additional simulation experiments.

There are two major additions in Goodman's version of the *E. coli* model. The first incorporates a pool for lactose and adds induction of lactose processing to the previously developed repressive controls on enzymes. This apparatus makes possible a "shift to lactose" experiment, described below. The second addition provides a more realistic interface between the cell and its environment. The original Weinberg model employs set internal concentrations of nutrients and does not consider supply and transport phenomena. Goodman adds chemical pools for the external input of glucose, lactose, amino acids, and nucleosides. The mRNA and protein pools consequently have four additional subpool components for the transcripts and enzymes needed to transport the nutrients into the cell.

Because he wished to study cell behavior for much longer time intervals than Weinberg, Goodman lengthened the basic time step during simulation from about one second to approximately one minute of real time. To compensate for the loss in accuracy and stability resulting from so drastic a change, some refinements appear in the simulation strategy. A multistep numerical integration routine is employed wherever oscillation is a problem, such as in glucose transport. A related technique "lags" the processing of some variables, so that the action taken during any single time step is half that required by the current and previous time steps combined. Finally, a "time sampled" simulation method is used for pools that experience a turnover several times their normal volumes during the one minute time step. Such turnovers can swamp the steady state values, leading to intolerable errors and/or oscillations. The solution involves a refinement of temporal resolution, yielding a prediction of long term behavior based on a sample corresponding to a second or less. Flux during the sampled period is normally a small and not unduly perturbing fraction of pool volume.

In his simulation experiments, Goodman monitored all pool concentrations closely and compared them carefully with laboratory data (some of which had become available after Weinberg's work). Growth in the three media proved satisfactory, as did shift up behavior. But shift down experiments revealed some deficiencies in the model, for which Goodman proposed specific solutions. In a new type of experiment, the cell was shifted from the glucose medium to one containing only lactose as an energy source. After one "debugging" operation, simulated bacteria at three different stages in the life cycle all responded to the shift with pool changes very similar to those seen in laboratory cultures.

Both the Margolis-Cooper and Weinberg-Goodman models of the bacterial cell are results of careful attention to the relevant details of the real system, combined with judicious and productive simplification procedures. Implementation of these models, respectively in the discrete event and continuous simulation paradigms, demonstrates a number of instructive practices. The simulation experiments suggest that the models have replicative validity and considerable predictive potential.

7.2 GENETIC ALGORITHMS FOR ADAPTIVE SEARCH

Problems of search are fundamental to biology and computer science. The cell seeks an appropriate balance of enzymes; evolution looks for organisms best suited to a given environment. The Michigan *E. coli* model was used to simulate these search process. Many algorithms devised by computer scientists also search for something, be it the root of a function, an item in a list, or an improved arrangement of data.

Holland (1975) has extracted and formalized the characteristics of search common to natural and artificial systems. He has devised a general adaptive search procedure based on the mechanisms of evolution. The *reproductive plan* is proposed as a good search procedure for a wide range of difficult and uncertain environments. Several computational studies have bolstered this theoretical prediction and revealed the character of reproductive plans in practice. Although much of this work is only remotely connected with modelling and simulation of biological information processing systems, the research does provide further evidence for the computational power of natural genetic algorithms.

7.2.1 Reproductive Plans

Search among any reasonably large set of alternatives usually benefits from a systematic search procedure, or *plan*. Each point in the search space, or environment, represents one possible solution to the given problem. The number of points may be finite or infinite and, in the latter case, the space discrete or continuous. If the search problem is implemented on a computer, the precision of the machine imposes a finite, discrete character on the space, which might therefore contain only a sample of the points in the actual environment. Associated with each point is a numerical measure of the goodness of the solution it represents, called its *utility* or fitness. If the utility is only implicit in the point's location, a utility function must be evaluated to generate the explicit value.

An adaptive (search) plan uses information about the environment in an attempt to shorten the duration and improve the results of the search. The information may include utilities for search points already investigated. Only this component of environmental "feedback" has been used in plans considered further. There are many kinds of adaptive plans. Function optimization techniques often use the rate of change (gradient) in utility to "climb" to the top of a "peak." The heuristic search techniques of artificial intelligence are also examples of adaptive plans. The limiting case is the search plan that ignores environmental feedback and simply enumerates points in the search space, in a predetermined or random order. If an adaptive plan makes any constructive use of feedback, its search performance should be better than the average behavior of an enumerative plan working on the same problem.

The performance of search plans (adaptive or otherwise) can be evaluated according to two criteria. An *effective* plan is one that will eventually find the best point in the search space. Effectiveness is necessary but not sufficient for good search performance. Even enumerative plans are effective. A major obstacle to effectiveness often occurs in environments where there are pronounced local optima ("false peaks") that are not globally the best.

When one of two effective plans usually finds the global optimum more quickly than the other, it is more *efficient*. Like effectiveness, efficiency is environmentally conditioned. One plan may be much faster than another in a simple search space, but fall apart when presented with a discontinuous, multimodal, or even just multidimensional environment. A plan which is efficient in a broad range of environments is called *robust*. Holland (1975) has shown that his reproductive plans are robust.

A reproductive plan works with a population of points in the search space, applying artificial evolution in an attempt to generate the global optimum. Each member of the population is called a *structure* and consists of a linear string of values for a set of attributes relevant to the search problem. Converting search space points to structures is not always straightforward; the "representation problem" is a current area of research in reproductive plans. Further, a plan usually works best if the attributes are binary, with 1 signifying the presence of a feature. Computational studies of reproductive plans for function optimization, for example, usually treat each bit position of each argument value as a separate attribute.

When a structure is viewed as a chromosome, its attributes are loci, and their values, alleles. When viewed as an individual in a breeding population, a structure can be copied to produce offspring. The essence of a reproductive plan is twofold: (1) reproduction according to fitness, and (2) application of *genetic operators*, including crossover, mutation, and (in some studies) inversion.

One version of the basic plan is the following.

Step 1. The generation counter is initialized. An initial population of structures is placed in the search space, usually at random points.

Step 2. The current population of structures is evaluated in the environment. The ratios of the resulting utilities are used to induce a probability distribution on the population.
Step 3. The required number (see step 5) of parents is selected according to the probability distribution from step 2. Exact copies of these structures are made.
Step 4. Genetic operators are applied probabilistically to the offspring, in pairs for crossover.
Step 5. The offspring replace a set of structures in the parent population. The number replaced can range from one to the entire population.
Step 6. The generation counter is incremented. The reproductive cycle begins again at step 2, using the new population.

The target structure in a given search problem will have some particular combination of alleles. Individually these alleles may or may not contribute positively to the utilities of other structures. Because it participates in multiple nonlinear interactions with the occupants of other loci, an allele's contribution depends on its context. When rewarding a successful structure, an adaptive plan should apportion credit to all the combinations of alleles contributing to the structure's utility. Reproductive plans apportion credit in this fashion, both automatically and in parallel for all allele combinations at once.

This *intrinsic parallelism* is the main reason for the power of reproductive plans. The concept of *schemata* provides insight into this aspect of the adaptive search process. A schema is a subset of structures with common values on selected attributes. If structures have just six binary alleles, the following are examples of schemata: all structures with 1's at positions 2 and 5; all structures "beginning" with 0 and "ending" with 1; all structures with 0 at position 3; and, of course, both "no structures" and "all structures." The *length* of a schema is the difference in the position numbers of its two most remote defining loci; in the above case, schemata lengths run from 0 through 5. A given structure is said to represent, or be an instance of, all schemata to which it belongs. Every structure of the above type is an instance of 2^6 schemata. In general, the number of schemata represented by a structure equals the number of alleles per locus raised to a power equal to the number of loci.

The utility of a schema is just the average utility of its instances in the current population. Each evaluation of a single structure affects the utility of all the schemata it represents, in parallel. The intrinsic evaluation of a population's schemata apportions credit to all possible combinations of alleles. The genetic operators also affect schemata in an intrinsically parallel fashion. A single crossover can generate new instances of many existing schemata, as well as instances of previously unrepresented schemata. Inversion alters the length of schemata, changing the degree of linkage between loci. And mutation guarantees that all schemata will be represented sooner or later, in a sufficiently large population.

The preceding qualification forms the basis for nearly all the work in the next section. Holland's theoretical results concerning the robustness of reproductive plans are usually based on a mathematical model in which the population size has no finite upper bound. Practical use of this search technique requires understanding of how finite and tractably small populations (on the order of 100 individuals) deviate from theoretical norms. These computational studies employ a form of simulation, although the "real system" is not a biological one, but rather an abstract one derived in part from biological mechanisms.

7.2.2 Simulation of Genetic Algorithms

Genetic algorithms, as implementations of reproductive plans have come to be called, have been studied by roughly a dozen investigators. Many, though not all, have been Holland's students. This discussion provides historically ordered samples of experimental results. No work is presented in its entirety; some important contributions are not mentioned at all. Portions of this presentation are based on that of Mercer and Sampson (1978).

Bagley (1967) may have been the first to employ the term genetic algorithm as it is used here. His work antedated by several years the version of Holland's framework presented above. Like several later investigators, Bagley was primarily concerned with adaptive search in environments with nonlinear interactions among dimensions.

Bagley devised a class of search tasks based on the simple game of hexapawn, in which each player moves three pawns on a three-by-three subset of a chessboard and tries to be the first to reach the opposite side. A multiparameter evaluation function was used to determine the next move for both the adaptive system and the environment (opponent). In an environment of "depth" n, the opponent based its decision on a fixed weighted combination of all parameter subsets of size n. Values of n greater than 1 may entail nonlinear interactions. The potential complexity of such interactions increases with n.

Bagley compared the performance of genetic algorithms with that of the "correlation" weight-adjustment schemes popular in early artificial intelligence programs. Because it can adjust weights independently, in pairs, and so on, the correlation algorithm has a "level" defined in the same sense as the environment's depth. The genetic algorithm, on the other hand, operates at all levels in parallel by means of its intrinsic evaluation of schemata. Bagley found that a "match" of environment depth and correlation algorithm level was critical to performance; a level-1 algorithm was inferior to a level-2 scheme on a depth-2 environment, but superior to the more complex plan in a depth-1 environment. At environmental depths greater than 3, even matched correlation algorithms tended to perform quite poorly.

Genetic algorithms were not significantly influenced by environmental depth, an indication of their robust character. In comparisons with (matched) correlation algorithms, genetic algorithms evinced equal or superior search performance. In his genetic algorithm simulations, Bagley usually employed populations of a few hundred chromosomes. His genetic operators included mutation, crossover, and a limited form of inversion.

In a series of experiments based more directly on the theoretical framework of Section 7.2.1, Cavicchio (1970) studied adaptive search in the context of a pattern classification task. An environment was composed of two "alphabets" of 16 hand-printed characters (*A* through *P*). Each character was represented as a 25 by 25 binary matrix, with ones in positions corresponding to grid squares touched by the character. An adaptive plan was trained on the first alphabet, with the correct pattern class supplied in each instance, and tested on the second alphabet. Performance was related to the number of correct classifications.

An individual chromosome contained several hundred genes or *detectors*, each of which inspected a small number (2 to 6) of points in the matrix. During testing, each detector would report whether or not the binary entries at its points matched in both the input pattern and a given stored pattern. The total number of matches against each stored pattern served as the basis for the classification decision.

To establish a context for the performance of genetic algorithms, Cavicchio first tested a set of enumerative plans, where detectors were generated at random. In an environment designated "difficult" (and used in most subsequent experiments), the enumerative plans scored an average of 17 on a scale of 0 to 100. Correlation plans improved performance to an average approaching 40, with a best score of 52. In the many variations of genetic algorithms that Cavicchio simulated, average scores ranged between 60 and 80. In this task domain, then, the reproductive plan seems to have a distinct advantage.

The extensive computation required to evaluate an individual's utility limited the population size in Cavicchio's experiments to about 20 individuals. Loss of genetic variance is a common cause of ineffective search in such small populations. Cavicchio developed several techniques to counteract this effect. A *selection scheme* limited the number of offspring a good parent could contribute to the next generation, in an attempt to avoid complete displacement of potentially valuable alleles by currently useful ones. A *preselection scheme* allowed offspring to replace their parents when they were superior to the parents; it was thought that displacement of a genetically similar individual, instead of a random one, would help maintain variance. Less successful techniques included the use of auxiliary populations and mutation pools in an attempt to simulate the effects of dominance in populations of diploid organisms.

One of Cavicchio's most innovative developments was a parameter modification scheme, or *meta-plan*. The probability of each genetic operator's usage was independently adjusted during search, on the basis of the utility of offspring created by its use. This scheme was designed to increase efficiency by allowing more large modifications (through crossover) early in the search, and to increase effectiveness by enhancing the contribution of small modifications (through mutation) later. Cavicchio wanted the crossover rate to decrease as the plan converged on an optimum, to allow a larger proportion of the offspring to be modified solely by mutation, thus facilitating a local search. Although it did not work entirely as intended, the parameter modification scheme did improve search somewhat; it also served as a point of departure for the meta-plan studies of Mercer and Sampson (1978) reported below.

De Jong (1975, 1980) undertook a systematic study of genetic algorithms in the context of function optimization. His test functions (environments) spanned a considerable range of search difficulty, from two-dimensional continuous unimodal functions to many dimensional discontinuous multimodal ones. A chromosome was a string of binary alleles, each corresponding to one bit in one of the coordinates of a point in the search space.

As alternatives to the average or maximum final utility, De Jong defined two measures of genetic algorithm performance more suited to his interests. *Online* performance reflects the cost of adaptive search when every putative solution must be used in "production" until another one is tested. This sort of situation would exist in a process control environment, for example, if a search for optimal parameters were carried out using the operating plant itself as a source of feedback. *Offline* performance would be relevant if the parameter search were carried out in a simulated model, to allow the actual plant to use the best solution so far discovered.

In his initial simulation experiments, De Jong studied the effects of four basic genetic algorithm parameters: population size, mutation rate, crossover rate, and "generation gap" or the proportion of parents replaced in a reproductive cycle. Small generation gaps and low crossover rates were both found to yield moderate improvements in early search performance, but at the expense of greater loss of variance in the population. This loss could be reduced by increasing the mutation rate, which in turn adversely affected online performance. As might be expected, larger populations (200 instead of 50 or 100 individuals) tended to have inferior short term performance; they also consumed more storage space.

De Jong then turned his attention to several schemes specifically designed to improve performance and/or maintain genetic variance. The *elitist policy* guaranteed that the best individual so far encountered was not replaced. This modification generally improved performance, substantially so on the easier unimodal test functions. The *expected value plan* attempted to maintain variance by reducing allele loss due to sampling error. This approach, which required that the number of offspring actually contributed by a parent be close to theoretical expectations, also had salubrious effects, particularly when combined with the elitist policy. Finally, a *crowding factor* model was studied, to see if replacing individuals genetically similar to the new offspring would help maintain variance. The result was an improvement of offline performance at the expense of online performance. The expected value and crowding factor methodologies have points of similarity with Cavicchio's selection and preselection schemes, respectively.

De Jong's last experiments compared the performance of his best genetic algorithms with that of two standard function optimization techniques. As might be expected, the latter methods outperformed genetic algorithms only on the test functions that possessed the characteristics for which the optimization techniques were designed. Once again, genetic algorithms demonstrated their robust character.

Brindle (1980) furthered the study of genetic algorithms for function optimization, using De Jong's findings as a point of departure. In all of her experiments, Brindle employed a factorial design for comparison of algorithms, with sufficient numbers of within cell replications to warrant the use of analysis of variance techniques; significance was assessed at the 5% level. Search performance was evaluated by De Jong's online measure and, more appropriately for function optimization, the single best point so far encountered. Loss of population variance was measured by (1) the proportion of loci homogeneous for one allele and (2) the frequency, over all loci, of the more common allele at each locus.

To study the relative behavior of varieties of genetic algorithms, Brindle designed a set of test functions, each of which contained a single global maximum among 2^{30} points. The number of local maxima varied from 8, for a damped sinusoid, to 2^{28}, for a 6-element Fourier sum. Of greater relevance to genetic algorithm performance than this Euclidean modality is a function's modality in Hamming space, since the minimum search step of a genetic algorithm (inverting a bit) represents a move of unit Hamming distance. Thus, one function that had 2^{28} Euclidean peaks but was Hamming unimodal proved especially susceptible to search by genetic algorithms. The damped sinusoid with only 8 Euclidean peaks, on the other hand, was so difficult that an algorithm comparable to De Jong's best was actually inferior to random search, at least with respect to the best overall performance measure. This type of performance was found to be significantly improved when the De Jong population of 50 individuals was increased to 200.

Using her new "optimal" basic algorithm, Brindle studied the effects of previously unexplored variations on reproductive plans. Several schemes for parent selection were compared, with a cheap and simple deterministic method turning out the be the best under most circumstances. In a radical departure from previous genetic algorithm implementations, Brindle undertook a series of experiments to evaluate the consequences of using diploid structures combined with a variety of dominance schemes. Although it had been endorsed by several previous investigators as a means of reducing loss of population variance, diploidy with dominance proved of little value in Brindle's function optimization application.

To compare her best genetic algorithm with a highly regarded "steepest descent" numerical search method, Brindle modified and augmented her set of test functions, including some that were much like those used in De Jong's similar comparison. The numerical method selected by Brindle compared much more favorably to the genetic algorithm this time, showing noticeably inferior long term performance only on some functions having both high Euclidean and low Hamming modality. Brindle concluded that, although there remained areas for profitable application of adaptive search by genetic algorithms, the utility of these techniques in optimization of functions that are difficult in Euclidean space was not as great as previously believed.

Cavicchio's parameter modification scheme (along with a similar suggestion by Bagley) inspired the *reproductive meta-plan* of Mercer and Sampson (1978). The basic idea was to use a genetic algorithm to control the probabilities of genetic operator application in the first-level reproductive plan. Cavicchio's pattern classification task was used as the basic search problem, mainly to facilitate comparison with the performance of his meta-plan.

The reproductive meta-plan's chromosomes were strings of genetic operator application rates. The meta "genetic" operators could mutate these rates and combine, by averaging, rates from two meta-parents. Each meta-parent determined lower level operator rates for a subset of detector-string individuals (the "meta-environment"). The meta-plan performance criterion was somewhat arbitrarily chosen to be the relative number of offspring, produced by parents under the meta-parent's control, that were good enough for inclusion in the next generation.

Mercer and Sampson carried out a limited set of experimental studies with a small meta-population (5 individuals). The results mostly revealed an improvement in performance relative to Cavicchio's meta-plan, which was in turn better than his basic algorithm. Whether such improvements can ever be substantial enough to justify the computational overhead of a meta-plan remains to be seen.

PART THREE

Information Processing in Development

Chapter Eight

Morphogenesis in Developing Organisms

The development of a multicellular organism is a process which is successfully repeated from generation to generation in the species. Since the "memory" of a species, its continuity, is supplied by its genetic material, the process of development itself must be encoded in the genome and thus available to every cell. In the individual organism, development involves orderly changes in multiplying cells, culminating in the complex structure of specialized cells and tissues which make up the mature state. Specialization of a cell implies an altered interpretation of the instructions for cell activities that reside in the genome. If the genome is regarded as a collection of stored programs, then development is a form of information processing with selective autonomous execution of these programs.

The object of the developmental biologist is to understand how the process of development is produced through interpretation of the genome. The ultimate goal is to break the code whereby sequences of nucleotides are related to each cell's behavior and future. There are two basic approaches to the problem. In the first, called the *molecular approach*, the focus is at the level of DNA and the individual gene. If a particular gene sequence may be identified with a particular event in development, or particular features of the organism, then the complete encoding of the developmental process may be obtained by exhaustive means. Since the amount of genetic material in multicellular organisms is immense, such an approach has limitations. Furthermore, the design of experiments to identify genes encoding particular phenomena is difficult.

The second approach to the study of development is that of the classical embryologist. It may be termed the *phenomenal approach*. The focus is on the events of development. The differences between cells in a developmental system are observed; influences and processes which appear to be involved in producing these differences are identified; and, finally, the implementation of these influences at the molecular level is investigated. An operational picture of development may thus be constructed.

Ultimately the two approaches should both produce the molecular encoding of the developmental process. The phenomenal approach, however, has led to concepts and models that contribute to an understanding of the dynamic nature of development. In this chapter, some of these concepts and models are introduced and experimental studies from which they arose discussed. The next chapter contains examples of simulation techniques used to study models of the dynamic aspects of morphogenesis.

8.1 CONCEPTS AND MODELS

8.1.1 Determination and Differentiation

An adult organism consists of many different cell types produced during development. These cells have undergone visible changes in morphology. They have different shapes and sizes and synthesize special products. The stage at which a cell's final type becomes overt is known as *differentiation*. It is preceded by a covert process known as *determination*, which selects a specific fate for a cell by singling it out from all the possibilities for which the genome is competent. Before determination begins, the cell is capable of becoming any of the cell types of the mature organism. As determination proceeds, the cell's future becomes increasingly more restricted, until it is committed to its final cell type. There is considerable experimental evidence for this process of "channeling" the cell through determination; examples are found in the work on the fruit fly described in Section 8.2.

Determination is said to be a covert process. But there must be some biochemical basis, however subtle, for a state of determination in a cell. It is therefore difficult to separate determination entirely from differentiation. Determined cells are already "differentiated" from others in some (undetectable) way. In general, however, use of the term differentiation is restricted to the later events, in which the final cell type is unveiled.

The singling out of particular pathways in development appears to be triggered by biochemical signals. These may take the form of intercellular interactions (for example, when the proximity of cells of one type to cells of another type initiates determination in one or both types; this process of induction is discussed below), hormonal signals, or environmental influences. Determination thus involves factors extrinsic to the genome.

Once determination has begun, the determined states of cells have been shown to be heritable, stable, and irreversible under the normal conditions of the developing organism. Heritability is demonstrated when cultures of determined cells maintain their state of determination and, in a favorable environment, resume their original pathway toward differentiation. Since it is the genome that progeny cells inherit, the particular state of determination must be the result of a modification of expression of the genome, a modification which is maintained through cell division. Experiments which artificially destabilize determination show that with a regular, low frequency, cells in an experimental medium can be made to switch their state of determination to that of cells on another pathway in the development of the organism. This phenomenon has been interpreted as indicating that the selection of each pathway is mediated by a single gene.

Once cells are fully determined (capable of developing completely without further interactions), it remains for them to undergo overt differentiation. This final stage involves selective gene expression, not loss of genetic material. Differentiaton is almost always completely stable and irreversible.

Multicellular development thus consists of a series of determinational events followed by differentiation into varied cell types. Both determination and differentiation appear to be mediated by selective gene expression, which in turn is affected by, or effected in response to, environmental or extragenetic biochemical cues.

8.1.2 Developmental Programs

The development of a multicellular organism involves a sequence of determinational and differentiative events that occur in a precise temporal order. The ordered sequence of such events is known as its *developmental program.*

The fundamental question concerning the developmental programming of an organism is what type of mechanism controls the sequencing of events. There may be some immutable specification of the program coded in the genome itself, as if it were written on a tape that is read sequentially as development proceeds. Alternatively, events may be causally linked, so that the occurrence of event B at time $t2$ is invariably triggered by the occurrence of event A at time $t1$.

The evidence points to the second type of control. Events have been shown to depend on critical levels of proteins produced by genes activated at earlier stages in development. In some cases, the developmental program has been interrupted and observed to resume when these levels have been re-established.

Various models for developmental programs have been based on this evidence for the causal relationships between events. In the well known

Britten-Davidson model, expression of batteries of genes is controlled by *regulatory genes*, which may reside in those portions of the DNA that contain repeated sequences. Expression of a regulatory gene produces activation (or, alternatively, repression) of the battery of genes under its control. Some genes in the battery are "producer" genes (those whose activities express cell behavior); and some may be other regulatory genes controlling other batteries. A cascading sequence of controlled events could thus be realized.

Garcia-Bellido, a prominent investigator of fruit fly development, has also suggested at least two classes of genes. *Cyto-differentiation* genes control cell behavior, whereas *selector* genes control developmental pathways. This model also provides for the inclusion of factors extrinsic to the genome in developmental events. A molecule specific for some gene combines with an "inductor molecule," a signal from the cell's environment, and can thereafter no longer promote transcription of its selector gene and contingent pathways.

The inclusion of the inductor molecule in such a model is important because the distribution of its "message" may not be uniform over all cells of the organism. Thus this factor, extrinsic to the genome, allows for the possibility of nonuniform genetic response, of spatially distributed differences arising among the cells of the organism.

The theoretical biologist Kauffman has developed a quantitative language for developmental programming in the form of a switching code to model developmental events. A symbol is used to represent a pathway selected in the developmental process. The alphabet of symbols depends on the number of choices a cell has at a developmental event. In the fruit fly system on which the model is based, there appear to be binary branches in these pathways and hence a (0,1) alphabet. The length of a binary string indicates the number of events which have been noted for a particular cell or tissue. The particular combination of symbols represents the history of pathways taken in the developmental process.

The success of this model in describing phenomena of fruit fly development supports the assumption that pathways are selected by the operation of a single gene acting as an on-off switch. As in the Garcia-Bellido model, the state of a switch may be influenced by factors both intrinsic and extrinsic to the genome. A spatially inhomogeneous setting of a switch is presumed to be caused by a nonuniform distribution of an extrinsic message.

The developmental biologist thinks of the establishment of such a nonuniform message distribution as *pre-patterning*. A collection of cells known as a *field* is the domain of the pattern, which imparts positional information to the individual cells by means of the inhomogeneity. The resulting nonuniform genetic response is known as *pattern formation*. These concepts are illustrated for a specific organism in Section 8.2.

The developmental program of an organism is thus the organized sequence of determinational events it undergoes while reaching maturity. The sequence appears to be controlled temporally by causal links between events. Individual events are probably controlled by single genes that regulate the expression of other genes. The action of the controlling genes is affected by factors extrinsic to the genome so that spatial ordering of developing tissues becomes feasible.

8.1.3 Induction

It has been noted that influences external to the genetically specified current behavior of a cell are implicated in the fulfilment of its genetic program. The well known phenomenon of *induction* is one example of how interaction with cells of another behavior (cells on a different developmental pathway) triggers further developmental events.

Induction was discovered in the 1920s when a group of amphibian embryo cells called the "organizer" was identified. The organizer possesses the capability of inducing development of the neural tube (a precursor of the central nervous system) when implanted in various host tissues. Furthermore, absence of the organizer prevents development of the tube in its normal position.

The development of most organisms involves phases of mass reorganizations of cell structures in the system. Most observable differentiations begin immediately after one such phase, known as gastrulation (Section 8.3.1). It is thought that the novel juxtaposition of previously remote tissues, which results from gastrulation, stimulates events through inductive interactions. It has been demonstrated in several systems that if normal interaction is prevented by experimental means, the developmental event does not occur.

Induction provides one of the intercellular, spatially distributed factors in the development of an organism. It has, however, proved difficult to find the actual basis for the inductive signal. On one hand, many sublethal toxic chemicals have been found to "induce" development in target tissues, which could imply an "unmasking" effect rather than transmission of a truly instructive message. On the other hand, there has been no success in isolating naturally produced diffusible substances that can definitely be shown to act as inductive agents.

8.1.4 Regulation of Gene Expression

Developmental programs whose determinational decisions are reached by interaction of genetic and intercellular spatial factors have as their domain the genome of the organism. In a given cell at a given time the result is a particular state of expression of the genome, revealed phenotypically or as a state of determination. The search for mechanisms of selective gene expression was stimulated by the discovery of the various types of RNA and their roles in transcription and translation, and by Jacob and Monod's elucidation of the *lac* operon in *E. coli* (see Section 4.5.1). One implication of the operon model is that the expression of structural genes is under the control of regulatory genes. In eucaryotes, the vast increase in the amount of DNA is not mirrored by a corresponding increase in cell products. It is thought that this implies a complex, combinatorial network of regulatory genes like that proposed in the Britten-Davidson model.

Two kinds of experimental findings address the actual mechanisms of gene regulation and provide direct evidence for control networks. First, biochemical phenomena in a cell occur at certain stages, which implicates these phenomena in regulation. For example, experiments with sea urchins have shown that the genome is not actively involved in development until about the time of gastrulation. The mRNAs in active translation prior to gastrulation were present in the unfertilized egg. When isolated and allowed to act as templates, these messages produce, among other proteins, histones. Since histones have a marked affinity for DNA, they may be agents of gene regulation when the genome becomes active.

In another case, it has been observed that multiple copies of the genes for making rRNA are formed during the development of an unfertilized egg. The result is massive synthesis of rRNA in the egg. This phenomenon of *selective amplification* might operate in other situations, such as in differentiated cells specialized for production of large quantities of particular substances. But such is not the case. In blood cells, for example, there is only one copy of the globin gene per cell. Except for rRNA synthesis, selective gene amplification has not been shown to be an agent in gene regulation.

The second type of experimental finding comes from mutation studies. A mutation in a particular gene can have an effect on more than one structure in the adult organinsm. If the mutation affects more than one structure, this may be because of subsequent interactions in the genetic program or because the mutation is in a regulatory gene and affects all the structural genes in its battery.

In the "steel" mutant mouse, for example, the mutation becomes evident in terms of sterility, anemia, and white color. Such disparate effects are most easily explained in terms of a battery of different genes controlled by the same regulatory gene, as in a Britten-Davidson control network. The visible effects of one class of mutations are restricted to clearly defined regions of the organism. This phenomenon stimulated Garcia-Bellido's formulation of the control model described above, since a regulatory gene in this model may have a spatially defined domain of action as a result of nonuniform distributions of the inductor message. A mutation in the regulatory gene itself would then be visible in its domain of action. Mutation studies of this kind provide evidence for a complex network in the regulation of gene expression in multicellular organisms. Actual identification of the molecular mechanisms responsible for regulation remains an area of active investigation.

8.2 PATTERN FORMATION IN IMAGINAL DISCS

Drosophila melanogaster, the common fruit fly, is an insect easily studied experimentally, by both phenomenal and molecular techniques. It is therefore a system which has provided fertile ground for experiments involving the concepts and models introduced above.

Fertilization of the *Drosophila* egg is followed by 12 to 13 mitoses over a period of about 3 hours. A multinucleate mass is thus formed. By the 9th cleavage, nuclei migrate toward the cortex (exterior of the egg). After the 13th mitosis, division pauses and membranes form to separate the nuclei into distinct cells. The hollow shell of cells thus formed is known as the *blastoderm*. Some 20 minutes later, gastrulation occurs and cellular division resumes. The organism is now in its larval phase.

Beginning in the blastodermal phase, small groups of cells called *imaginal discs* become isolated from the surrounding tissues. Each consists of a single invaginated layer of epidermal cells. These discs are destined to become external features on the adult fly, such as eyes, wings, legs, and genitalia. The discs begin to be determined in the early larval phase, during which there are three *instars* (periods between molts) over about four days. The insect then becomes a pupa for another four days and finally undergoes metamorphosis into the adult fly. During metamorphosis the imaginal discs evert and become their predetermined structures.

8.2.1 Experiments with Discs in Culture

Because of the isolation of discs from other larval tissues, they can be removed and cultured in various media. By this means it has been shown that discs cultured in the adult female abdomen grow but do not differentiate, whereas those implanted in male abdomens neither grow nor differentiate. A hormone produced by the adult female is needed for disc growth; this hormone must also be present in the normal larval phase of both sexes. Disc differentiation, however, must be facilitated by some factor present only in the larvum. When a disc which has been growing in a female abdomen is reimplanted in a larvum, it resumes differentiation toward its original goal. Its original state of determination is preserved and bequeathed to progeny cells during implantation and culture.

After repeated culture of discs in many successive adult abdomens, it has been observed that occasionally, upon re-implantation in a larvum, the discs differentiate into a structure other than the one for which they were originally determined. The repeated culturing process has somehow destabilized the determined state of the disc tissue. The cells are said to have undergone *transdetermination*. Tabulation of transdetermination frequencies between disc pairs indicates that the stability of determination is a quantifiable property.

In the Kauffman model (Section 8.1.2), the observed frequencies of transdetermination for each disc are encoded as a string of binary digits. Transdeterminations occur when the code for one disc is transformed into the code for another disc by resetting one or more of the binary switches. Changes from 0 to 1 are more frequent than from 1 to 0. Transdeterminations requiring more digit changes occur less often. The ability to explain such data in terms of a switching code supports theories of determination in which developmental decisions are mediated by the presence or absence of transcription of a single gene.

8.2.2 Gradients and Positional Information

During the blastodermal phase, *Drosophila* divides into longitudinal segments of differentially determined tissue, within which the imaginal discs are formed. Within the imaginal disc itself, an invariable geometry of subdivisions into genetically distinct groups of cells is observed. This subdivision of discs occurs throughout the larval stage. It is therefore clear that, when a developmental decision is made that separates one group of cells entering one state of determination from another group entering another state of determination, the domain for each state is decided in geometric or "positional" terms.

For instance, the blastoderm initially separates into two genetically distinct groups of cells, organized geometrically into the anterior and posterior halves of the organism. Similarly, at a certain stage of determination of the wing disc, the central portion of the disc becomes determined to form the distal portion of the wing, while the surrounding tissue becomes determined to form proximal features. Developmental models for *Drosophila* must therefore include a means by which the cells' determinational states may be made regionally specific so that a cell's new state of determination depends on its position in the organism. The cell must then be responsive to some sort of *positional information.* The Garcia-Bellido model would supply this positional information in the form of a nonuniform distribution of inductor substance.

The molecular basis for positional information has been the focus of much recent work in developmental biology. There is evidence from *Drosophila* studies that information in the form of a smoothly varying field, or *gradient*, of some kind is interpreted by discs during development. Surgical experiments on discs show that fragments which are implanted and allowed to grow are capable of regenerating missing features "downstream" from them in this gradient, but not those upstream. Such gradients of regenerative potential are thought to be most probably effected by spatially nonuniform concentrations of some diffusible substances that have yet to be detected.

Mathematical models for concentration distributions, in the form of partial differential equations for reaction-diffusion systems, have demonstrated the possibility of nonuniform distributions arising from initial uniformity in growing systems. Some of these models have been studied using simulation and are described in the next chapter.

The response of cells to a field of positional information is a pattern in the multicellular structure of differentially determined cells. One instance of such pattern formation is seen in the phenomenon of *compartmentalization* within imaginal discs. During larval development the discs undergo successive, geometrically invariant subdivisions, in a fixed temporal order, into genetically distinct compartments. (This process is a continuation of the subdivision process initiated in the blastoderm.) The significant finding with respect to compartments is that certain mutations are compartment specific. The mutation is visible only in those parts of structures on the adult that come from one compartment. This discovery suggests that each compartment is under the control of a single gene (the selector gene in the Garcia-Bellido model), in which the original mutation must have occurred. Compartmentalization thus provides evidence for the kinds of control structures for development posited in the models of Section 8.1.2.

Studies of other developmental systems have revealed pattern formation phenomena which parallel those in *Drosophila*. For example, many insects undergo segmentation in the larval phase. Compartmentalization has been observed in other organisms. There is evidence for gradients of regenerative potential in newt limbs and cricket legs. There is evidence for positional information in the form of an anterior-posterior gradient in *Hydra*, and a proximo-distal gradient in the developing vertebrate limb. Models of some of these systems have been studied by simulation (see Chapter 9).

8.3 MORPHOGENETIC MOVEMENTS

In early embryogenesis there are periods of extensive, well coordinated movements, of both individual cells and two dimensional sheets of cells, which rearrange the basic form of the embryo. These movements are called *morphogenetic* because they are fundamental to the structure and future development of the organism. The main phases of movement are known as *gastrulation*, *neurulation*, and formation of the extra-embryonic membranes.

8.3.1 Gastrulation

The blastodermal phase of development involves cell division without extensive mutual rearrangement of cells. Subsequently the embryo undergoes rearrangement into three basic sheets of cells: *ectoderm* (exterior skin), *mesoderm* (middle layer), and *endoderm* (interior lining). The ectoderm is destined to form the epidermis and the nervous system. From the endoderm the gut wall and associated glandular structures will develop. And the mesoderm will become all other internal structures (muscle, blood, bone, kidneys, and so on). Details of this reorganization into three layers vary among species.

The simplest form of gastrulation which has been experimentally analyzed in detail occurs in echinoderms and amphibian embryos. In the latter, as shown in Figure 8.1, the blastoderm begins to flatten at one location on its exterior surface. Cells in this area then begin to push into the hollow interior of the blastoderm. These first cells, initially isolated, move apparently at random inside the blastoderm, but then adhere around their site of entry. Other cells on the exterior then begin to move inside through the same point, but remain attached to their neighbors, so that the whole sheet of cells forms a hemispherical invagination into the blastoderm. This invagination then elongates and becomes attached to the opposite interior wall of the blastoderm; the resulting pit is the primitive gut.

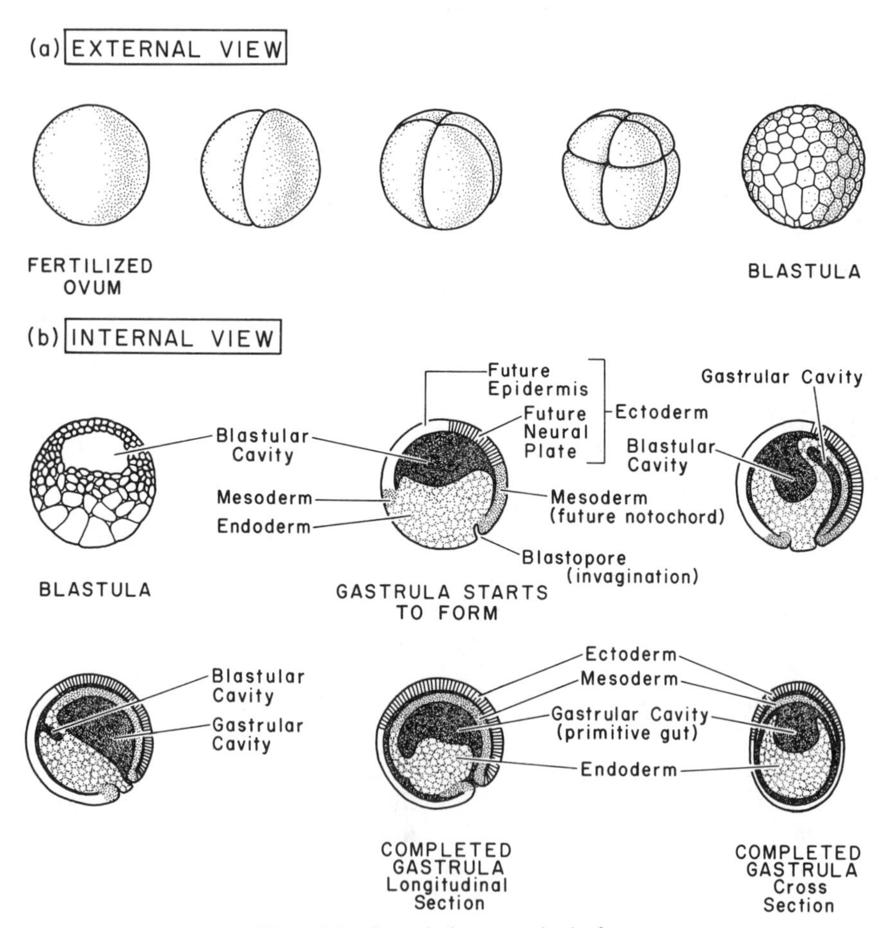

Figure 8.1 Gastrulation stages in the frog.

The cells most remote from the point of invagination end up covering the exterior surface of the embryo after other cells have migrated inside. These exterior cells are the ectoderm. Endodermal cells, initially closest to the entry point, move inside first and fill the lower half of the gastrula. The upper surface of their mass forms the floor of the gut. The mesodermal cells move in next and form the walls and roof of the gut. A column of specialized mesodermal tissue forms within the mesodermal layer. Called the *notochord*, this column will become the backbone, giving rigid support to the gastrula. Endodermal cells then spread upward from the gut floor to line the gut.

Gastrulation in other organisms involves other forms of movement of cell masses. In all cases, however, sheets of cells move, preserving their two dimensional integrity, and cause juxtaposition of formerly distant cell types. "Models" of gastrulation tend to be purely descriptive. Experiments concerned with the self-sorting of aggregates of cells, described below, are motivated by the desire to explain phenomena such as gastrulation in terms of individual cell properties. But there are difficulties in basing models for mass movements of cells on local cell interactions.

8.3.2 Neurulation

In vertebrate embryos, gastrulation is followed by a period of movement in the outer ectodermal layer above the notochord (on the dorsal side). Toward the end of gastrulation the embryo starts to elongate along its axis and the dorsal surface flattens out into the *neural plate*, as shown in Figure 8.2. As this plate flattens, the lateral edges rise up to form the *neural folds*. These folds define the division of ectodermal tissue into prospective neural tissue and the rest of the ectoderm. The plate is broad at the anterior (head) end and narrows posteriorly. Eventually the folds bend inward over the neural plate and fuse together. Exterior cells fuse to form the covering epidermis, while interior cells fuse to complete the *neural tube*. The middle of the tube closes first, followed by the large head cavity (the future site of the brain) and the posterior portion.

The neural tube forms the basis of the central nervous system. The beginnings of the optical system appear as evaginations of the brain cavity. When the neural folds fuse, there is a band of neural tissue left behind above the site of fusion called the *neural crest*. These cells will migrate from their position between the neural tube and the epidermis to form the components of the peripheral nervous system (Section 10.1).

8.3.3 Cell Sorting Phenomena

The primary objective of cell sorting studies is the explanation of morphogenetic movements observed in laboratory experiments. The basic finding in such studies has been that mixtures of disassociated cells of different types have a capacity for self organization into masses of cells of similar type. Thus a mixture of ectodermal and endodermal cells, for example, will sort out into a layer of ectoderm covering endodermal cells. Not only do the tissues sort out, but they often obey the order of stratification found in nature. Pioneers in the field saw this phenomenon occurring in two phases. First there is directed movement of cells which have become developmentally distinct (endodermal

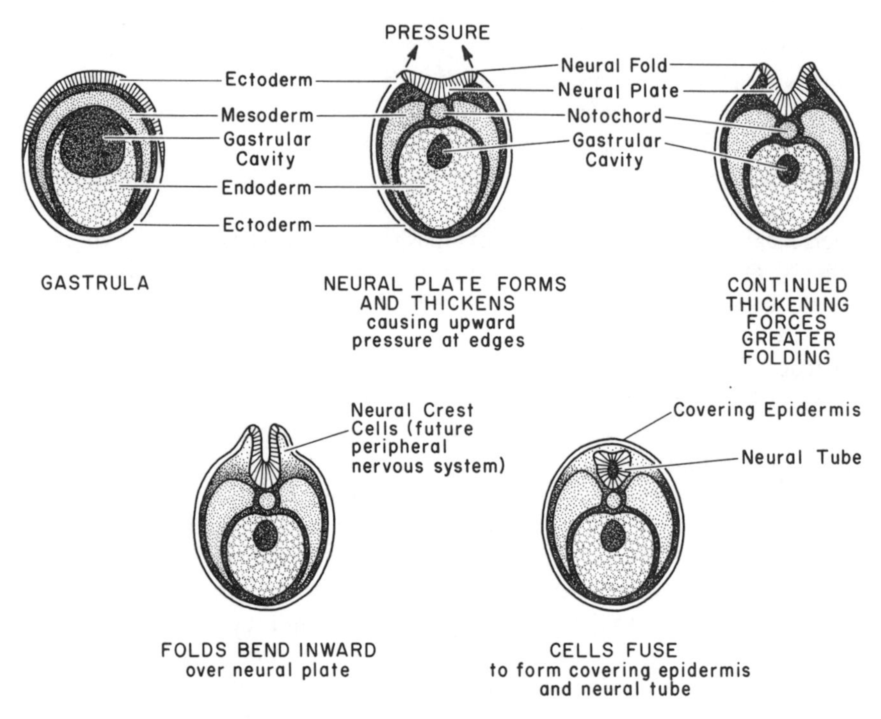

Figure 8.2 Neural tube formation.

cells move outward, ectodermal cells move inward). Second, the cells adhere selectively to one another, thus forming coherent masses or sheets of tissue.

The phenomenon of selective adhesion has been explained in two ways. In the first theory, like cells adhere because of surface properties that are mutually recognizable. This view arises from studies in which aggregates of cells both from different tissues and from different vertebrate species sort out in terms of tissue type and not species (for example, an aggregate of intermixed mouse and chick cells of the same tissue type is produced). The interpretation is that mutual recognition is a property of the tissue type and is species independent.

The second view arises primarily from the early work of Steinberg. It is proposed that there is some measure of adhesion between any two cells, but in some cases a greater strength of adhesion between cells of the same type. Sorting out is seen as a transition of the system to a configuration of maximal thermodynamic stability, where like cells minimize some form of energy in adhering to one another. In such a system, any mixture of motile cells will move toward a predictable and stable final configuration. Tissues with the strongest mutual adhesion will form the internal component of a mixture. In particular, with more than two cell types an onion-like configuration of layers of tissue would be expected. Such configurations have been observed experimentally.

Further evidence for the Steinberg hypothesis is provided by instances where mixtures of cells from the same tissue in different species are observed to sort by species. In such cases there is presumed to be a significant difference in mutual adhesiveness; in cases where sorting by species does not occur, it is assumed that the mutual adhesions happen to be too close to promote sorting. Such results suggest that adhesion is a continuous variable, not a binary switch involving absolute recognition or rejection.

Some controversy exists concerning the "central clumping" or onion configurations obtained in some studies. Other experimenters, notably Trinkaus, do not observe such final states, but instead obtain many "clumps" of the internalizing cell type distributed throughout the other tissue. The differences are suggested in Figure 8.3. This question is addressed by some of the simulation oriented models described in the next chapter. These models attempt to grasp the basic difficulty of modelling large scale structure solely in terms of local cellular interactions.

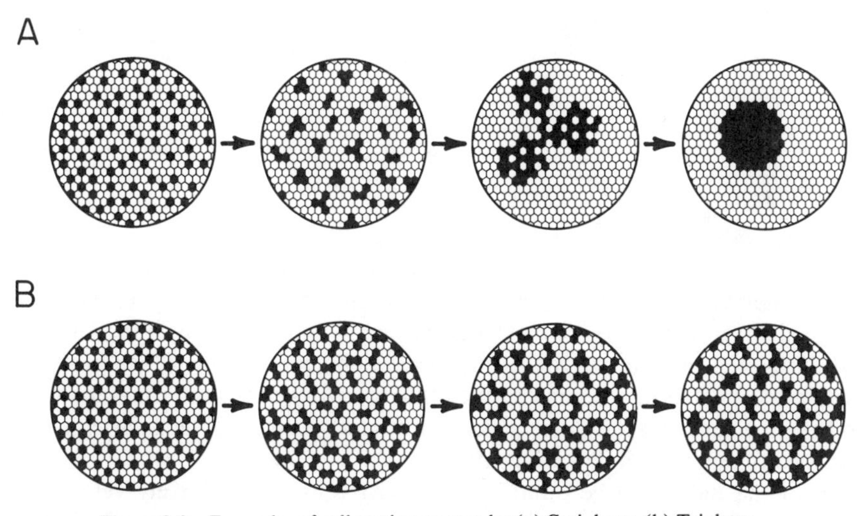

Figure 8.3 Examples of cell sorting as seen by (a) Steinberg, (b) Trinkaus.

In the larger perspective of morphogenetic movements, the phenomena of development cannot be explained in terms of the sorting out observed in the laboratory. In the developing organism, the layers of the gastrula form by sheets of cells rolling in through an opening, not by vertical sorting of individual cells into layers. Models for morphogenetic movements, like those for developmental phenomena of a genetic nature, which seek to relate individual cell behavior to the spatially and temporally organized development of the entire organism, must bridge the gap between purely local behavior and multicellular structures and processes.

Chapter Nine

Simulation of Morphogenesis

Models of development vary widely, with respect to the organism under study, the developmental events or stages selected, and the level of resolution chosen for modelling. Yet there are elements that many models share as they attempt to address central issues in multicellular development. Models which have been elaborated in a form designed to facilitate study by computer simulation tend to have special characteristics, and to encounter special problems, arising from the need to represent multicellularity.

Both the individual cell and structures of cells can be nicely represented using the cell space formalism, which makes it one of the most widely used techniques in morphogenetic modelling. Within this class of cell space models there are two types, which may be termed *mobility oriented* and *genetically oriented*. Models whose main focus is cell movement have difficulty coping with large amounts of cell-specific information, since each time a cell moves, all its contingent information must become associated with a new position in the array. At the other extreme, genetically oriented models may prohibit cell movement altogether, focusing instead on the dynamic information content in each cell.

All models using the cell space formalism have difficulty representing growth by cell division. Addition of a cell in the interior of the tessellation must be accommodated by extensive cell movement. An alternative to the cell space, often more suited to representing growth, is the linked list data structure, which has been employed by some modellers.

Both cell space and linked list models usually have a discrete time base. In multicellular structures, this means that continuously occurring processes in a field of base-model cells must be represented as discrete changes in the parameters identifying individual cells, occurring at particular times when the cell is being "processed." Most models represent these changes in terms of a fixed iterative processing order among the cells at each system-wide time step. A few models use special simulation techniques to obtain effects which resemble those of parallel and/or asynchronous activity in groups of cells.

Within the spectrum of developmental models, then, certain common techniques can be recognized, namely, those needed to represent spatial and temporal structure in iterative terms. Confronting these implementation issues illuminates three basic problems of developmental modelling: (1) how to represent both mobility and genetic information; (2) how to model simultaneous and continuous changes; and (3) how to incorporate cell divisions (growth). A complete morphogenetic model should solve all these problems; but, as will be seen below, most existing simulation work has had more modest goals.

9.1 MODELS OF CELL SORTING

All of the models considered in this section are mobility oriented and attempt to explain the experimental phenomena in which cells sort out according to type (Section 8.3.3). The *real systems* in all this work are thus cell mixtures in laboratory preparations. The *experimental frame* typically limits attention to variables associated with cell type and position, ignoring such factors as the nutrient medium. Validation of a *base model* within such a frame can usually be attempted only by visual comparison of cell patterns produced by simulation with those photographed in the laboratory. The major simplifying assumptions used to obtain a discrete-discrete *lumped model* allow cells to occupy points on a grid (usually two dimensional) at successive points in time.

9.1.1 Goel

Goel and his co-workers were among the first to employ the cell space formalism in constructing models for simulation studies of the Steinberg differential adhesion theory. In the first model developed by this group (Goel et al. 1970), cells of two types ("b" and "w") are arranged in a two dimensional array. A cell can interact with its four nearest and four diagonal neighbors. Changes in a given configuration are governed by an energy function, $E = N(bb)A(bb) + N(bw)A(bw) + N(ww)A(ww)$, where $N(xy)$ is the number of x-y cell confrontations and $A(xy)$ is the strength of x-y adhesion. The model is based on the supposition that maximal E corresponds to minimal surface free energy of the stable patterns, as predicted by the Steinberg hypothesis. A cell may move to a neighboring position if such an action can increase the value of E.

In the FORTRAN implementation of the model, cells in a 20 by 20 array were processed serially in random order, once each per global time step. Calculation of the E function was based only on the local (eight cell neighborhood) value for each cell. This decision was made both for computational expediency and on the grounds that individual cells cannot detect large scale phenomena. Global pattern changes arose from combinations of elementary moves, as individual cells attempted to improve their local energy values.

Steinberg's optimal patterns, including the onion pattern, could not be obtained in simulations starting from random initial configurations. Rather, stable non-optimal configurations showing various degrees of clumping were encountered. Such configurations resulted from entrapment in local maxima of E; further elementary moves would decrease E locally. The situation was only slightly improved when such small local decreases in E were permitted. Three dimensional simulations (of a 10 by 10 by 10 array) showed similar cross sections, so the results were not artifacts of the restriction of the model to two dimensions. The conclusion was that maximally stable patterns could not be achieved by strictly local cell motility.

Leith and Goel (1971) investigated modifications to the original model in further attempts to produce Steinberg's maximally stable patterns. The changes included different motility rules (for example, restriction of movement to nondiagonal neighbors only), different scanning techniques, and use of a hexagonal cell tessellation. The desired patterns were achieved when the criterion for cell movement was that a nondecreasing E be calculated over an extended neighborhood for each cell. This "extended zone effect" was interpreted by Leith and Goel, and by later modellers of cell sorting, to imply the ability of cells to sense and respond to types of cells several diameters away.

It should be noted that the need for an extended zone in the simulation does not necessarily mean that the original base model cannot also produce Steinberg's results. For example, the effect of sequential processing order was not analyzed. And the effect of representing the energy function in discrete, local terms was not evaluated. The possibility must be considered that the energy function may vary steeply, and can therefore be represented with sufficient accuracy only by including many cells in the calculation. This requirement for an accurate local value of a quantity does not imply any actual ability of real cells to appreciate remote events.

Later work by Rogers and Goel (1978) involved an interactive APL implementation of a revised model. New features of this version included the ability to vary parameters such as adhesion coefficients, rates of interaction, and initial configurations. In addition, "simultaneous" exchanges between all candidate cells for movement were simulated. An important new provision was that any two cells on a common interface could change places, rather than just immediate neighbors; the result of a directed series of local exchanges could thus be effected in a single "move." Many cell sorting phenomena were observed in simulations of this version of the Goel model, including sorting to Steinberg's patterns, rounding up of uneven tissue masses, and engulfment of one tissue by another.

9.1.2 Antonelli

The work of Antonelli and his colleagues seeks to explain cell sorting in purely local terms, without resort to extended influences. Antonelli, Rogers, and Willard (1973) present an analysis of sorting with two cell types. The Goel energy function is applied over an array of hexagonally tessellated cells. According to the *exchange principle*, two cells of different types cannot both increase their adhesive energy in an interchange of adjacent positions. But if the sum of both changes in energy is non-negative, a move is termed allowable. (Some moves that would be allowed by Goel, because only one cell of the pair is considered, are not allowable under this principle.) Moves that result in no energy change are called "O-moves" and allow modelling of random motion of isolated cells. For each cell, the optimum allowable move is chosen from among the six (or fewer) possibilities.

The model contains the following additional features. There is a Hamming type distance measure (correlation metric) to characterize configurations in terms of connectivity. Also possible is a constraint referred to as simultaneity, which means that each cell is given only one chance to move during the random sequential scan of a global time unit. In the nonsimultaneous case, a cell may move more than once. In simulation experiments comparing the two alternatives, "qualitatively similar" final configurations were attained, but more quickly under the nonsimultaneous condition.

The final states obtained by simulation were actually "steady" rather than "stable." Small oscillations, caused by the presence of O-moves, continued indefinitely. O-moves also allowed single cells to join clumps; occasionally such accretion led to fusion of clumps. The larger the incidence of O-moves, the faster the approach to steady state, but the final configurations were again qualitatively similar.

No central clumping was obtained by simulation. The authors suggest the onion phenomenon is an experimental artifact, perhaps because cell mixtures are centrifuged. Their results more closely resemble the observations of Trinkaus (see Section 8.3.3) than the predictions of Steinberg. The absence of central clumping, however, does not necessarily point to an error on Steinberg's part; a more likely explanation could once again involve the discrete numerical approximation inherent in the simulation.

Antonelli et al. (1975) applied the same model to more than two cell types. In their simulation experiments, the authors attempted to deal with the following four phenomena.

Transitivity. If cell type *a* segregates internally to type *b*, and type *b* internally to *c*, then if *a* also segregates internally to *c*, the sorting process exhibits transitivity. Mathematically, the exchange principle neither predicts nor denies transitivity. A stochastic version of Steinberg's theory, with adhesion proportional to the product of densities of adhesion sites on confronting surfaces, yields inequalities that ensure transitivity.

Pattern reversal. The type of the internally segregating component can depend partly on the relative quantities of the different cell types. Some biological experiments suggest a time dependent variation of adhesion values in which the initially more strongly self-adhering cells become the less adhesive type. Whether the first segregation is then reversed depends on cell type proportions. In simulations using the exchange principle, and approximating a continuous variation of adhesion values by an abrupt change during a run, the Antonelli group obtained pattern reversal for some relative proportionality values.

Engulfment. The authors could not simulate engulfment using the exchange principle because they could not ensure a preference for isotypic adhesions over heterotypic ones; mixing occurred rather than movement around the foreign mass. It was suggested that either extensions to the exchange model were required or that engulfment, like central clumping, should not happen in well controlled experiments.

Duality. With three or more cell types, results can be difficult to interpret because of "figure-ground ambiguity"; in a final configuration with many clumps, the internally segregating component is not always easy to identify. Biological results suggest a gradual movement of the external component outward, which the exchange principle could not simulate. Extension of the motility rules was again suggested.

9.1.3 Rogers and Sampson

The Rogers and Sampson (1977) model is not derived directly from Steinberg's differential adhesion theory but is based on a "hit-and-stick" rule for cells undergoing random motion. There are two types of cells, A and B, with A the internally segregating type. The cell space is a hexagonal tessellation, scanned in random order each time step. B cells are ignored during the scan. Solitary A cells move to neighboring positions, or remain where they are, according to a random walk with uniform probabilities of 1/7. Moving cells aggregrate as new neighbors belonging to A-cell clusters are encountered. Type A cells in clusters of size two or larger are allowed to generate provisional moves individually; the resulting overall configuration is allowed if two conditions are met. First, cluster *size* must not decrease, so that clusters do not fragment. Second, cluster *connectivity* (number of cell-face confrontations) must not decrease, so that clusters do not become less compact. Colliding clusters associate permanently.

APL programs were written to simulate the above basic model. Clusters did form from initially random configurations, but were not as compact and convex as those seen in the biological experiments of Trinkaus. Two new motility rules were therefore added to the original model. First, a cell joining a cluster was allowed one additional move if that move would increase connectivity. Second, movement in large clusters (size greater than five) was confined to individual chances to improve connectivity by movement of single cells. (Allowable overall moves in large clusters had been insignificant in number anyway.) The authors justified these modifications in terms of a "settling in" effect, whereby clusters tend toward more stable, compact configurations. Such tendencies cannot be modelled using the hit-and-stick principle alone.

In simulations using the modified model, biologically realistic clustering was obtained in populations of up to 50 type A cells at 20-25% density. Starting configurations included randomly distributed isolated A cells, randomly distributed A cells in which small clumps already existed, and a central cluster or "target" surrounded by isolated cells. It was concluded that the clusters ultimately obtained from isolated cells were more natural than the larger ones obtained from the latter two kinds of initial conditions.

A continuous model, in which Brownian motion theory predicts the numbers of clusters of given sizes at successive times, was compared with a tabulation of such numbers from the authors' discrete version, for small cluster sizes averaged over multiple runs from initially isolated distributions. Close agreement was obtained when coefficients were adjusted to reflect the progressive slowing of aggregation as larger clusters appear.

Rogers and Sampson contrast their strictly local model, which shows results similar to those of Trinkaus, with Goel's extended zones of influence, and with Antonelli's exchange principle in which it is possible for single cells to escape from clusters.

9.1.4 Hurum and Sampson

A model which investigates the potential contributions of gradients to cell sorting phenomena has recently been developed by Hurum and Sampson (1984). By giving different interpretations to the gradients, several different biological theories of cell sorting can be accommodated, including Steinberg's differential adhesion hypothesis, differential cell contractility, gradients of adhesive factors, and chemotaxis.

A given cell is the source of one or more gradients, with types depending on the type of source cell. A gradient decreases in strength with distance from its source. Each global time step consists of four successive phases.

1. For each cell x and gradient type i, $g(i,x)$, the sum of gradient i values from all other cells within the range of x, is calculated.

2. $G(i,j)$ is the coefficient of influence of gradient i on cell j. The sum, $E(x)$, of all different gradients on cell x (which is of type j) is computed by adding the products $g(i,x)G(i,j)$ over all i.

3. Every adjacent pair of cells is tested for possible exchange. The work required to exchange cells x and y is denoted $W(x,y)$ and computed as the net difference (positive or negative) in E that would result from the interchange.

4. If $W(x,y)$ is greater than a preset value (close to zero), then the potential exchange is stored in a binary tree ordered by the magnitude of $W(x,y)$. When all pairs have been tested, the tree is traversed. Those exchanges providing the greatest energy improvement are implemented first; and no cell is allowed to move more than once.

Hurum and Sampson implemented their model in Pascal and provided interactive selection of neighborhood space (hexagonal and 4- and 8-cell rectangular), two types of distance metric (Euclidean or number of intervening cells), various gradient functions, ranges and coefficients, sequential or "simultaneous" (binary tree) cell exchange, three variations of the basic model (including one using vectorial gradients), and various output options. In addition, it was shown how minor modifications could make both of the "rate limiting" steps (formation of gradients and testing of cell pairs for exchange) suitable for parallel computation.

The limits of sorting out and engulfment using only nearest neighbor exchanges were studied by simulation. The 8-neighbor cell space gave the best results, provided that corner neighbors were treated differentially from side neighbors. Linear gradients proved to be better than those varying polynomially or exponentially with distance, a finding at least partly attributable to the inherent spatial discretization. Simultaneous cell exchange produced more symmetric and usually more compact sorting than did sequential exchange. Neither the choice of distance metric nor that of model variation significantly affected simulation results.

For both sorting out and engulfment, emergence of concentric patterns required that the heterotypic gradient coefficient have a value intermediate to those for the homotypic coefficients of the two cell types, in keeping with the results of Goel. Local clustering was unavoidable when either the internally segregating cell type was dilute or when the gradient range was below a critical value. Otherwise, good sorting out patterns were obtained. Engulfment was more difficult to demonstrate; but partial engulfment could be obtained with favorable combinations of certain parameter settings. These results showed that the information about the environment provided by gradients was partially effective in overcoming the limitations imposed by the nearest neighbor exchange motility rule.

9.1.5 Matela and Fletterick

A radically different approach to modelling cell masses has been developed by Matela and Fletterick (1979, 1980). Instead of cell space techniques they employ a topological approach. Graph theoretic notation is used to encode cellular structure. Cells are represented by points on a plane, joined by "edges" if the cells are in physical contact with one another. Cells can have variable size and shape and be deformable. The entire cell mass is represented by a triangulated map of points joined by edges.

Cell sorting can be modelled using this formalism by identifying two cell types and postulating differential adhesion. The approach is thus yet another embodiment of Steinberg's hypotheses.

An exchange model is established as follows. In a given group of four cells, suppose edges exist forming the perimeter of the quadrilateral, but only one edge crosses the interior of the quadrilateral (only one opposing pair is in contact). This interior edge can be lost, and the other opposing pair joined, if the latter pair's bond is preferred in adhesive terms to the original one. Once an exchange is made, the pattern of quadrilaterals changes and new exchanges can be considered. The authors used table lookup to decide on the preference for a new bond.

Simulation studies demonstrated central clumping from arbitrary starting configurations, suggesting that the lack thereof in most of the locally oriented cell space simulations discussed above may be a consequence of discrete spatial representation. The authors intend to address the difficulties involved in relating the topological representation of a cell mass to its actual physical structure, thereby extending the capabilities of their model to other developmental phenomena.

9.2 CELL PROLIFERATION MODELS

9.2.1 Herman

Growth in multicellular organisms can be modelled using the theory of L-systems. An L-system is a cellular automaton in which components of the array can be replaced by other arrays. Because of the geometrical complexity of such replacements in two and three dimensions, most biological applications of L-systems have been in modelling structures with a one dimensional organization.

Both the components and the state set of an L-system model are discrete. The state of a cell usually refers to a genetic state which is heritable and changes according to a transition function. The function (which may be nondeterministic) is uniform over the array and takes as arguments the present state of the cell and the states of its neighbors. One dimensional L-systems have counterparts in formal language theory. The set of states is an alphabet; and developmental possibilities may be generated by applications of a set of production rules to initial state configurations. Because of this formal equivalence, many predictions about L-system models can be derived theoretically.

L-systems have been used by Baker and Herman (1972) to model the regulation of cell states in a one dimensional, filamentous organism (a species of blue-green algae). In this organism, a special type of cell (heterocyst) is observed at regular intervals on the elongating filament. Other multiplying cells differentiate into heterocysts at locations which maintain the regular spacing. In the L-system model, a nonheterocyst cell's state changes in response to a signal from heterocyst cells transmitted, one cell at a time, through the line of cells. The signal transmission is modelled by a difference equation.

The later work of Herman and Liu (1973) applies the techniques of L-systems to development in simple two dimensional organisms. Their FORTRAN program CELIA simulates iterative copies of linear filaments. The pigmentation patterns on the shell of a species of tent snail have been duplicated by simulation results using CELIA. The shell has an active growing edge. New cells exhibit pigmentation in the form of continuous lines in a peaked triangular pattern, with darkly pigmented triangles occurring and terminating simultaneously along the same line, that is, with their bases along the growth edge. Herman and Liu model these patterns using an L-system filament to represent the growing edge at each time step. Deposition of pigment occurs when the concentration of a pigment precursor is above threshold. The simulation results can be changed by varying input parameters to produce the patterns found on other types of shell.

While the studies of Herman et al. are limited by the one dimensionality imposed by their model of cell proliferation, they demonstrate the importance of cell interactions (here, in the form of linear gradients with thresholds) in determining patterns of cell states.

9.2.2 Hogeweg

If morphology in a growing organism is accounted for by cell divisions controlled by strictly local cellular interactions and by genetic information stored in individual cells' DNA, then, once an initial configuration is selected, final form is fully determined by encoded "rules" and neighborhood interactions. Hogeweg (1978) employs interactive simulation using cell spaces and L-systems to study models having these characteristics.

She considers the effects of discretization in generating lumped models suitable for simulation. Models using a discrete time base have an implicit form of global control when cells are seen as undergoing synchronous changes during each global time step. To avoid this artifact of the lumped model, SIMULA/67, in which asynchronous processing can be simulated, was chosen for model implementation.

Hogeweg uses a discrete event simulation technique, in which a "next event" strategy is employed. A list is maintained of those cells that can possibly change state in the current time step. This list contains all cells that changed state in the previous time step, together with their immediate neighbors. Inactive cells are not maintained explicitly during processing, but can be regenerated by the emergence of an active neighbor.

Hogeweg demonstrated her discrete event strategy using SIMULA/67 for a cell space model. She specified active cells by state and geometric position. The position was used to link this information to an entry in a hash table, which permitted next states to be computed using the desired transition rules and the neighbor states obtained through uniform operations on the hash table. A prototypic "process" was defined for an active cell; a particular cell's position and state corresponded to local variable settings in this process. Each "specialization" of the process existed as a separate copy of the code, and could be activated or delayed using SIMULA/67 features. Active cells could thus be (conceptually) processed in parallel, or asynchronously, or under "local synchrony" by controlling the updating time for cells' new states. At any point, cells which did not change state during one process invocation were deleted from the hash table. Note that use of a hash table means that the cell space is still of bounded size; growth is not actually unrestricted.

The discrete event strategy in SIMULA/67 was also used for an L-system model. Here the structure was given not by geometric position, but solely in terms of immediate neighbors (corresponding to a linked list, not an array or hash table). In determining cells' next states, those of neighboring cells had to be investigated by access through list structures, not by uniform array operations. With list structures, problems of boundaries to growth were circumvented by means of truly dynamic data acquisition. Final configurations were presented for three versions of the L-system model, based on different processing strategies and exhibiting structurally different branched forms.

Hogeweg stresses that these studies are of the development of cellular forms, not real organisms, and therefore are of heuristic interest to biologists. Her contributions lie in the use of a programming environment capable of simulating nonsequential processing, the use of data structures suitable for dynamically changing cell populations, and the use of a next event scheduling strategy. The last is of particular value when a majority of cells is inactive at any one time.

9.2.3 Ransom

A model for cell division has been designed by Ransom (1977a, 1977b, 1981) to study the development of clonal patterns in the imaginal discs of *Drosophila* by means of simulation. A *clone* is the set of descendants of a single cell (or small number of primordial cells). If the cells of the clone are given a detectable, heritable marker early in development, then the shape of the clone can be followed through the normal processes of development, and insight gained into the establishment of structures in the disc and the resulting organ.

Ransom's lumped model is of the discrete-discrete variety. Edinburgh IMP was chosen as the programming language. Spatial discretization consisted of embedding cells in a hexagonal tessellation. Cells were actually stored in a one dimensional array whose index mapped to an address on the hexagonal plane. Cells were characterized by a digit representing a clonal marker. Temporal discretization consisted of global time steps, in which every individual cell could attempt to divide once. Cells were processed in a random serial order by selection from a one dimensional array of candidate cells. The array was reduced and maintained in compact form after each cell division, which necessitated the movement, on average, of half the parent cells of that generation per cell division. This aspect of processing could thus become quite time consuming for simulation of a sizeable system.

Cell division in the hexagonal space is accomplished by a "pushing" algorithm. The nearest empty location along the six directions in the tessellation is identified, and all cells along that line moved outward. This movement entailed locating and moving affected entries in the one dimensional array.

Simulations of growth in eye-antennae and leg discs were undertaken. The oriented divisions (preferred directions for division) observed in the real eye could be reproduced through constraining the cell mass by an expanding membrane, represented at each time step by predefined cell positions in which division was forbidden. Unconstrained growth led to radial divisions of the kind observed in actual leg discs.

9.2.4 Duchting

Duchting (1978) has presented a two dimensional cell space model for competitive growth of two cell species with different division rates. Simulations of the effectiveness of removing some of the faster growing cells were studied, with a view to modelling tumor growth and treatment. The initial version of this model was rather simplistic. In particular, cells in the 10 by 10 rectangular array encountered severe edge effects.

Later work by Duchting and Vogelsaenger (1983) extends the model to a three dimensional (40 by 40 by 40) cell space containing a nutrient medium and arbitrarily placed tumor cells. Each cell goes through a series of cell-cycle phases (the length of each phase being determined stochastically) culminating in a readiness to divide. The cell divides if it is within a critical distance of the nutrient medium. It does so by using a pushing algorithm to find and fill the nearest empty space. Each cell has an assigned probability of mortality and, at any time step, can be removed or can die because it is too far from the nutrient medium. Thus natural cell death can be modelled, as can surgical removal of tissue and death by radiation treatment or chemotherapy. In addition, necrosis of tumor interiors can modelled in terms of lack of nutrient.

The model was implemented in FORTRAN IV. Simulation of growth therefore involved considerable data relocation in arrays. Output for the simulation studies was provided in the form of plots of cross sections through the tumor spheroid in any desired plane. Cells were represented in these plots by symbols standing for their current phase in the cell cycle.

The authors thus developed a means of modelling tumor growth and interruption of growth by various methods of treatment. Based on their preliminary results with this system, Duchting and Vogelsaenger suggest that simulation experiments can partially replace extensive biological tests in the study of cancer.

9.3 MODELS OF PATTERN FORMATION

In Chapter 8 it was seen that pattern formation in developing organisms can be explained in terms of a genetic response by individual cells to a field of positionally specific information established dynamically in the growing structure. The studies described in this section attempt to model gradient or prepattern formation and the resultant genetic changes.

9.3.1 Wilby and Ede

The formation of the skeletal structures of the avian wing has been the focus of simulation studies by Wilby and Ede (1974, 1976). They use a two dimensional rectangular cell space to model the cross section down the limb and laterally through it. The model uses a pushing algorithm, as described above, to allow cells to divide toward the nearest empty space. In addition, realistic shapes for the wing are generated by making cells respond to an imposed mitotic gradient down the wing and by allowing cells to move slightly down that gradient.

In addition to experiencing an imposed mitotic gradient, cells can respond to another gradient and differentiate into cartilage cells. This gradient is generated by special cells in an "initiator region" and diffused passively by other cells. If the initiator region lies along the anterior edge of the array, and if particular parameter settings are chosen in the difference equations used to represent discretized diffusion, then periodic patterns of concentration are seen in the simulation results. These patterns may model iterated bone structures such as digits. The system parameters needed to produce such results, however, are chosen arbitrarily.

9.3.2 Meinhart and Gierer

Meinhart and Gierer have attempted to model the actual biochemical mechanisms responsible for pattern formation (Gierer and Meinhart 1972, Meinhart 1978). They use two coupled, nonlinear reaction diffusion equations, for activator and inhibitor species, to model pattern formation in organisms that can be regarded as one dimensional. The models assume initial gradients in the densities of cells which are sources of the two chemical species. Then, if a reaction diffusion system is set up in which the activator (which catalyzes both itself and the inhibitor) has a lower rate of diffusion than the inhibitor (which inhibits only the activator), patterns of concentration of the two species are established with an accentuated peak at one end of the array. The authors suggest that such patterns could act as prepatterns for long term differentiation of additional cells into sources of the species. Results of computer implementation of the model (by unspecified numerical techniques), when compared to quantitative data from hydra, suggest that the model can account for certain duplication phenomena.

Use of an initial gradient in the density of already differentiated source cells, in order to obtain a prepattern for further cell differentiation, suggests that the model is one of pattern "reinforcement" rather than pattern "formation." Establishment of the initial density gradient has not been explained. Less initial disparity is necessary in the work of Meinhart (1978), where the field is polarized by a single source of activator and inhibitor at one end. Such a model is used to discuss segment formation in insect embryos. Simulation results for insect embryos successfully demonstrated realistic one dimensional patterns. In addition, certain surgical experiments on real embryos can be simulated using Meinhart's model. Interrupted patterns were repaired when the size of the original system was restored and the reaction diffusion model allowed to resume operation.

9.3.3 Descheneau

An environment for modelling development has been created by Descheneau (1981). A pool of component models comprises a "model base" as envisaged by Zeigler (1984) in his recent theory of multifacetted modelling. Particular components of the base are used to model different phenomena of development and can be combined and allowed to interact dynamically when a particular developmental system is to be simulated.

Because it is impossible to identify all genes in the vast DNA of multicellular organisms, and because a true appreciation of development as a dynamic process, by means of such a "bottom up" treatment is difficult, Descheneau incorporates the notion of a *difference model*. Such a model focuses on differences between cells arising during developmental events. Genetic models which use this type of technique were seen in Chapter 8, including induction (in which properties unaffected by the inductive event are ignored) and positional information models (in which characteristics of cells unrelated to position are ignored).

Difference models have the power to vary their scope (some or all cells) and proceed from simplicity (identical cells) to diversity (differentiated cells). Each cell is characterized by a set of local parameters relevant to the difference model being considered. This state set may include concentrations of various chemical substances, various kinds of forces such as pressure, or internally regulated values such as cell cycle clocks. In addition, a cell has a genetic state. Genetic information is characterized by a *genetic activation code.*

During a developmental event, a cellular population is divided into mutually differentiated subpopulations, or at least into subpopulations of different determination. This difference is coded by the allocation of binary digits at a developmental event. In a binary subdivision, subpopulation A would be given 10 while B would be given 01. The 1 symbolizes the activation of behavior peculiar to each cell type. The behavior of a cell is characterized by the operation of this code on the set of state variables being maintained for the system under study. Code digits accumulate as developmental events occur and differences between cells increase. The state of a cell is also dynamically affected by processes indicated by its genetic activation code. Such processes may be discrete (e.g., cell division at intervals of time) or continuous (e.g., chemical flow).

The modelling framework is of mixed type (continuous time, continuous and discrete variables and processes) and is multilevelled in that the perspective may range from individual cells to large scale cellular structures. Simulation is done using ALGOLW, which allows dynamic acquisition of storage for information defining new cells during growth. Structures of cells can be represented by linked lists. Such data structures are related to a cell space by maintaining addresses in the lists and by hash tables. ALGOLW also has efficient numerical operations for simulating processes modelled by differential equations.

Simulation of development in a given experimental frame consists of the invocation of relevant components from the implemented model pool. Typically, models for cell division, intercellular processes, and the particular mechanisms for recognition and imposition of developmental events are chosen.

Descheneau describes simulation of various models of cell division, including clocked division, division on the attainment of a threshold value of a metabolite, and division in response to a combination of deterministic and probabilistic factors. Her model uses an efficient algorithm for the cell movements arising from divisions, the list structures allowing pointers to be changed rather than requiring transfer of data for each moving cell. One of the major difficulties encountered in other developmental models, that of combining mobility and significant genetic data for individual cells, is thus avoided by the use of appropriate data structures.

Features for intercellular communication of the reaction diffusion type are also included. The continuous processes are discretized by numerical approximations to these nonlinear differential equations. Patterns of concentration (as theoretically predicted on the basis of linearized stability theory) are achieved in rectangular two dimensional structures.

Determination, as modelled by changes in the genetic activation code, arises dynamically as a local event in each cell, in response to critical situations encountered during growth and intercellular communication in cell structures. In hypothetical structures, growth rates and modes of communication interact to determine the patterns of positional information which provoke developmental events.

Differentiation, as modelled by changes in behavior resulting from changes in the activation code, was simulated for two real systems. Segmentation in the blastoderm of *Drosophila* was achieved in response to patterns of positional information simulated as arising during cell divisions and reaction diffusion. Theoretical models of the type discussed in Section 8.2.2 can thus be successfully implemented. The development of bone structures in the mammalian limb was simulated using a reaction diffusion model as the sole pattern forming factor in a growing chick wing. The ability of the reaction diffusion model to explain the establishment of a nonuniform prepattern without initial inhomogeneities (such as the source cells of Gierer and Meinhart or the intitiator region of Wilby and Ede) was demonstrated dynamically.

Descheneau's major contribution for the developmental biologist is the provision of a framework into which particular models can be inserted, and then combined with selected models for other developmental phenomena. The modelling resource is capable of operating at various levels of resolution. And, because it is based on the concept of a difference model, it maintains a top down perspective on development.

PART FOUR

Information Processing in Neural Systems

Chapter Ten

Neural Architecture

Processing of information in the nervous systems of higher animals involves a variety of encoding and signalling mechanisms, to be treated in the next chapter, that occur in a structurally complex milieu, the subject of this chapter. Although many of the macroanatomical descriptions in the first three sections below apply to vertebrates generally, the treatment here emphasizes neural architecture in mammals, usually with specific reference to man. The microstructures (cells and synapses) considered in the last two sections, however, are similar in the nervous systems of most organisms.

The *central nervous system* (CNS) consists of the brain and spinal cord. A collection of other neural and associated structures comprise the *peripheral nervous system*. Similar types of components in these two general divisions of the nervous system have traditionally been given different names. Information transmission lines, bundles of neural fibers which serve mainly to carry data from one place to another, are called *tracts* in the CNS and *nerves* in the periphery. Information processing nexi, regions of interconnection containing neural cell bodies, are known as *nuclei* in the CNS and *ganglia* elsewhere. Areas consisting mostly of nuclei have the characteristic grayish hue of neural tissue, hence the term "gray matter." Tracts and nerves are wrapped in varying amounts of fatty sheathing material whose paler color has given rise to the term "white matter."

Direction and orientation in the nervous system are specified by a number of standard anatomical terms, which usually come in pairs. The more frequently encountered ones include: *medial* (toward the middle) and *lateral* (toward the side), *anterior* (toward the front) and *posterior* (toward the rear), and *dorsal* (toward the back) and *ventral* (toward the abdomen). The primary longitudinal axis of the CNS is always considered anterior-posterior, regardless of whether the spinal cord runs front to back (as in quadrapeds) or top to bottom (as in bipeds). Since there is complete bilateral symmetry of structure in the nervous systems of vertebrates, many of the diagrams below show one side only or illustrate different components on the two sides.

10.1 THE PERIPHERAL NERVOUS SYSTEM

The nerves and ganglia that transmit information between the CNS and the rest of the organism are structured rather differently around the brain than along the spinal cord. Direct communication between brain and periphery is

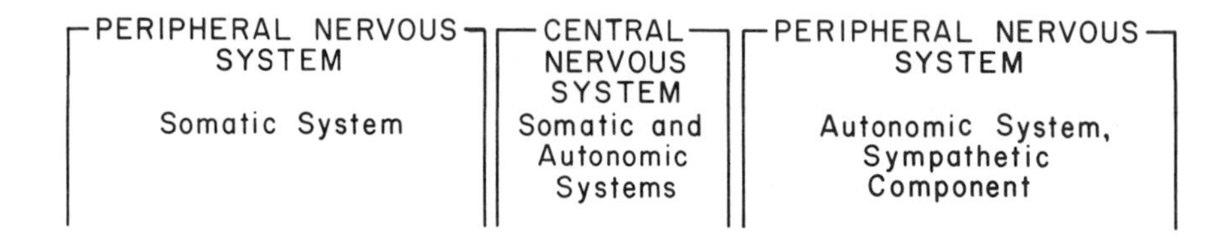

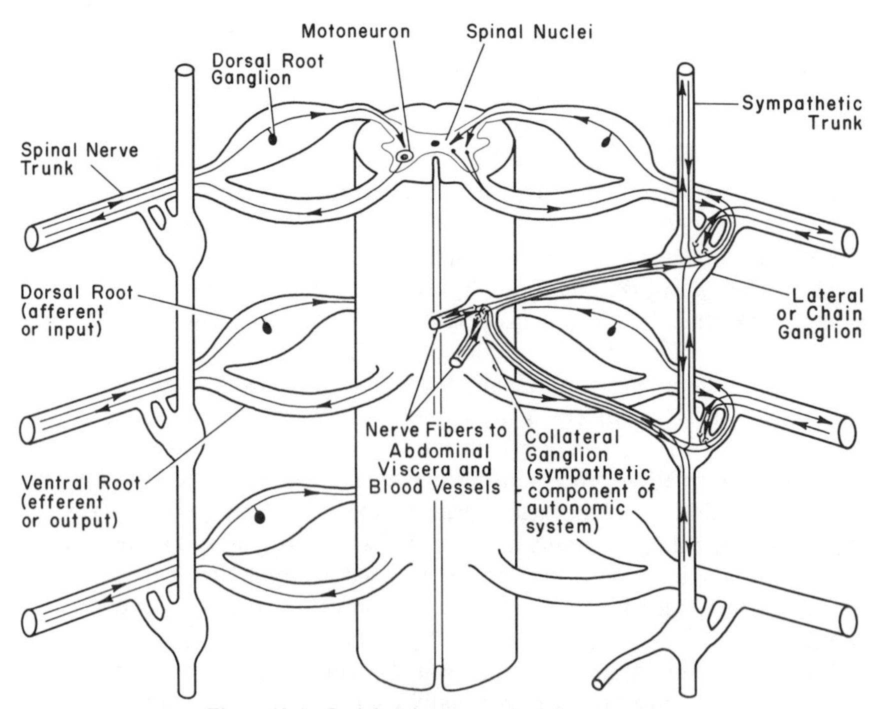

Figure 10.1 Peripheral nervous system components.

mediated by the *cranial nerves* (Section 10.1.2), whose organization is dominated by data pathways for the specialized sensory systems in the head, such as vision.

Access to the spinal cord is limited by the bones of the vertebral column. Associated with each intervertebral gap is a pair of *spinal nerves* and associated ganglia. The structural organization at each level is essentially the same and is shown diagrammatically in Figure 10.1. Note that the spinal nerve trunk divides to enter the cord as both dorsal and ventral roots. There is also a branch to the system of lateral and collateral ganglia, which are components of the *autonomic nervous system* (Section 10.1.3).

10.1.1 Spinal Nerves

In man, the 31 pairs of spinal nerves and their associated spinal cord segments have been assigned to five groups. From top to bottom, there are 8 cervical (C1-C8), 12 thoracic (T1-T12), 5 lumbar (L1-L5), 5 sacral (S1-S5), and 1 coccygeal (Coc 1) nerve pairs. So ordered, these nerves mediate communication with successively lower, but overlapping, portions of the body. By careful mapping of the loss of sensation and/or motor control, neurologists can often determine which nerves have been affected by disease or injury.

A human spinal nerve is a cable containing thousands of individual fibers. Roughly two thirds of these carry information toward the CNS from *sensors* in the skin, muscles, and internal organs. Most of these *afferent* (input) fibers are attached to cell bodies in the *dorsal root ganglion* (see Figure 10.1) and enter the spinal cord as the dorsal root. *Efferent* (output) fibers emerge primarily from cell bodies in the ventral part of the spinal cord. These fibers carry information away from the CNS, to muscles and other *effectors*. The sensory and motor structures at the termini of spinal nerves are considered further in Chapter 13.

There appears to be some degree of organization within a spinal nerve. Larger nerves have up to several hundred distinct "bundles" of fibers. A bundle contains both afferent and efferent fibers from some localized region within the general body area served by the nerve. Within a bundle, fibers are probably clustered according to function. Among the afferent fibers, for example, it would be more likely to find those mediating pain together, rather than randomly interspersed with those mediating temperature sensation. Thus the principles governing organization of well-mapped regions in the CNS (see, for example, the discussion of the visual cortex in Section 13.3) appear to have some application in the periphery as well.

10.1.2 Cranial Nerves

Man has 12 pairs of peripheral nerves that connect directly with brain structures. The cranial nerves have been given both names and the numbers I through XII. Unlike most nerves, the olfactory (I) and optic (II) nerves have exclusively afferent fibers, serving to carry information from sensors in the nose and retina, respectively. The path and information content of the optic nerve are discussed in Section 13.3. The oculomotor (III), trochlear (IV), and abducens (VI) nerves are two-way channels used primarily in the control of eye movements.

The trigeminal nerve (V) controls the muscles of mastication and handles most of the cutaneous sensory input (touch, pain, temperature) in the head region. Perhaps the most versatile of the cranial nerves, the facial (VII) is involved in control of the muscles of facial expression, the secretion of tears and saliva, touch reception in the external ear, taste reception on the front part of the tongue, and sensors in the blood vessels of the face. Many of these functions are shared with the glossopharyngeal (IX), vagus (X), and accessory (XI) nerves, a group which is also involved with control of the musculature in the larynx and pharynx.

The acoustic nerve (VIII) has two major independent divisions. The cochlear division carries information from auditory receptors in the inner ear to the brain and some gating control information in the other direction. The vestibular division of the eighth nerve carries information from the semicircular canals that contributes to the maintenance of balance. Finally, there is the hypoglossal nerve (XII), which functions primarily in the control of tongue musculature.

This brief survey has revealed that some cranial nerves are highly specialized, such as the optic nerve, whereas others have even more functions than the general purpose spinal nerves. It may be the case that, during the evolution of the special senses, those channels carrying large amounts of important information, such as vision, audition, and (for many higher mammals) olfaction, eventually acquired "private lines." Other less vital or simpler special functions, like crying and tasting and chewing, could have been grafted onto one or more spinal nerve type channels.

Analysis of the distribution of function among cranial nerves should also take into account the fact that mere coincidence of location of two or more central nuclei could have caused their perhaps unrelated peripheral communication functions to be assumed by a single cranial nerve. The same applies from the other end; the auditory and vestibular senses, for example, presumably share the eighth nerve because of the proximity of their sensors. As in any complex, multimodal communications system, compromises between homogeneity of function and coincidence of position can lead to unexpected cabling arrangements.

10.1.3 The Autonomic Nervous System

To this point, the discussion of the peripheral nervous system has emphasized the so-called *somatic* components, namely the special and cutaneous sensory inputs and the two-way channels for control of skeletal (voluntary) muscles. The muscular and glandular structures of the systems for respiration, circulation, digestion, elimination, and reproduction (the viscera, for short) are regulated by the *autonomic* components of both the central and peripheral nervous systems. Centrally, autonomic and somatic components are usually indistinguishable on structural grounds. And the two systems frequently share peripheral nerves. But since the most distinctive structural feature of the autonomic system is an additional set of peripheral ganglia (along with some special nerves), it is appropriate to consider briefly the entire system at this point.

The autonomic nervous system has two subdivisions, based on a dichotomy in the distribution of peripheral components and a corresponding difference in influence on visceral function. The *sympathetic* division influences nearly all the viscera, typically mobilizing systems for response to stress. Among the effects of sympathetic stimulation are pupil dilation, blood vessel constriction, sweating, increased heart and respiratory activity, decreased digestive activity, and release of adrenaline. Anatomically, the sympathetic system output to the periphery occurs between levels T1 and L3 (inclusive) of the spinal cord; thus it is also known as the thoracolumbar division of the autonomic nervous system.

Sympathetic ganglia are of three types, two of which were seen in Figure 10.1. There is one lateral ganglion near the ventral root at each output level; extensive vertical interconnection of these ganglia, outside the spinal cord, has led to the name "chain ganglia." These ganglia actually extend to the upper cervical and lower sacral levels, where they do not connect to the spinal cord. There are four major collateral ganglia in the sympathetic division, and a variable number of small accessory ganglia, near the cord and closely associated with spinal nerve trunks.

The other division of the autonomic nervous system is the *parasympathetic*, which distributes information to fewer visceral structures (rarely, for example, to blood vessels or sweat glands), by way of ganglia which tend to be near or within the target organ. The CNS connections to the parasympathetic division are from the brain (cranial subdivision) and the sacral portion of the spinal cord (sacral subdivision). The vagus nerve is a major channel in the cranial subdivision, influencing the heart, lungs, and upper digestive system. Pelvic viscera are influenced by the sacral subdivision.

The effects of parasympathetic activity include pupil constriction, blood vessel dilation, lacrimation, salivation, decreased heart and respiratory activity, and increased digestive activity. Unlike sympathetic influences, the parasympathetic effects are not a coordinated response to an environmental occurrence. Rather, they are ongoing processes that serve to maintain desirable physiological operations under normal environmental conditions.

10.2 SPINAL CORD AND BRAIN STEM

Figure 10.2 shows the major structural features of the vertebrate CNS as they appear in man. The structures above the spinal cord have traditionally been classified into three regions: the *hindbrain* (rhombencephalon), containing the medulla, pons, and cerebellum; the relatively small *midbrain* (mesencephalon); and the richly elaborated *forebrain*, made up of the diencephalon (thalamus and hypothalamus) and telencephalon (cerebral hemispheres). Another convenient CNS subdivision is the *brain stem*, which contains all brain structures except the cerebellum and cerebral hemispheres.

10.2.1 Spinal Cord

In addition to being the only path for information flowing between the spinal nerves and the brain, the spinal cord does considerable processing of that information; further, it autonomously governs a great variety of local reflex channels (see Section 13.1). As shown in Figure 10.3, the cross-sectional organization of the spinal cord has a central butterfly shaped region of gray matter, the spinal nuclei. The surrounding white matter consists of ascending and descending spinal tracts, some of which are labelled in the figure. Because increasing numbers of spinal nerves supply input and receive output during the ascent toward the brain, the number of fibers in the tracts increases, making the white matter noticeably larger at higher levels. The small hole at the midline is the central canal, a remnant of the embryonic neural tube. The canal connects with larger ventricles in the brain, a system through which cerebrospinal fluid circulates.

Neuroanatomists have identified some 15 distinct nuclei in the spinal cord gray matter, the majority of which are named for either their placement in cross section (e.g., dorsolateral, ventromedial) or their vertical distribution (e.g., lumbosacral). In three dimensions, most nuclei are elongated cylinders, although only a few run the entire length of the cord. Extensive study has allowed assignment of a primary type of afferent or efferent information to each nucleus. The types, including the various cutaneous senses, control of particular muscle groups, and autonomic output, are usually closely associated with the information flowing along nearby tracts.

The spinal white matter contains more than 20 tracts for carrying information to and from the brain. Although the majority of these tracts run the entire length of the cord, the individual fibers of some tracts may not extend more than a few segments. There is thus some localized transfer of information vertically, as well as horizontally, within the spinal cord. As already suggested, these local pathways interface with spinal nuclei to govern various reflex systems.

The naming of spinal tracts largely follows a simple convention. The first part of the name relates to the position of the tract in cross section. The second,

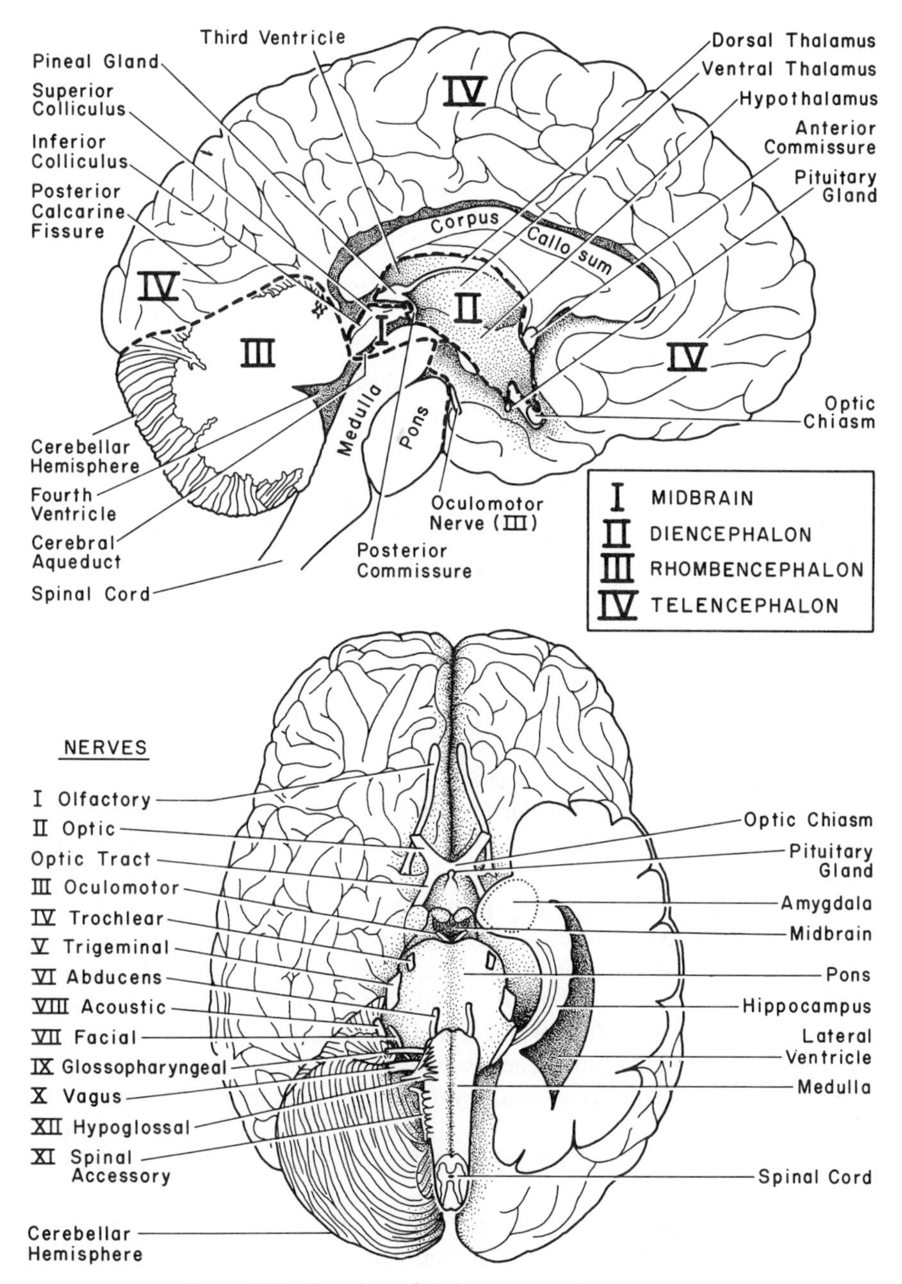

Figure 10.2 Two views of the human central nervous system.

usually compound, part of the name identifies the origin and destination of the tract, having "spinal" first for ascending and last for descending tracts. Thus the lateral and ventral spinothalamic tracts (see Figure 10.3) both carry information up to the thalamus, whereas the lateral and ventral corticospinal tracts both bring information down from the cortex of the cerebral hemispheres.

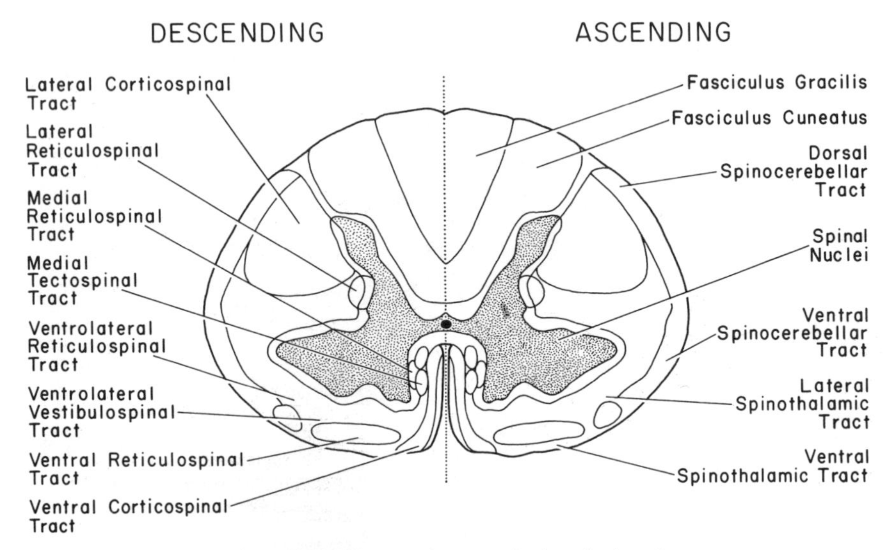

Figure 10.3 Some major tracts in the spinal cord.

The roles of some individual tracts in particular sensory and motor systems will be explored in Chapter 13. For present purposes it is sufficient to note the primary tracts involved in the four major categories of information flowing in the spinal cord: (1) cutaneous sensation via the fasciculi gracilis and cuneatus (fine touch), lateral spinothalamic (pain), ventral spinothalamic (pressure and crude touch), and spinocerebellar tracts (generalized touch, for coordination); (2) information from muscle, tendon, and joint receptors via the fasciculi gracilis and cuneatus, spinovestibular, and spinocerebellar tracts; (3) skeletel muscle control via the corticospinal tracts (fine, voluntary movement) and the tectospinal and vestibulospinal tract systems (crude and/or involuntary movement, coordination); and (4) visceral control via the reticulospinal tracts.

10.2.2 Medulla, Pons, and Midbrain

The transition from spinal cord to brain stem is not an abrupt one. Many tracts continue upward for some distance, in approximately the same positions as in the cord. But others terminate in nuclei that give rise to new pathways. And the progressive enlargement of the neural column is accelerated by the appearance of the cranial nerves and their associated nuclei. As the complexity multiplies, this account necessarily becomes sketchier and more selective.

Sitting atop the spinal cord is the *medulla oblongata* (medulla, for short), within whose walls the central canal gradually opens into the wider passage of the fourth ventricle. Cranial nerves IX through XII (accessory, vagus, glossopharyngeal, hypoglossal) enter the brain through the medulla, their fibers terminating in and originating from nuclei there. As a result, the medulla is heavily involved in the coordination of somatic and autonomic functions in the oral cavity, as well as in the modulation of respiration and heartbeat (via the vagus nerve).

Most functions on one side of the body are managed in the opposite side of the brain. Tracts thus usually cross to the opposite side (decussate) at some point. Some of this crossing occurs in the spinal cord where the fiber enters or exits as part of a nerve. Two major systems, however, cross in the lower medulla. In the motor decussation, descending fibers destined to become the lateral corticospinal tract shift to the side of the body they will influence. Just above the motor decussation lie the nuclei gracilis and cuneatus, terminations of the corresponding fasciculi. A new common tract, called the medial lemniscus, arises as fibers from these nuclei cross in the sensory decussation. It may be coincidence that the systems for fine touch discrimination and fine motor control both decussate in the lower medulla. There is no evidence that they interact there.

Pons, meaning bridge, is the name given to the structure above the medulla because of the prominent arching bulge on the lower two thirds of its ventral surface. The bulge is a massive band of fibers connecting pons and cerebellum. The rest of the pons, not much different from the medulla in general character, provides upward continuation of tracts and interface with cranial nerves V through VIII (trigeminal, facial, abducens, acoustic). The pons and medulla share the primary relay system for auditory input (see Section 13.2); nerve VIII enters the brainstem at their border.

The fourth ventricle, having reached its widest point at the boundary between medulla and pons, progressively narrows until (in the midbrain) it is known as the cerebral aqueduct. In the diencephalon, the channel broadens again, forming the third ventricle. The uppermost parts of this internal circulatory apparatus are the paired (first and second) lateral ventricles, one inside each cerebral hemisphere.

In the midbrain appear the first visual information processing structures, along with second-stage auditory stations. In lower vertebrates, the midbrain is a primary control center for these senses. In mammals, especially man, most control has evolved into the highly developed cerebral cortex, leaving mainly reflexive and unconscious mechanisms in the midbrain.

On the dorsal surface or tectum (roof) of the midbrain are four small bumps known pairwise as the superior (anterior) and inferior (posterior) colliculi. The inferior colliculus is an important processing station for auditory information. The intricately structured superior colliculus (seven layers have been distinguished) is involved in human visual reflexes and other aspects of oculomotor coordination. The primary nuclei of the third and fourth (oculomotor and trochlear) cranial nerves are also found in the midbrain.

The midbrain also contains major nuclei of some efferent systems. Notable for their involvement in muscle tone, movement stabilization, and spatial orientation are two regions named for their characteristic coloring; the red nucleus in the intermediate level (tegmentum) of the midbrain and the substantia nigra in the bottom (ventral) level. Finally, there are tracts which use the midbrain as a passageway, having little or no connection with its information processing functions. One example is the medial lemniscus, carrying somatic sensory data from nuclei in the medulla to the thalamus.

10.2.3 Diencephalon

The two major components of the diencephalon perform an astounding variety of integrative and relay functions. The thalamus processes most of the information flowing between the cerebral hemispheres and lower centers. The tiny hypothalamus coordinates somatic, autonomic, and endocrine functions in the regulation of reproduction, sleep, emotional reactions, temperature, fluid balance, and weight.

The thalamus is composed of about 30 pairs of nuclei. Most of these, including all the ones that are reasonably well understood, are in the *dorsal thalamus*. Notable structures in the *epithalamus* include the mysterious pineal gland and the *posterior commissure*. The latter is a tract with the sole function of conveying information between the two sides of the brain. Interlateral communication has been part of the routine connection pattern at lower levels. At and above the thalamic level, however, the two sides of the CNS are well separated physically, requiring special arrangements for lateral transfer of information. Epithalamic nuclei are probably involved mostly in olfactory information processing and do not connect forward to the telencephalon.

The *ventral thalamus* is apparently involved in motor coordination. It also contains the uppermost regions of the *brainstem reticular formation*, a densely interconnected column of gray matter near the midline, running down into the medulla. More a system than a clearly demarcated structure, the reticular formation makes far ranging afferent and efferent contacts and has been implicated in the coordination of numerous bodily functions.

The nuclei of the dorsal thalamus can be divided into two general groups. The *specific* sensory and motor relay nuclei have well identified and clearly localized connections with higher structures. Included in this group are the medial and lateral geniculate nuclei, which are the thalamic centers for processing auditory and visual information, respectively. These centers, along with those relaying somesthetic information (posterior ventral nuclei) are considered further in Chapter 13. Other specific nuclei are involved in efferent control systems, both for skeletal muscles and autonomic functions. Still others are called "integrative" specific nuclei, since they connect with similarly designated areas in the cerebral cortex.

The *nonspecific* or generalized thalamic nuclei, a phylogenetically older group of structures, are widely interconnected among themselves, with specific thalamic nuclei, and in some cases with higher centers. The nonspecific system is thought to be involved in maintenance of function in the thalamus itself, among other things.

Although about a dozen nuclei have been distinguished in the *hypothalamus*, assignment of specific functional roles to them remains tentative at best. Two exceptions are the supraoptic and paraventricular nuclei, which are known to contain neural cells capable of secreting chemical substances into the posterior lobe of the nearby pituitary gland. Since the anterior lobe of the pituitary gland is the master control point of the endocrine system, the hypothalamus is evidently the key structure in neuro-endocrine coordination.

For its more neurological functions, the hypothalamus connects extensively with the midbrain, thalamus, and phylogenetically more primitive regions of the telencephalon. Although the organization of information processing in the hypothalamus is mostly unknown, stimulation or destruction of specific hypothalamic regions in experimental animals can lead to abnormal eating, drinking, sleeping, or emotional behavior.

10.3 CEREBELLUM AND CEREBRUM

From the standpoint of physical appearance, the cerebellum and cerebrum (telencephalon) have a lot in common. Both have paired hemispheres, with an extensively folded outer layer called the *cortex*. Within the cerebral and cerebellar hemispheres, the organization of gray and white matter is the reverse of that in the spinal cord and brain stem. The outer cortex contains most of the neural cell bodies (gray matter), with the tracts (white matter) inside. Deeper within their hemispheres, both cerebellum and cerebrum have additional nuclei and tracts.

10.3.1 Cerebellum

Because of its accessibility and its interesting external geography, the cerebellum has long been an object of neuroanatomical study, during the course of which its apparent subdivisions have been subjected to a variety of elaborate nomenclatures. But extensive terminology is not really needed, since one cortical ridge (folium) has the same internal organization as any other. Different cortical areas correspond systematically to different body regions. The "circuit diagram" of cerebellar cortex is discussed in Section 11.3.

Cerebellar input and output travel by way of three major bands of fibers, connecting with the pons, the midbrain, and the medulla. The outward flow of information comes from the four pairs of interior cerebellar nuclei, whose behavior is in turn influenced by outputs from the cortex. Cerebellar output travels to numerous brainstem nuclei involved in the motor systems, where it serves to maintain normal muscle tone and posture, and to smooth the coordination of precision movements. All cerebellar activity goes on below the level of human consciousness.

To carry out tasks like those just mentioned, the cerebellum needs input information from all of the body's muscle, tendon, and joint sensors, from all higher (voluntary) motor centers, from the visual and auditory systems, and even from autonomic sensors (notably those in the vestibular system). A dozen major input paths to the cerebellum have been identified and associated with the three bands of communication fibers mentioned above. Examples are the dorsal and ventral spinocerebellar tracts, which converge on the cerebellar cortex by way of the medullary and midbrain fiber bands, respectively.

It is interesting that most of the information carried by these spinal tracts terminates on the same side as the sensors from which it originates, at least in man. The dorsal tract does not cross, whereas much of the ventral tract crosses the midline twice. There may be a reason for some cerebellar information processing to deviate from the general principle of management from the opposite side. But no good explanation is likely to emerge soon, since no one has yet provided a satisfactory rationale for the principle itself.

10.3.2 Cerebral Hemispheres

This section briefly describes the interior nuclei and subcortical fiber systems of the telencephalon, leaving a more detailed treatment of the cortex to the next section. The major subcortical telencephalic nuclei include the caudate nucleus, the lenticular nucleus, and the amygdala. The term *basal ganglia* is sometimes applied to all three structures and sometimes to the caudate-lenticular complex, a functional group involved in motor control. The latter usage is employed here. This system has extensive connections to the cerebral cortex, diencephalon, and brain stem.

The amygdala sits below the basal ganglia, at the bottom of the hemispherical cavity. Actually a complex of many smaller nuclei, the amygdala connects with the olfactory system and with nearby cortical areas, notably the hippocampal cortex (see below). The amygdaloid areas in each hemisphere are connected through the *anterior commissure*. Some of the functions of the amygdala and its associated structures are autonomic in character.

The white matter of the cerebral hemispheres can be divided into three types of fiber systems: (1) the *projectional* systems carry information between the cortex and lower centers; (2) the *commissural* systems carry information between the two hemispheres; and (3) the *associational* systems carry information from one part of the cortex to another, in the same hemisphere.

The major projectional system is known as the *internal capsule*, a massive sheet of fibers with more than a dozen component tracts. Many of these tracts connect specific regions of the cortex with their thalamic counterparts. Included in this group are the (somato)sensory radiations, the auditory radiations, and the visual radiations. Other components of the internal capsule interconnect the cerebral cortex and the lower brain stem. The corticotectal tract, for example, runs from visual cortex to the superior colliculus and other midbrain nuclei of the visual system. There is also the corticospinal tract, which descends through the internal capsule on its direct path to the spinal cord.

Two commissural systems, the posterior and anterior, have already been mentioned. But most interhemispheric information transfer goes by way of the *corpus callosum*. This massive band of fibers (see Figure 10.2) crosses the roofs of the lateral ventricles, interconnecting the hemispheres along their entire anterior to posterior extent. Cutting the corpus callosum in an experimental animal produces surprisingly little change in behavior. The same is true in humans where the callosum has not developed or has been cut to prevent transmission of epileptic seizures from one side of the brain to the other. For a long time the function of the corpus callosum was unknown. More recently, careful studies of "split-brain" animals and patients have revealed the subtle but important bilateral coordination made possible by information transmission in the commissure.

Association fibers connect one region of the cerebral cortex with another region on the same side. One long association fiber bundle is the inferior longitudinal fasciculus, which connects visual and auditory areas, presumably facilitating audiovisual associations. The shortest association fibers, prevalent everywhere, are simply U-shaped loops that dip below the cortex and connect adjacent regions.

10.3.3 Cerebral Cortex

The most evolutionarily advanced piece of neural tissue known, the human cerebral cortex has an average depth of 2.5 cm and an area of some 2000 cm^2. About two thirds of this surface is buried in the *fissures* (sulci) created by folding. The ridges between the fissures are called *gyri*. Many of the more prominent fissures and gyri appear in approximately the same places in all brains. These landmarks have been given names, some of which are shown in Figure 10.4. In addition, the figure shows how a few major fissures serve as the basis for division of the cortex into four *lobes*, frontal, parietal, temporal, and occipital. Other regions of importance, to be discussed subsequently, are also identified in Figure 10.4.

About 10% of the cortex, located on the medial side in the lower frontal region (not seen in Figure 10.4), is of a phylogenetically more primitive type.

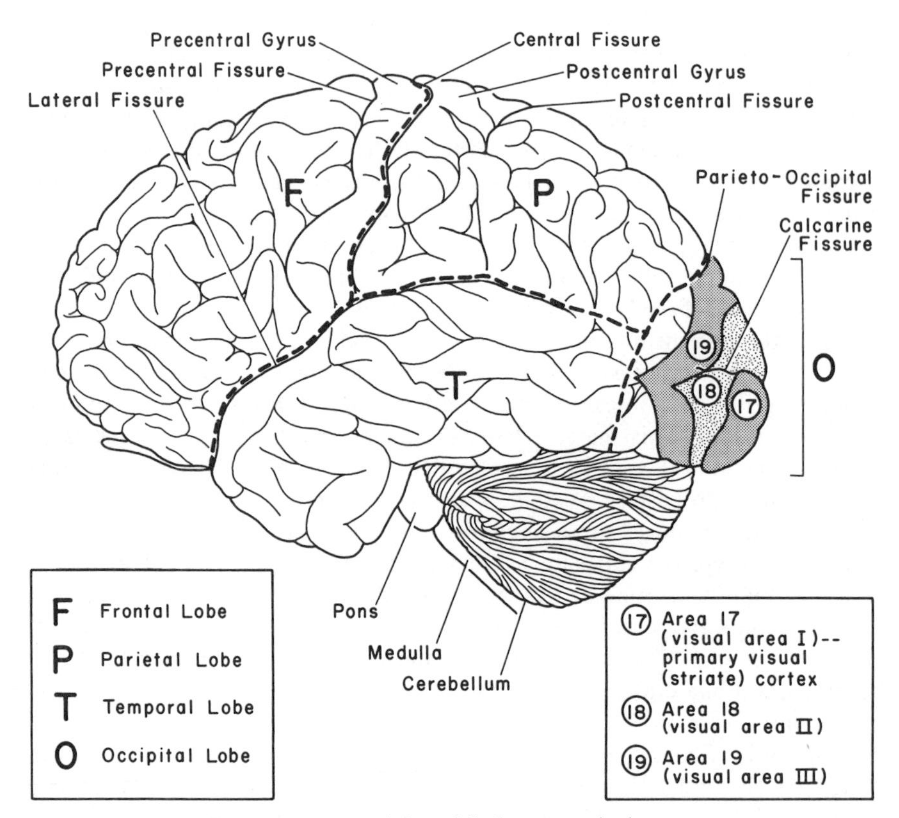

Figure 10.4 Lateral view of the human cerebral cortex.

Included in this portion is the *hippocampus*, a region with close ties to the olfactory system and amygdala. The terms "rhinencephalon" and "limbic lobe" have been applied to collections of structures in this area which seem to function together in the control of emotional behavior. The hippocampus is also involved in short-term memory, at least in man.

The remaining 90% of the cerebral cortex is called *neocortex* and has a regular layered structure. The six levels can be distinguished by the kinds and relative densities of neural cells and interconnecting fibers they contain. The fourth level is specialized for input of information to the cortex, the fifth for output. Differences in the appearance and relative thickness of the various layers have led some investigators to map large numbers (e.g., 53) of individual regions on the surface. These maps are useful for reference purposes, although the regions seldom correspond closely to identifiable functional areas.

In fact, the degree to which specific functions can be assigned to particular regions, as was frequently done in the discussion of lower structures, varies considerably in the cerebral cortex. At one extreme are the primary sensory areas, for vision (at the occipital pole), audition (in the temporal lobe, near the end of the lateral fissure), and somesthesis (along the postcentral gyrus). Within such regions there is a close correlation between cortical location and the placement of associated sensory phenomenon. It is possible to draw a *somatotopic map* of the body on the postcentral gyrus, for example. There is also a somatotopic map in primary motor cortex, which is located along the precentral gyrus and is the main origin of the corticospinal tracts.

The retinotopic and tonotopic maps, found respectively in visual and auditory cortex, are fully discussed in Chapter 13. In all such maps, relative position is preserved, while areas of greater importance (the center of the retina, the fingertips) receive proportionately more space. Clearly these areas of the cortex evince localization of function in the extreme.

On the other hand, there are regions of the cortex for which no specific function whatever has been discovered. These are sometimes called "association areas," although the term rather begs the question. These are clearly not the elusive memory storage areas, since destruction of nonspecific cortical tissue is never accompanied by the loss of particular memories or skills. There is instead a cumulative, generalized deficit with increasing tissue destruction. Clearly, a great deal remains to be learned about how information is organized for storage and retrieval in the brain.

Another aspect of the localization issue is the allocation of functions to the two hemispheres. In a perhaps surprising departure from the principle of bilateral symmetry, the human hemispheres are to some degree specialized. The cortex of one side, the left in most people, is primarily responsible for language skills and formal reasoning. A few "language areas," where damage interferes with some aspect of communication, have been identified (see Section 11.3.3). The other hemisphere is oriented more toward spatial reasoning and temporal pattern perception.

Although some of the above conclusions about functional localization in the human cortex are generalized from experimental work in higher mammals (mostly cats and monkeys), such cannot be the case for language functions. In fact, most of the human cortex has been mapped as a necessary adjunct to brain surgery. Before unhealthy tissue is removed, it must be determined if the consequences (e.g., loss of speech) are too severe. Minute electrical stimulation can be applied to the area while the patient is conscious. Provocation or blockage of language activity is an indication of "speech center" involvement. Similarly, stimulation in a primary sensory area causes the patient to "see," "hear," or "feel" localized stimuli. These probes are not unpleasant since the brain, where all pain is actually perceived, has no pain receptors of its own.

10.4 CELLULAR COMPONENTS OF THE NERVOUS SYSTEM

Like all biological tissue, neural tissue is made up of cells having the general characteristics outlined in Chapter 3. Only a small number (perhaps 10% in the

CNS) of neural tissue cells, the *neurons*, have an unarguable information processing function. Several varieties of *glial cells* comprise around half the mass of the nervous system.

10.4.1 Glial Cells

The neuroglia ("glia" for short) include up to four distinct cell types. The *microglia*, however, are not native to the nervous system, being derived from the embryonic mesoderm. Entering the CNS from the bloodstream, microglia function as scavengers, not unlike white blood cells. The remaining types of glia share the ectodermal origin of neurons (Section 8.3.2).

The *oligodendroglia* of the CNS, and their peripheral counterparts, the *Schwann cells*, have sheathing functions. Schwann cells wrap around individual fibers of nerves in spiral layers, with little cytoplasm in between the membranes. The result is a sheath composed mainly of the lipid *myelin*. Between Schwann cells the neural fiber is exposed in small gaps called *nodes of Ranvier*. This arrangement is used to accelerate information transmission, as discussed in Section 11.2. The smallest diameter peripheral fibers have less elaborate Schwann cell sheaths that do not affect the speed of information transmission; this small minority of fibers is referred to as "unmyelinated." In the tracts of the CNS, fibers are sheathed by oligodendroglia. Again, the largest fibers are myelinated. But the proportions are the opposite of those in the periphery; most central fibers are unmyelinated.

The last type of glial cell is the *astrocyte*, named for its frequently star-shaped appearance. There are two varieties of astrocytes, the fibrous, usually found in nerve trunks and central white matter, and the protoplasmic, commonly associated with neuron cell bodies in gray matter. It is not known if there is a functional difference corresponding to this distributional one.

In fact, very little is known about the function of astrocytes. Their frequent simultaneous association with a blood vessel and a neuron has led to the long standing hypothesis that these glia provide metabolic support for neurons. But there is no good evidence for this role, nor any indication that such support is required, since neurons stripped of their glia continue to function in an apparently normal fashion. This last observation also causes problems for the many other theories of astrocyte function, ranging from "structural support" and "insulation" to specific interaction with neurons in the processing and storage of information. Although they must be important, it would be premature to assign any information processing role to neuroglia (except acceleration of transmission in myelinated fibers).

10.4.2 The Neuron

Like other eucaryotic cells, each of the half trillion or so neurons in the human CNS has an external membrane, a cytoplasm replete with ribosomes and organelles, and a nucleus containing a normal complement of chromosomes. Neurons have specialized by limiting or emphasizing certain properties common to all cells. First of all, neurons enter a somewhat unusual state of the cell cycle, and do not divide. When all neurons are fully differentiated, around the time of birth in man, there can never be any more; and many will be lost from this initial population. There must be a reason. Perhaps new neurons would foul up established circuits. Whatever the advantages of nonproliferation, cumulative loss of neurons is a likely contributor to senescence.

The other specializations adopted by neurons are more clearly relevant to their function. Transmission of information over distance requires cells that span distance. Neurons have carried to an extreme the filamentous extensions of membrane surrounded cytoplasm that many cells evince. Some human neurons have fibers more than two meters in length. Another specialization relates to secretion, again a feature of most cells. Since mammalian neurons usually influence one another by emission of chemical substances, there are specialized secretory structures in the termini of neural fibers.

Finally, transmission of information requires a change of state that can travel in space. Although all cell membranes respond to electrochemical stimulation, neural membranes do so in ways specialized for encoding and transmitting messages. To summarize, neurons are just like other cells except that they (1) do not divide, (2) may extend over great distances, (3) secrete special chemicals to influence other neurons, and (4) reliably propagate membrane disturbances.

This introduction to the neuron concludes with a look at the structural components of a typical or "model" neuron. Neurological research increasingly suggests that this model neuron is a rarity. Yet aspects of its structure are found in all neurons. The model has three major components. (1) The cell body, or *soma*, contains the nucleus and much of the cytoplasm. The soma can receive input from other neurons over most of its surface. (2) The *dendrites* are fibrous extrusions, often intricately ramified, which can considerably extend the area of the neuron's input surface. In the model neuron, dendrites always transfer information toward the soma. (3) The output component is the *axon*, potentially a very long fiber, carrying information to other neurons, or target organs. The area where the axon exits the soma, known as the *axon hillock*, is often important in the encoding of neural signals.

10.4.3 Varieties of Neurons

The extent to which the cellular architecture of neurons may follow or diverge from the model just presented can be demonstrated with a close look at a few specific types of neurons found in all higher vertebrates and sketched in Figure 10.5.

The *spinal motor neuron* (Figure 10.5a) has been extensively studied, mainly because of its accessibility, and closely resembles the model neuron. The cell body of this efferent is located in the ventral gray matter of the spinal cord. The extensive dendritic tree has no special pattern, but allows for inputs from widely distributed sources. The long axon emerges from the cord, joins the spinal nerve, and eventually terminates in skeletal muscle. Although unbranched for most of its length, the motor neuron axon usually exhibits two kinds of *collaterals*. At the muscle, the axon ramifies into many terminal collaterals. There are also *recurrent* collaterals, axon branches near the cord boundary that return to the region of the cell body. By means of recurrent

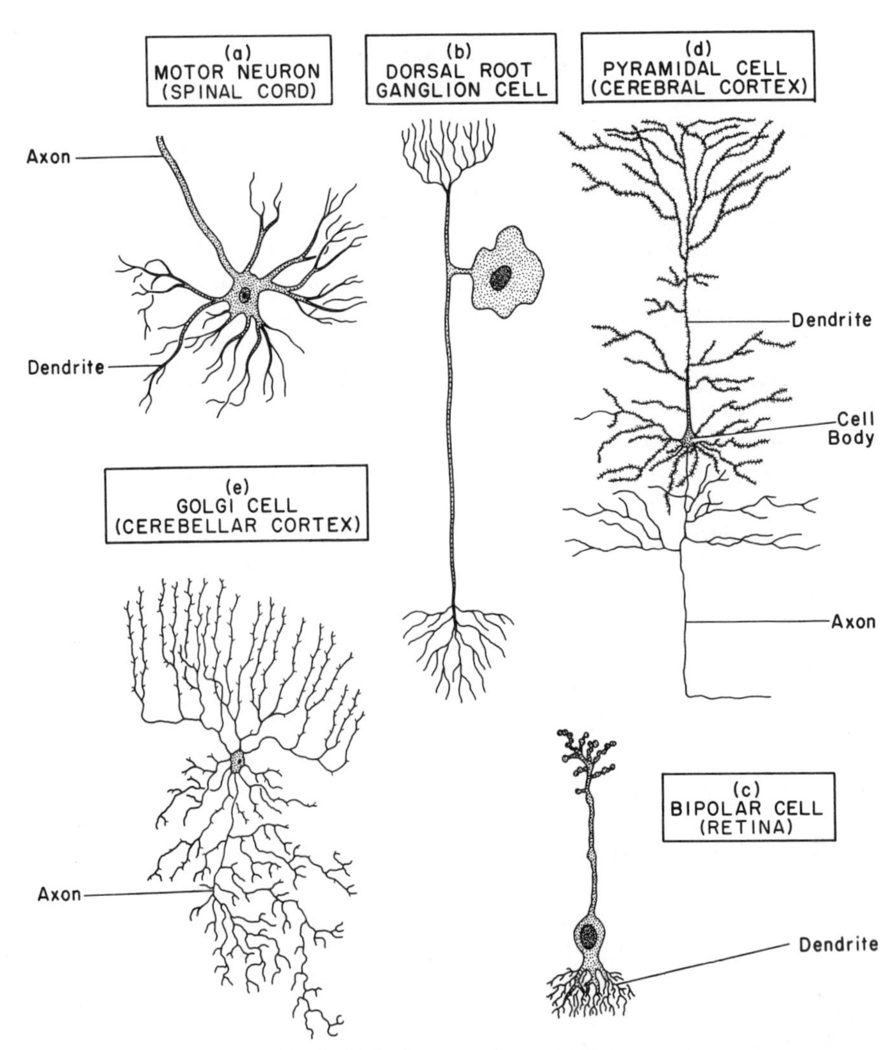

Figure 10.5 Some vertebrate neurons.

collaterals, spinal motor neurons indirectly limit their own activity, in a negative feedback control loop.

An afferent spinal nerve component, the *dorsal root ganglion cell* (Figure 10.5b), provides a first instance of the frequent difficulty in distinguishing axon from dendrite. This is a *unipolar* cell (having one major process) whose cell body is located in the dorsal root ganglion of the spinal cord (see Figure 10.1). The single long fibers (which include the two-meter ones mentioned above) are homogeneous in appearance and transmission mechanism, from their origins in peripheral sensors to their destinations in the spinal cord. These fibers develop as dendrites do, but acquire the transmission mechanism of the model axon. They carry information both toward and away from the cell body. The dorsal root ganglion cell may thus be regarded as a neuron with no dendrites and an unusual axon, or as a neuron with no axon and bizarre dendrites. The limitations of the model terminology are becoming apparent.

Almost the opposite situation is found in the *bipolar cell* (Figure 10.5c) of the mammalian retina. Here the cell body is located centrally between two processes. Since the direction of information transmission is fixed, there should be no problem distinguishing axon from dendrite. Yet both processes have similar structure and support only those signalling mechanisms characteristic of dendrites. There is no axon hillock. The model must be stretched to allow either dendrites as outputs or rather strange axons. These evident exceptions to model neuron architecture are not rare; both dorsal root ganglion cells and retinal bipolar cells number in the millions in man.

The last two examples of neural structure in Figure 10.5 return to the organizational plan of the model neuron. They have been included to suggest the wide variety of process patterns present in even conventional neurons. Figure 10.5d shows a *pyramidal cell* of the cerebral cortex. The basal dendrites spread out in the same cortical level (usually the third or fifth) as the cell body. The apical dendrites extend upwards to terminate in the top level. The axon, after giving off many horizontal and recurrent collaterals, descends out of the cortex.

Figure 10.5e shows a *Golgi cell* from the cerebellar cortex. With its highly ramified axon and extensive dendritic tree, the Golgi cell may connect to on the order of 10^5 other neurons, including the numerous granule cells. These latter cerebellar neurons may be the most populous cell type in the human brain; there are an estimated 40 billion in the two cerebellar hemispheres.

10.5 SYNAPTIC STRUCTURES

Neurons can influence each other in many ways. Mere physical juxtaposition may permit ion-flow based signalling mechanisms (described in the next chapter) to interact. Most strong influences involving reliable information transmission, however, occur at *synapses*. Although they manifest considerable structural variation, synapses can be grouped into two general types, electrical and chemical. In higher animals chemical synapses have received far more attention and appear to be the primary mode of interneuronal communication. Electrical synapses play a greater role in invertebrates; these structures are considered only briefly below and not at all in the next chapter's discussion of synaptic transmission mechanisms.

Regardless of type, synapses may be classified according to the communicating portions of the *presynaptic* and *postsynaptic* neurons. Perhaps the most standard arrangements are *axosomatic* and *axodendritic*, in which the axon of the presynaptic cell sends information to the cell body (soma) or a dendrite of the postsynaptic cell. However, the uncertainty described above concerning the nature of some fibers can make these terms difficult to apply. Also, there are unambiguous instances of *dendrodentritic* and *axoaxonic* synapses. Even stranger arrangements will be mentioned below.

10.5.1 Electrical Synapses

Coupling that permits flow of electrical current between cells is a common arrangement in many tissues, such as heart and liver. In these cases the cell-to-cell junctions are very tight, with an actual fusion of the exterior layers of the membranes. Tight junctions between neurons are probably rare and may not have specific signalling functions.

The so-called gap junctions, on the other hand, are usually electrical synapses when they occur between two neurons. At a gap junction, there is an intercellular distance on the order of 10^{-9} meters (roughly a tenth the distance at a chemical synapse). Some observers have reported channel-like structures within the gap. The gap junction offers little resistance to the flow of current, in one or both directions. Some well-studied electrical synapses have rectifying properties, transmitting information in one direction only. Others are bidirectional. No structural basis for the difference is known.

Electrical synapses may have been the earliest form of interneuronal communication. Compared to chemical synapses, they are fast and reliable. But electrical synapses lack much of the flexibility that will be seen in chemical signalling. It would appear that a combination of the two synaptic mechanisms could be advantageous in some circumstances, since a few such "mixed junctions" have been discovered.

10.5.2 Chemical Synapses

Chemical transmission of information from a neuron to another cell was first discovered, and remains best studied, at the interface between spinal motor neuron and skeletal muscle. This *neuromuscular junction* has many characteristics in common with synapses between neurons, the major difference being that the effect of transmission is the contraction of the postsynaptic cell rather than further information processing.

A diagram of the essential features of most chemical synapses is given in Figure 10.6. The *synaptic cleft* separates the presynaptic membrane from the postsynaptic one by about 10^{-8} meters. The presynaptic fiber often terminates in a knob (or "end foot") like that shown, although synapses can also occur, without special enlargements, where two fibers cross.

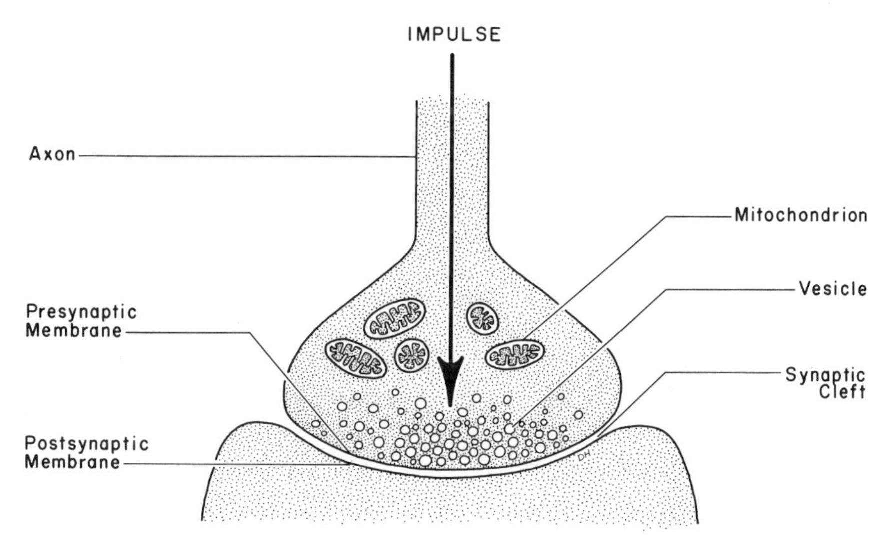

Figure 10.6 A chemical synapse.

Within both the presynaptic and postsynaptic membranes of a chemical synapse, specialized structures may be seen. The postsynaptic membrane can have thicker regions, below which may be seen filamentous strands. The presynaptic membrane sometimes shows symmetrical thickening. The interior of the presynaptic terminal usually has large numbers of mitochondria, typical of regions with considerable biological activity. But the most characteristic structures at a chemical synapse are the presynaptic *vesicles*. These are membrane surrounded bags, each containing thousands of molecules of the transmitter substance. Large numbers of vesicles sit in close proximity to the inside of the presynaptic membrane.

The shape of the vesicles, along with a number of other structural differences, has led to a general classification of chemical synapses. *Type I* synapses have wider clefts, greater (but less symmetric) membrane thickenings, large spherical synaptic vesicles, and usually an axodendritic position. *Type II* synapses have narrower clefts, lesser (but more symmetric) membrane thickenings, smaller and more flattened vesicles, and usually an axosomatic position. There is reason to believe these differences are associated with a functional distinction: Type I synapses seem to be mostly of the sort that tend to elicit activity in the postsynaptic neuron, whereas Type II synapses are mostly inhibitory in character.

10.5.3 Specialized Synapses

The relatively small, single point of information transfer shown in Figure 10.6 has been called a "simple" (chemical) synapse. Probably the vast majority of synapses in the human nervous system are of this type. But investigators are discovering increasing numbers of more elaborate synaptic organizations. The mixed chemical and electrical junction has already been mentioned. Multiple, serial, and reciprocal synapses complete the present inventory of synaptic structures.

Multiple synapses occur when a single presynaptic fiber or structure has two or more adjacent terminals that influence postsynaptic membranes of one or more postsynaptic cells. This is actually the standard organization at the neuromuscular junction, where the terminating axon of a motor nerve forms an end plate consisting of dozens of closely spaced presynaptic terminals arrayed along the muscle fiber. Another example of a multiple synapse is at the terminal of a mossy fiber input to the cerebellar cortex (see Section 11.3), where some 20 granule cells participate in hundreds of synaptic interactions with the single fiber. A structure of this complexity is sometimes called a synaptic *glomerulus*. A simpler multiple synapse occurs in the retina (see Section 13.3), where a single receptor cell terminal influences several postsynaptic neurons.

Serial synapses occur when adjacent regions of one structure or fiber are respectively postsynaptic to one cell and presynaptic to a third. In *reciprocal synapses* the serial relation returns to the cell of origin, leading to adjacent synapses with influences in opposite directions. In mammals, serial and reciprocal synaptic arrangements have been found together in both the olfactory bulb (the termination of cranial nerve I) and at several points in the visual system (see Section 13.3).

Chapter Eleven

Neural Information Processing

The discussion of neural architecture in Chapter 10 began with major components of the nervous system and descended to the level of synaptic structures. This chapter's functional treatment begins with information processing at the synapse and ascends to behavior in small networks of neurons. Larger functional units are discussed in Chapter 13.

Today's understanding of neural communication mechanisms is based on decades of painstaking laboratory work. Three major classes of techniques have been developed to probe the nervous systems of experimental animals and human patients and volunteers.

First, there is destruction of neural tissue (also known as ablation or lesioning). Removing a portion of the nervous system, or of an individual neuron, sometimes yields insight into its functional role. What can be learned from this relatively crude technique is limited because one neural area, however small, may be indirectly involved in the normal operation of many subsystems, some of which are physically or functionally quite remote. A lesion in the brain stem, for example, could destroy part of the cerebrospinal tract, leading to the (incorrect) conclusion that brain stem centers are the origin of fine control of voluntary motor activities.

A second investigative technique makes use of electrical or chemical stimulation of neural tissue. Chemical stimulation is often used in attempts to establish that particular substances are transmitters at chemical synapses.

Electrical stimulation allows correlation of bodily movements and sensations with activity in particular portions of the nervous system. As with ablation, it is not necessarily the case that primary centers are located at the point of stimulation. But at least the nervous system remains intact, reducing the likelihood of such malfunctions as may result from tissue destruction.

Perhaps the most valuable technique is recording of neural signals, a method often used in conjunction with stimulation. Like electrical stimulation, recording requires electrodes. The size of these devices is directly related to the number of neurons contributing to the recorded signal. At one extreme, large metallic disc electrodes can be attached to the scalp. The resulting signal is the *electroencephalogram*, or EEG, which reflects generalized activity in large chunks of neural tissue. Variations in the EEG waveforms associated with different regions of the brain are important in the diagnosis of disorders and the analysis of normal activities, like sleep.

Smaller electrodes can be inserted directly into the nervous system, permitting the investigator to monitor activity in a population of hundreds or thousands of neurons. The behavior of such small neural communities is important in the study of *evoked potentials*, neural signals which occur in response to controlled stimulation of a sense organ or sensory pathway. Analysis of evoked potentials is making an increasing contribution to the understanding of brain functions in higher mental activities like perception and memory.

The greatest advances in the understanding of neural signals have come from the development of *single unit* recording techniques, using microelectrodes made of fine tungsten wire or of narrow glass capillary tubes containing an electrolytic solution. A single unit may be a neuron cell body, individual fiber, or synaptic region. The recording can be either extracellular, a relatively simple way of monitoring a cell's activity, or intracellular. The delicate insertion of a microelectrode into the interior of a neuron allows measurement of the fundamental electrical changes underlying neural signalling. When coupled with sophisticated electronic equipment, such as feedback controlled current generators, intracellular electrodes permit systematic experimentation with the signalling apparatus.

11.1 SYNAPTIC TRANSMISSION

There are two quite different neural signalling mechanisms, each with its own advantages. The propagation of charged ion fluxes, discussed in Section 11.2, has, among many other signalling functions, a role at electrical synapses. This mechanism generally satisfies requirements for both (1) effective spatial and temporal integration of inputs to a neuron and (2) rapid and reliable transmission of the result of that integration, sometimes over relatively long distances. At the synaptic junctions between neurons, speed is less critical; and unchanging passive integration of influences may not be sufficient. Transmission of information by molecules, whose configurations can interact with those of others present in the postsynaptic membrane, contributes such features as amplification, inversion, and temporal variation to the neural signalling repertoire. The rest of this section deals exclusively with chemical synapses.

Synaptic information transmission has six basic steps. After its (1) *synthesis*, a chemical transmitter is packaged by (2) *storage* in presynaptic vesicles. Arrival of a signal at the presynaptic membrane triggers (3) *release* of transmitter, which then moves across the cleft by (4) *diffusion*. At the postsynaptic membrane, transmitter molecules (5) *interact* with receptors, until the transmitter succumbs to (6) *inactivation*.

11.1.1 Nature, Synthesis, and Storage of Transmitters

The first neurotransmitter to be identified and studied, and the one whose function remains best understood, is the chemical agent at skeletal neuromuscular junctions. Known as *acetylcholine*, or ACh for short, this transmitter is now known to be the agent for all parasympathetic neurons and preganglionic sympathetic neurons, as well as at many central synapses. Junctions where ACh is a transmitter are called *cholinergic*.

Like most subsequently identified neurotransmitters, ACh is a small molecule, containing about two dozen atoms. It is synthesized from two precursors, choline and acetate, in a two step reaction requiring the participation of ATP and the enzyme choline acetyltransferase. This synthesis, which occurs in the presynaptic cytoplasm, employs raw material resulting from the breakdown of ACh, a complementary reaction discussed in the next section. After synthesis, ACh is packaged into vesicles. The vesicles themselves are thought to be manufactured originally by the membrane system in the soma of the neuron. Like the transmitter, however, vesicular membrane is probably subject to extensive recycling at the synapse.

The other chemical transmitter in the peripheral nervous system, *norepinephrine* or NE, operates at the terminals of postganglionic sympathetic neurons to stimulate visceral muscles and glands. NE is also an agent in the CNS. Junctions where NE is a transmitter are called *adrenergic*. A close relative of the hormone adrenaline, NE (sometimes called noradrenaline) has peripheral effects that synergize with the system-mobilizing actions accompanying release of the hormone from the adrenal glands.

NE is synthesized from the amino acid tyrosine in several steps. The intermediate *dopamine*, probably itself a neurotransmitter, is packaged in vesicles before the final reaction step converts it to NE. Although there are agents which destroy NE, a substantial proportion of it appears to be recycled without modification (see below).

Although there are some other peripheral transmitters in invertebrates, ACh and NE are the only ones operating in the peripheral nervous systems of mammals. The existence of other central transmitters has long been suspected, but their identification made extremely difficult by the relative inaccessibility and complexity of central neural architecture. Since so many chemical substances can be isolated from synaptic regions, the criteria for identifying a transmitter must be quite strict. In addition to the substance itself, there must also be present systems for its synthesis and inactivation. Further, local application of the substance (via a micropipette) must produce the same results as synaptic activity. And these results should vary in the same way when subjected to experimental modification by drugs or electrical stimulation.

There is now at least partial evidence for many central transmitters in addition to ACh and NE. Two of these substances, GABA (gamma aminobutyric acid) and 5-HT (or seratonin), have been implicated with near certainty. An amino acid not used in polypeptides, GABA is a known transmitter at neuromuscular junctions in crustaceans. Several other amino acids are probable transmitters. Seratonin is a monoamine, as are ACh and NE. Both monoamines and amino acids have a positively charged nitrogen atom, which suggests that this may be a feature of many transmitters.

11.1.2 Transmitter Release and Inactivation

The arrival of a signal at the presynaptic membrane initiates a sequence of events culminating in the dumping of thousands of transmitter molecules into the synaptic cleft. There is a *synaptic delay* of about half a millisecond between arrival of the presynaptic signal and its measurable postsynaptic consequences. Since transmitter diffusion and receptor interaction can be responsible for only about a tenth of this delay, release of transmitter is evidently a time-consuming process. It is also a poorly understood one.

One known requirement is the presence of a sufficient concentration of extracellular calcium ions. The signal modifies the presynaptic membrane in a way that facilitates calcium influx. Calcium is somehow involved with the mobilization of vesicles, their migration to the membrane, and/or their fusion with the membrane to release their contents (a form of exocytosis, see Section 3.1.3).

Individual vesicles participate either fully or not at all in a given release event. Since the number of transmitter molecules per vesicle is approximately constant, the number released is always some multiple of this basic *quantum*. Quantal release of transmitter was discovered through investigation of continual very small postsynaptic responses at the neuromuscular junction. Even in the absence of presynaptic signals, there is sporadic release of the contents of individual vesicles, presumably caused by a relatively low concentration of intracellular calcium. Quantal transmitter release determines the units for encoding of information at a synapse. Facilitation of a synapse's effectiveness, for example, cannot occur by increased vesicular content. Somehow, a given level of presynaptic activity must begin to cause the release of more vesicles.

Once the released transmitter molecules have carried out their primary function of interacting with receptors (discussed in the next section), they must be inactivated. Transmission of information requires both appearance and disappearance of signal elements. There are three major ways of getting rid of transmitter that has done its job; different combinations of these mechanisms are used for different transmitters.

The easiest way to terminate a signal is to allow diffusion to carry the transmitter molecules away from the postsynaptic area to other regions of the intercellular matrix. Although it is probably a factor in transmitter inactivation at most synapses, diffusion has some disadvantages. First, the transmitter cannot easily be recovered for reuse by its presynaptic terminal of origin; since relative transmitter availability may be a synaptic parameter with information processing relevance, any mechanism tending to average that availability should be minimized. Second, diffusion may be too slow, particularly if the synaptic region has the sort of enclosed geometry frequently provided by surrounding glial cell membranes. If there is rapid access to nearby regions, however, they may contain postsynaptic membranes which could be inappropriately influenced by the diffusing transmitter.

The second inactivation mechanism involves reuptake of transmitter through neural membranes in the region of the synapse. In addition to being a mechanism for transmitter removal, reuptake by the presynaptic membrane has appealing recycling efficiency since the molecules that re-enter the presynaptic membrane are immediately available for packaging into vesicles. Reuptake is the primary inactivation mechanism at adrenergic synapses.

Finally, transmitter molecules can be inactivated by modification. The paradigmatic example of this process is the cleavage of ACh into its component acetyl and choline fragments by the enzyme acetylcholinesterase (AChE). Extracellular AChE is prevalent in regions of cholinergic synapses. In fact, its presence is one of the defining criteria for cholinergic transmission. There are lesser amounts of AChE within synaptic terminals. It would appear that the vesicles have a protective as well as a quantizing function. Enzymatic modification is a secondary inactivation mechanism at adrenergic synapses.

11.1.3 Postsynaptic Receptors

The intermolecular allosteric interactions which mediate postsynaptic information processing in the nervous system are not different in principle from those within all cells or those in the extensive hormonal signalling systems of higher animals. Yet merely finding, much less understanding, those membrane elements which recognize and respond to neurotransmitters has proven one of the most difficult neurophysiological endeavors.

Initially, it should be emphasized that much information processing flexibility at synapses resides in the postsynaptic receptors. A given neurotransmitter can have opposite effects at different synapses in the same organism. In invertebrates, there are individual synapses that exhibit opposite responses to a single transmitter (see Section 11.3.3). Thus it would not be surprising to find several different receptors, or receptor arrangements, for a given transmitter in a given organism.

Cholinergic receptors have been isolated from regions where they are accessible and especially dense, such as neuromuscular junctions and the specialized electrical organs of some fish. These receptors are large protein complexes, made up of several subunits and combined in some cases with lipid or carbohydrate elements. Receptor organization must somehow convert the recognition of neurotransmitter molecules into the particular membrane permeability changes discussed in Section 11.2.

Even less can be said about adrenergic receptors. At the synapse-like terminals of postganglionic sympathetic neurons, adrenergic receptors are embedded in visceral muscles and glands. Here the receptor function may be to trigger the ubiquitous "second messenger" system of cyclic AMP, which is known to mediate responses to hormones by some visceral structures. It is even possible that cyclic AMP participates in a sequential response to the transmitter that terminates in membrane permeability changes of postsynaptic neurons.

11.2 NEURAL SIGNALS

11.2.1 Ions and the Neural Membrane

All neural signalling involves the flow of electrically charged ions through the neuronal membrane. Such flow temporarily alters the voltage drop, or *potential*, across the membrane. Like most cells, the neuron not involved in active signalling maintains a *resting potential*, with the interior of the cell 30 to 80 millivolts negative with respect to the extracellular fluid. Since resting potentials vary widely among different types of neurons, and can be different at different places within a given neuron, these values are potentially a major parameter in neural signalling.

Three kinds of ions are mostly responsible for the resting potential and its modification during information transmission. Positively charged potassium (K^+) ions are much more prevalent within the cell, at roughly 20 times their extracellular concentration. Positively charged sodium (Na^+) is reciprocally distributed, with about 10 extracellular ions for each intracellular one. Negatively charged chloride (Cl^-) ions have approximately the same transmembrane concentrations as sodium. Other negatively charged particles contribute to the resting potential but do not move across the membrane during signalling; these are multiatomic organic groups.

The flow of all charged ions is governed by three factors: concentration gradient, charge gradient, and permeability. All other things equal, any ion which can cross the membrane easily will flow in a direction that tends to give it the same concentration on both sides. This gradient runs downhill from the side with more ion to the side with less. The charge gradient impels the ion in a direction that tends to decrease the voltage drop across the membrane. All other things equal, a positively charged ion should move inward and a negatively charged one outward. Not only are all other things never equal, as discussed below, but also not all ions penetrate the membrane freely.at all times, making permeability the primary consideration in the distribution of some charged components (e.g., the immobile organic groups.)

Potassium permeability is always high. But under resting conditions this ion is subject to opposing gradients. To equalize its concentration, potassium must flow out of the cell, a movement that increases the voltage drop across the membrane. The compromise is a significantly higher internal potassium concentration. Similarly, chloride is pulled two ways, into the cell to equalize concentration and out of the cell to equalize charge. Only sodium would appear to have a clear mandate. By flowing into the cell this ion could reduce differences in both charge and its own concentration. Sodium, however, penetrates the resting membrane much less easily than the other two ions. Further, the small amount of "leakage" of sodium into the neuron is compensated by the active transport mechanism which continuously exchanges intracellular sodium for extracellular potassium (see Section 3.1.2). Variation in the sodium entry rate seems to be the basis for the wide range of resting potentials mentioned above.

Within the constraints of permeability, the various charge and concentration gradient tensions are resolved in a state of equilibrium having the resting potential and concentration inequities already mentioned. It is also of some interest to know the contributions of each ion to this equilibrium state. A physical chemistry equation (named for Nernst) relates the charge that exactly balances an ion's concentration differences to the logarithm of the ratio of those concentrations. Use of the Nernst equation gives this *equilibrium potential* for potassium as about -85 millivolts, somewhat below the resting potential because of sodium and chloride effects.

By contrast, an equilibrium potential of about +50 millivolts would be the voltage difference required to maintain the sodium concentration disequilibrium, if the ion were free to move. Put differently, if potassium and chloride ions were immobilized and the membrane made fully permeable to sodium, a new resting potential of 50 millivolts, interior positive, would result. The chloride equilibrium potential is intermediate between the resting potential and the potassium equilibrium potential.

The above deductions about the components of the resting potential have received extensive experimental confirmation. They may be summarized in the form of an equivalent electrical circuit, as shown in Figure 11.1. In addition to an overall membrane capacitance, a voltage source ("battery") and a *conductance* (reciprocal of resistance) is shown for each of the three ions involved in signalling. Information transmission is a consequence of alteration in the membrane structure in ways which temporarily increase some of the conductances, allowing the net potential to move toward the equilibrium potential(s) of one or more of the individual ions.

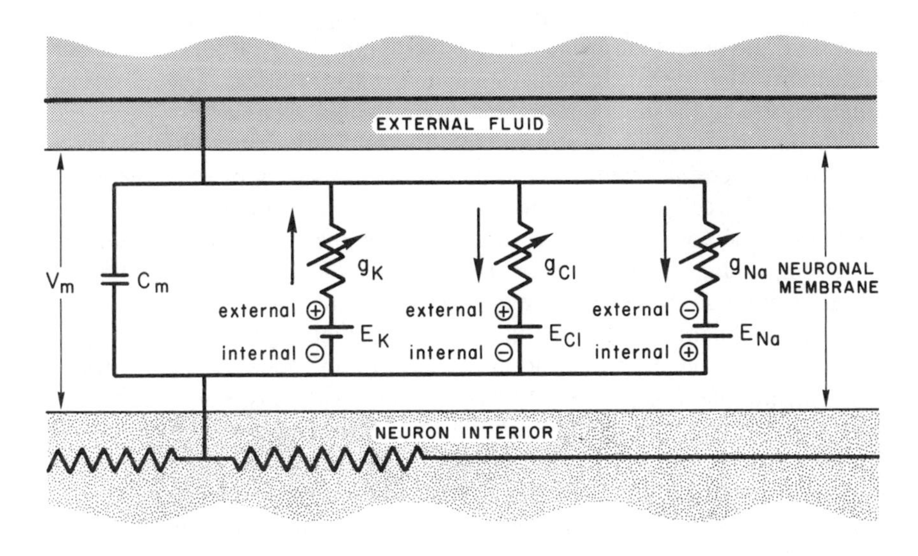

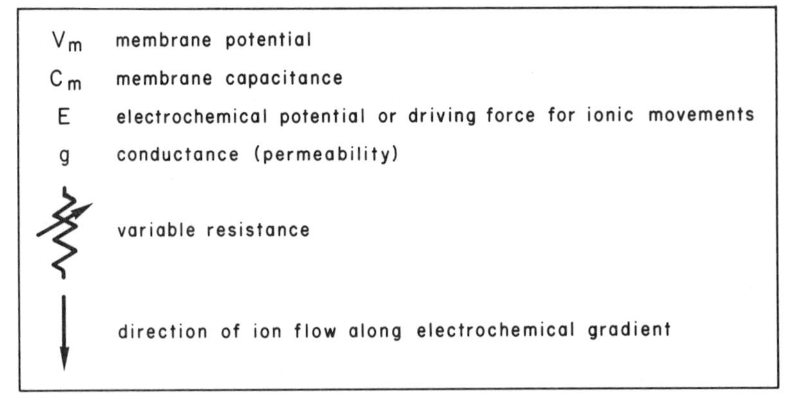

Figure 11.1 Equivalent electrical circuit of the neural membrane.

11.2.2 Postsynaptic Potentials

Depending on the type(s) of receptors embedded in the postsynaptic membrane, transmitter flow through a chemical synapse can produce two general kinds of effects. In one of these the membrane is *depolarized* (i.e., the voltage drop is reduced). Because the consequence of such a depolarization is usually a tendency to increase activity (and/or promote transmitter release) in the postsynaptic neuron, the phenomenon is termed an *excitatory postsynaptic potential*, or EPSP. The other type of effect involves *hyperpolarization* of the postsynaptic membrane, increasing the voltage drop. These *inhibitory postsynaptic potentials*, or IPSPs, are so named because they usually tend to depress message transmission by the postsynaptic cell. Comparable aspects of both PSP types are depicted in Figure 11.2

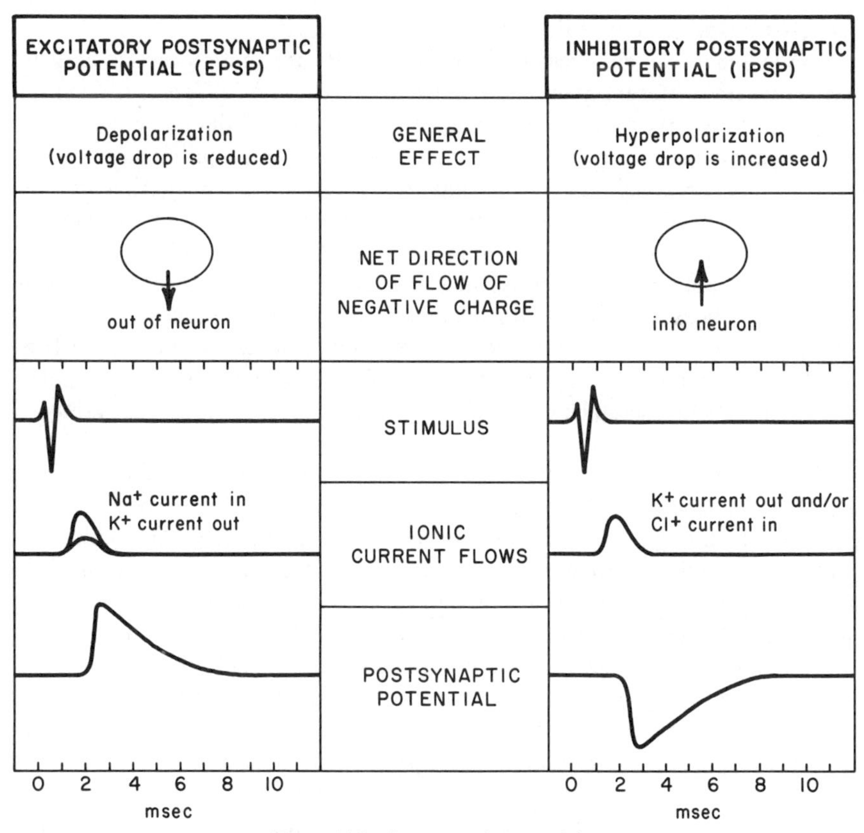

Figure 11.2 Postsynaptic potentials.

The amplitude of the deviation from resting potential during a PSP is variable. Factors such as the amount of transmitter release and the history of activity at the synapse presumably determine the PSP amplitude at any given time. The time course of most PSPs is on the order of 10 milliseconds, a faster change in potential being followed by a slower (usually exponential) decay back to the resting level.

The ion movements underlying PSPs are fairly well understood, although the ways in which receptors change conductances are still mostly a matter of conjecture. In an EPSP, there is a general increase in conductance for all ions. In fact, the equilibrium potential that would result from sustaining the membrane changes indefinitely is near zero. The primary consequence of this general increase in conductance is an influx of sodium, hence the depolarization. Upon transmitter inactivation, the membrane repolarizes to the resting level as a result of internal diffusion of the excess sodium, which is eventually actively transported out of the cell.

The conductance changes that mediate IPSPs are more selective and may involve several different mechanisms, perhaps because of different transmitters and/or different receptors. An increase in potassium conductance may be the most common basis for the IPSP, which often has an equilibrium potential close to that for potassium itself. Enhanced chloride flow is another probable mechanism, which may in some instances combine with potassium movement.

Conductance increases have long been regarded as the only basis for postsynaptic potentials. Recently, however, similar but often more prolonged effects have been identified with decreases in the conductances of particular ions. A drop in sodium conductance, for example, can resemble a slowed IPSP. Conductance-decrease synapses seem to employ the same kinds of transmitters as "classical" synapses. The scope of postsynaptic receptor functions is thus still emerging.

To have their intended effects on the postsynaptic neuron, PSP influences must not be confined to the postsynaptic membrane. These potentials are propagated through adjacent membrane regions, the original conductance changes expanding with ever decreasing amplitude. Because this signalling mechanism does decay in both time and space (in contrast to the action potential considered below) it has been called *passive* or electrotonic conduction.

Within a neuron, postsynaptic potentials can interact in both space and time. At any given synapse, the effects of one PSP may be decaying as another begins. The two potential changes will then add algebraically, in a phenomenon known as *temporal summation*. The two potentials need not have the same sign, although this is thought to be almost always the case in higher animals.

As they spread away from their postsynaptic membrane positions of origin, decaying PSPs encounter each other. The multiple additive interactions that can result are known as *spatial summation*. The surface of the neuron should evidently be regarded as a field upon which perhaps thousands of moving potential changes interact continuously in time and space. This is the basic computational process of nervous systems. The results of the computation depend on both the spatial geometry and temporal sequencing of the input signals.

11.2.3 Action Potentials

Postsynaptic potentials decay too quickly to be used for transmission of information, except over the relatively short distances within dendrites and soma or in dense networks of small neurons (e.g., the retina). For reliable signalling over longer distances, the axons of many neurons (and the dendrites of some) are endowed with a specialized membrane which supports a different kind of transmissible electrochemical change, known as the *action potential.*

Although action potentials can be elicited by artificial electrical stimulation at any point along the axon, they normally originate in the region of the axon hillock. At any given time, integration of recent PSP activity over the entire neural surface will have some (possibly null) net effect at the axon hillock. If this effect is a depolarization that differs from the resting potential by an amount at least equal to the cell's *threshold*, an action potential will be sent down the axon. In more prosaic parlance, the cell "fires." Thresholds vary considerably among different neurons and with time in a single neuron. Frequently, depolarization on the order of 10 millivolts is sufficient to trigger an action potential.

The action potential itself is uniform in character, consisting of a rapid reversal of the membrane voltage drop, making the cell interior roughly 40 millivolts positive, followed by a somewhat slower return to resting level. As can be seen in the diagram of Figure 11.3, this entire process takes only a few milliseconds. The ionic basis for action potentials has been carefully studied. The upswing phase is associated with a transient membrane change which allows a massive local influx of sodium ions. The peak is thus close to the sodium equilibrium potential. In recovery phase, the membrane's resting potential is restored through potassium efflux. The cumulative sodium-potassium imbalance is ultimately redressed by active transport.

Action potentials differ from postsynaptic potentials in at least three major ways. First, the action potential has been termed an "all or none" phenomenon; if it occurs at all it does so with full amplitude, unlike the graded responses of PSPs. This means that information carried in PSP amplitudes must be coded differently in action potentials. Second, action potentials propagate without diminution, using a conduction scheme quite different from the passive one that serves PSPs. Third, action potentials do not summate or interact the way PSPs do.

Action potentials are effectively isolated from one another by the *absolute refractory period.* During this time, which lasts on the order of one millisecond following the peak of the potential, the neural membrane is incapable of sustaining another action potential. The absolute refractory period thus imposes an upper bound of about one kilohertz on action potential frequencies in a single fiber.

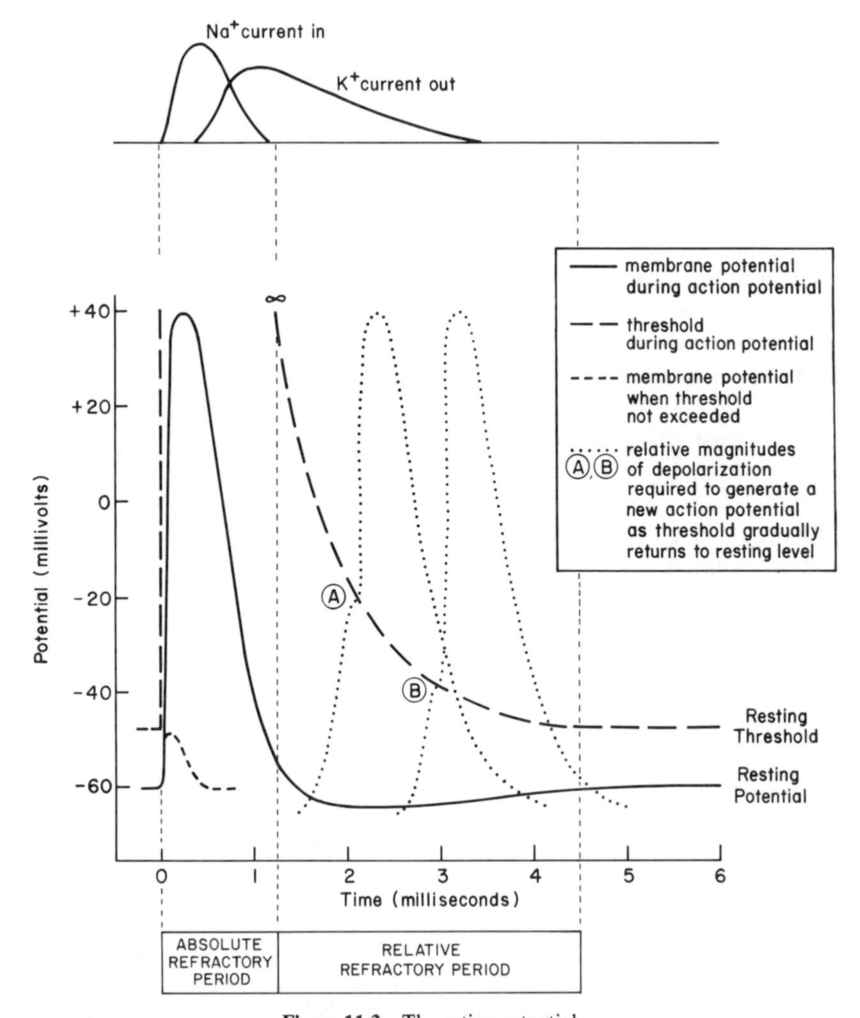

Figure 11.3 The action potential.

Because of the way in which the action potential travels, effective signalling would be impossible without the absolute refractory period. The action potential is a wave phenomenon; a region of disturbance, but nothing material, moves down the axon. Essentially, a new action potential is initiated in the tissue immediately ahead of the present disturbance site because of local electrotonic propagation of a suprathreshold depolarization. The tissue immediately behind the present disturbance also receives this depolarization; were it not in an absolutely refractory condition (the action potential having just "been there"), the entire axon would undergo continuous action potentials. The capability for unidirectional message transmission would be lost. This is not to imply that axons can support only action potentials that move away from the cell, although such are the normal physiological type. Electrical stimulation in the middle of an axon will trigger action potentials in both directions. The point of the above analysis is that each of these disturbances will travel in one direction only.

The speed of action potential transmission varies considerably. In smaller, unmyelinated fibers the range is about 0.5 to 2.5 meters per second, with faster transmission in fibers of larger diameter. Large myelinated axons take advantage of the "insulating" characteristics of the Schwann cells. The action potential "leaps" from one node of Ranvier to the next, a mechanism known as *saltatory conduction.* Transmission speeds in these fibers range from 10 to 100 meters per second, again being faster in axons of larger diameter. (Axon diameters in mammals range from about 10^{-7} to 10^{-5} meters.)

Another refractory period, immediately following the absolute one and lasting for a few milliseconds, provides the mechanism for encoding in action potentials information that was previously stored as PSP amplitudes. During this *relative refractory period*, the threshold drops smoothly from its value during the absolute refractory period (effectively infinite) to its resting value. Thus the amount of depolarization required to initiate a second action potential is, for a while, inversely proportional to the time elapsed since the first one. If a suprathreshold depolarization is maintained at the axon hillock, its magnitude will determine the frequency of the resulting action potentials, up to the maximum frequency imposed by the absolute refractory period. Amplitude encoded information is thus converted, by the mechanism of the relative refractory period, into frequency information, a process known to engineers as *pulse frequency modulation*, or PFM for short.

Some neurons appear to use the relative refractory period as a timing device for spontaneous or *pacemaker* output activity. When the resting potential of a cell is just sufficient to trigger an action potential, without any external input, then the duration of the relative refractory period will determine the frequency of spontaneous firing. This arrangement allows modulation of a "background" firing rate to be used for encoding information.

The neuron thus emerges as a rather sophisticated information processing element, interpreting a spatially distributed array of chemical transmitter quantities, integrating these influences, and translating the result into frequency modulated trains of action potentials. In central neurons, these trains ultimately cause release of transmitter substances at presynaptic terminals, thereby influencing the information processing in dozens to thousands of other neurons.

11.3 NEURAL CIRCUITS

The term neural circuit is used here to designate any group of up to several hundred interconnected neurons with a unified physiological function. The concept is intended to embrace both actual networks and hypothetical arrangements which have been proposed to explain aspects of neural information processing. A detailed inventory of neural circuit studies could thus fill several volumes. The much more modest goals of this section are, first, to present a few exemplary instances of well studied neural circuits and, second, to provide essential background for some of the simulation studies considered in Chapter 12.

11.3.1 Circuitry of the Cerebellar Cortex

The pervasive role of the cerebellum (Section 10.3.1) in the regulation of movement may account for its wide range of sensory inputs. Despite the diversity of input information, the computational circuitry in the cerebellar cortex is relatively simple and exceptionally regular. The connections within a one-millimeter slice from any folium are the same as those in the thousands of other such slices. The only obvious variable is the bodily region implicated by the somatotopic map. This rigid geometry has encouraged extensive study. The organization of cerebellar cortex is probably better understood than that of any other brain area in higher vertebrates. Comprehension of function has understandably lagged behind that of organization. The next chapter considers some models of cerebellar cortex that have been studied by simulation.

Figure 11.4 presents, both graphically and schematically, the organization of the representative folial slice mentioned above. There are five major types of neurons, the most distinctive of which is the *Purkinje cell*. This neuron provides the only pathway for information flow out of the cortex, its axon descending to the interior cerebellar nuclei. Interestingly, the synapses onto cells in these nuclei are all inhibitory, suggesting that the cerebellar cortex contributes to regulation of movement by selective dampening of information flow through the nuclei. The Purkinje cell axon also gives off collateral branches within the cortex, distributing its inhibitory influence to neighboring Purkinje cells and to other types of cortical neurons. The Purkinje dendritic tree is highly branched, with terminal ramifications extending up to the cortical surface. The dendrites are confined to a planar region, orthogonal to the longitudinal axis of the folium.

The remaining types of neurons shown in Figure 11.4 all mediate the distribution of input information to the Purkinje cells. The input pathways, of which there are just two, are described first. The *climbing fibers* carry information which comes mostly from a brain stem region called the inferior olive. After delivering collateral input to the cerebellar nuclei, these fibers ascend through the cortex, exerting extensive axosomatic and axodendritic synaptic influence on the Purkinje cells directly. There are also climbing fiber synapses which initiate control loops by influencing other types of cortical neurons. All climbing fiber synapses, in fact all inputs to the cortex, are excitatory.

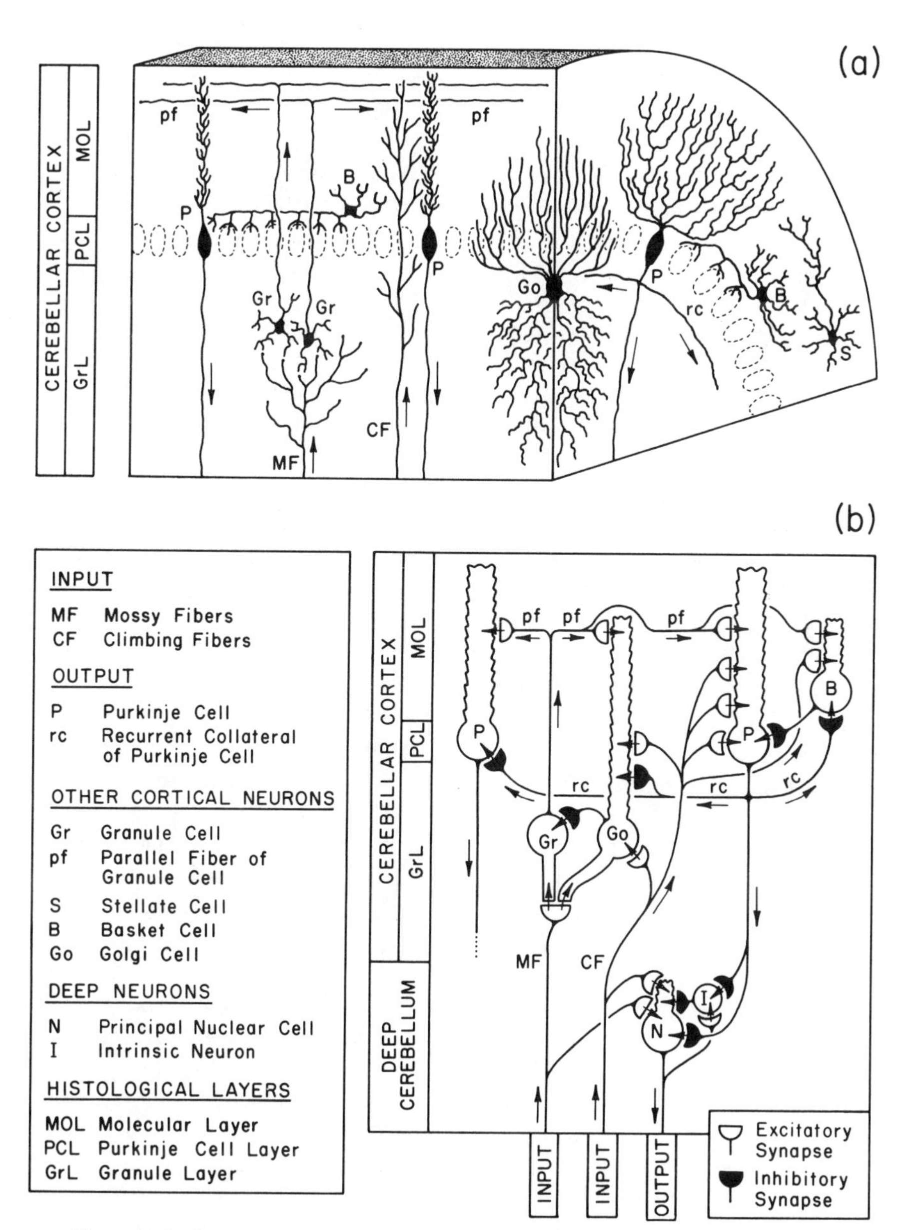

Figure 11.4 Cerebellar circuits: (a) three dimensional sketch, (b) connection diagram.

The second type of input pathway, the *mossy fibers*, exerts exclusively indirect influence on the Purkinje cells. The mossy fibers terminate near the bottom of the cortex in complex synaptic arrangements involving large numbers of cortical neurons. The origins of information carried by mossy fibers are diverse and not yet fully identified; primary sources include the vestibular system, the spinal cord, and various brainstem nuclei.

The *granule cells* relay mossy fiber inputs to the Purkinje cells. Small in size but large in number (there may be 40 billion in the human cerebellum), granule cells send their axons vertically to the top of the cortex, where the fibers bifurcate to run along the folial axis. These *parallel fibers* can extend two or three millimeters, causing EPSPs in the dendritic trees of hundreds of Purkinje cells. The excitatory input to Purkinje cells thus comes from just two sources, climbing fibers and parallel fibers, the latter delivering "preprocessed" mossy fiber input.

The major inhibitory influence on Purkinje cells comes from the *basket cells* and the closely related *stellate cells*. The latter are named for their star-shaped appearance, the former for the dense basket-like engulfment of Purkinje cell axon hillock regions created by the terminal arborizations of their axons. The possibility of strong inhibitory control at the point of Purkinje action potential initiation gives each basket cell a virtual stranglehold over the hundred or so Purkinje cells it influences. The stellate inhibitory connections appear to function identically in the circuitry, but are less influential and not structurally distinctive. Basket and stellate cells receive excitatory input from climbing and parallel fibers, which suggests that compensatory negative regulation of Purkinje cell output is a primary role for these two cell types. In addition, the basket cell receives inhibitory influence from the cells it "controls," via Purkinje cell axon collaterals. So the negative regulation is itself subject to negative feedback control.

The last type of neuron in the cerebellar cortex, the *Golgi cell*, has perhaps the most complex relations with other cells and fibers. A relatively large neuron situated near the middle of the cortex, the Golgi cell sends dendritic branches upward and an axon tree, unusual in the extent of its branching, downward. Excitatory inputs to the Golgi cell come from climbing fibers as well as from both mossy and parallel fibers. The Golgi cell is inhibited by a Purkinje axon collateral. The primary output of Golgi cells is inhibitory to granule cells, allowing both feedforward and feedback (via the parallel fibers) negative control.

The neuroanatomical studies which uncovered the cerebellar circuitry, combined with extensive electrophysiological experiments, have led to a general picture of how it all works. Several possible control circuits have been mentioned in the course of the above presentation. In addition, the two input pathways have radically different influences on Purkinje cell behavior. The ratio of climbing fibers to Purkinje cells is near unity; the synaptic connections are very strong. Thus the climbing fiber pathway provides powerful and precise control. By contrast, a single mossy fiber may, by way of the granule cell interface, influence on the order of 100,000 Purkinje cells; and each Purkinje cell may be subject to the influence of a comparable number of mossy fibers. The synaptic connections, however, are weaker, and the transmission delay potentially much longer. Some investigators believe that the mossy-parallel fiber system provides information about the "context" of movement. The climbing fibers, in this view, signal the "appropriateness" of the individual Purkinje cell's response to the context, serving possibly to modify the efficacy of the parallel fiber input. Simulation of a model related to this idea is considered in Section 12.2. Modifiable synapses are further discussed in the remaining sections of this chapter.

While much of the above is speculation, one feature of the geometry is fairly clear. Because of the progressively increasing transmission delay over the parallel fibers to more remote Purkinje cells, there is a mechanism for interconversion between space and time. Conceivably a Purkinje cell could "measure" the distance between two simultaneous movements by comparing the delays associated with the arrival of the information. Conversely, an event signalled at a particular mossy fiber will give rise to a spreading wave (or "beam") of Purkinje cell activity. Such spatiotemporal transforms have been suggested as information processing techniques in other areas of the nervous system. But seldom has there been seen elsewhere the degree of architectural uniformity found in cerebellar cortex, a feature that could enhance the precision of the transforms and might therefore be necessary in a center for coordination of sometimes very rapid movements.

11.3.2 Cell Assemblies

From a demonstrable and quite specific neural circuit, the focus now shifts to a class of hypothetical neural arrangements. Cell assemblies were initially an attempt to explain the nature of brain changes accompanying learning in experimental animals. For many years, cell assembly theory dominated the study of learning mechanisms, partly for want of plausible alternatives. As seen in the next section, however, there is today no shortage of candidates for learning and memory mechanisms; and the cell assembly has lost popularity. Nevertheless, as an interesting (at least historically) and relatively simple neural circuit, and as one which has been the object of several computer simulation studies, the cell assembly retains a prominent place in this discussion.

The concept of the cell assembly was introduced in the late 1940s by the McGill psychologist D. O. Hebb. In the original formulation, a cell assembly was considered to be a group of perhaps a few hundred neurons, with relatively strong interconnections, capable of functioning as a unit in perception and thought processes. Since attempts to excise or interfere with specific concepts or memories by means of controlled brain lesions were meeting with no success, the cell assembly was unlikely to be anatomically compact or minimally connected. Early additions to the theory were the notions that the neurons involved in an assembly could be widely distributed in (presumably) the cerebral cortex and have a sufficient redundancy of connections so that deletion of a few would not damage the assembly's functional integrity.

Hebb gave considerable attention to the manner in which cell assemblies might be expected to develop from the randomly connected neural circuits that were widely believed to be the organism's cortical endowment at birth. A process was needed whereby repeated common activity, a presumed concomitant of the repeated experiences leading to learning, could strengthen the connections among a subset of neurons. As a postulate for synapse modification Hebb proposed that, if the firing of one neuron is repeatedly influential in causing the firing of a second neuron, the first neuron will become increasingly important as a determinant in the firing of the second; that is, the (excitatory) synapse from the first neuron to the second neuron will grow in "strength."

Several physiological mechanisms could be imagined as a basis for Hebb's postulate. Although he favored an increase in presynaptic membrane area as the source of increased strength, Hebb did not tie his theory to any particular mechanism. One reason for the previously mentioned decline in the popularity of cell assembly theory has been the failure of many experimental investigations to find a physiological basis for Hebb's postulate, or even merely to observe it in action.

Whatever the mechanism, it would presumably require some time to effect the strength change, a period during which a temporary record of the commonality of activity would have to be maintained. Hebb believed this temporary record was held in the form of recirculating impulses among the neurons of the nascent cell assembly. Disruption of such "reverberation" would explain the loss of memory for events immediately preceding electroconvulsive shock or accidental brain trauma. For Hebb, then, the reverberatory activity was both a temporary record and the driving force producing the permanent physical change, a process known to psychologists as *consolidation.* Some descendants of the original cell assembly theory appeal to independent mechanisms for the temporary record and for consolidation; reverberation, too, has lost popularity.

Hebb discussed cell assembly development and coordination in the context of an experiment in which a rat learns, over many trials, to find food behind a door with a triangle on it, regardless of whether it is the left or right of two doors. Since the rat presumably has had little or no previous experience with well-formed triangles, the first several trials will be accompanied by the development of cell assemblies for the shape's components, perhaps the individual angles, while the animal's visual system scans and rescans the patterns on the doors. As the rat begins to recognize the triangle as a whole, the constituent assemblies will begin to establish relations among themselves and with other, previously uninvolved, neurons. The triangle should eventually come to invoke an ongoing sequence of activity among the constituent assemblies and a new "superordinate" assembly. Hebb called such activity a *phase sequence.*

The original theory was rather vague concerning the nature and mechanics of interassembly relations. The manner in which a currently active assembly would influence which other assembly would next be active, a process which has come to be known as *priming*, has been given several different treatments during the evolution of the theory. Priming is a critical issue; at stake is the explanation of how one thought leads to another in the absence of any directing external stimulation.

Most major versions of cell assembly theory, following Hebb's first one, can be seen to pivot on their particular resolutions of the priming issue. A somewhat more detailed account of the chronology from this point may be found (with references) in Sampson (1969). The second version of the theory (he called it "Mark II") came from Hebb's colleague P. M. Milner. The major new mechanism was inhibition, a phenomenon not widely accepted until the early 1950s. Although he introduced it as a damping mechanism (to prevent an "explosive" accumulation of excitation), Milner also thought inhibition would serve as an umbrella, under which a priming process (mediated by excitatory interassembly connections) could potentiate without actually activating some other assembly. This *excitatory priming* mechanism was based on the now abandoned physiological premise that subthreshold excitation of a neuron would make it easier to fire later.

The third version of the theory was developed as a consequence of some very early simulation studies, roughly contemporaneous with Milner's work. J. H. Holland and others implicitly proposed an *inhibitory priming* scheme, in which the next assembly to become active would be that one which had been *most* inhibited and therefore least subject to the fatigue effects of even minimal background activity. A similar kind of postinhibitory rebound has been invoked to account for other neural information processing phenomena.

A fourth version of cell assembly theory originated from a suggestion by I. J. Good that cell assemblies might contain *subassemblies*, smaller and even more coherently connected groups of neurons. On the assumption that assemblies could share subassemblies, priming would become a natural transferral of activity to that other assembly with the most subassemblies in common with the currently active one. The hierarchical notions implicit in both Hebb's phase sequence and Good's subassembly were developed by Sampson (1969) into a multilevel hierarchical decomposition of neural circuits, with consistent inter- and intra-level relations. Priming was seen to occur in parallel on three different time scales. Some preliminary simulation work with this model will be reported in the next chapter.

11.3.3 Notes on Learning and Memory

The ultimate goal of many students of neural information processing has long been to understand the neural circuitry, and even the molecular mechanisms, underlying distinctively human brain functions, such as language and abstract thought. Even to approach these issues requires good working knowledge of the only slightly less remarkable processes of learning and memory. Here experimental animals can be used, since whatever goes on in the human brain when the elements of language are learned and reproduced in appropriate contexts may have some relation to the simpler learning processes common to most animals. But even a rat has a very complicated brain. Paradoxically, some of the initial progress toward understanding vertebrate learning and memory mechanisms, which may in turn lead to an understanding of the neural bases of language and thought, is now occurring through study of the very much simpler neural circuits found in insects and crustaceans.

This discussion starts "at the top," indicating the sort of skimpy data that are available concerning specialized human neural circuits. The viewpoint then gradually descends to the level of cockroaches and snails, where important new discoveries about learning and memory are now occurring. A thorough treatment of all this would require more than one book of its own. What follows are a few arbitrarily chosen observations.

Although there is currently some dispute about the extent to which other higher primates can be taught to communicate the way people do when they talk to one another, the scope and versatility of human language behavior remains unrivaled. Almost nothing is known about the neural basis of this behavior. As suggested in Section 10.3, damage to certain regions on one side (usually the left) of the cerebral cortex interferes with language behavior, producing disorders known as *aphasias*. Two regions have been named after early investigators of aphasic syndromes.

Broca's area, located in the frontal lobe near the lateral fissure, appears to function primarily in speech production. Injuries in this region seem not to affect language comprehension significantly, yet can leave the patient almost inarticulate. By contrast, damage to Wernicke's area, which is situated in the upper temporal lobe, leads to a much more noticeable deficit in language reception than in production. The production-reception dichotomy is but the coarsest subdivision in the neurologist's taxonomy of aphasic behavior; and several other cortical areas have also been implicated. Notwithstanding its clinical importance, however, the correlation of brain areas with language deficits does not contribute much to understanding how verbal behavior is normally mediated by neural circuitry.

Another brain region whose circuits may have evolved to perform special functions in man is the hippocampus. Although other higher primates do not seem to be so affected, people with bilateral hippocampal lesions exhibit a profound learning and memory disorder. Such patients have normal memory for events prior to the brain damage, but cannot process new data. Only by overt rehearsal or other mnemonic tricks can such people retain small amounts of information, like a street address. The slightest distraction destroys the memory. New acquaintances must reintroduce themselves at every subsequent meeting. There is thus an evident role for the hippocampus in human learning and memory. The exact nature of this role is controversial; whereas many investigators believe that hippocampal injury interferes with consolidation, others think that retrieval processes may be disrupted.

In monkeys, experimental lesions in different regions of the cerebral cortex produce different disorders of learning and memory. One of the most distinctive of these is the loss of the ability to perform simple delayed choice tasks after bilateral lesions in small areas of the frontal lobes. A normal monkey can watch the experimenter place a food reward under one of two cups and quickly learn to pick up the proper cup after they have been hidden from view for a minute or more. Any such concealment leads to random performance in the animal with frontal damage. One hypothesis is that frontal regions help maintain in memory the proper temporal sequencing of events. As for hippocampal function in humans, however, the cellular circuitry involved in the memory function remains unknown.

Yet there are inarguable cellular changes in the brain when experimental animals undergo extensive learning experiences. Rats reared in an "enriched" environment, with broad spectrum sensory stimulation and opportunities to acquire a variety of motor tasks, have thicker cerebral cortices, with a greater density of glial cells, than their environmentally "impoverished" litter mates. Some studies at the molecular level even suggest that there are specific changes in glial and neural RNA base ratios in particular brain regions of rats that have been taught moderately complicated motor tasks.

There have been a number of theories concerning the "biochemistry of learning," involving roles for RNA, proteins, and even DNA. It would be surprising if such fundamental biological molecules did not participate at least indirectly in modification of neural circuits. Specific proposals connecting biochemical changes with alterations in synapse effectiveness (assuming such alteration is the physiological basis of memory) have usually not, however, received widespread support. And connecting biochemistry with behavior involves bridging an even wider gap.

Yet the gap can be narrowed in the relatively simple nervous systems of primitive organisms. Some invertebrates have on the order of 10,000 neurons. An individual ganglion may contain only a few hundred cells, some of which can be seen to have the same connections in every organism of the species. Such "identifiable neurons" permit repeated experiments with the same neural circuit. When these circuits involve only a few neurons, prospects for understanding the learning mechanism are vastly improved.

Such is the case in the abdominal ganglion of the sea snail *Aplysia*. One identifiable cell in this ganglion has cholinergic synapses to several other identified neurons. By simultaneously exciting a cell which accelerates heart rate and inhibiting a cell which constricts blood vessels, this neuron can exert a powerful synergistic effect to increase the flow of blood. A third synapse from this same neuron is interesting in that it mediates both excitation and inhibition, the "sign" of the influence being dependent on the frequency of arrival of presynaptic action potentials. Since the transmitter for all these effects is ACh, there are evidently some rather sophisticated configurations of postsynaptic receptors in this primitive organism. Such configurations may, in fact, be more sophisticated than those found in higher animals. When there are only thousands, instead of billions, of primitive circuit elements to work with, the elements may have to exhibit greater flexibility of function.

One difficulty with studying learning and memory in these simple and accessible neural circuits is that the organism may not evince much capacity for learning. *Aplysia*, however, has a simple reflex (the withdrawal of the gill when the siphon is touched), which manifests *habituation*, a form of learning in which the reflex becomes less pronounced and eventually disappears with repetition of the same stimulus. The circuitry for the gill withdrawal reflex involves a couple of dozen sensory neurons, six identified motor neurons, and three identified intermediate neurons. Experiments have revealed a probable molecular mechanism for short term habituation of the reflex. The sensory neurons excite the motor neurons less effectively because they release less transmitter at the synapses. Decreased transmitter release is apparently a consequence of lessened calcium ion flux.

Other studies with *Aplysia* have yielded insight into longer term habituation mechanisms (and their connection with the short term ones) and into the basis for the complementary learning phenomenon of *sensitization*, in which the normal reflex response is enhanced by associating the siphon stimulation with another stimulus that is unpleasant to the organism. The mechanism for short term sensitization appears to involve an enhanced calcium flux resulting from increased cyclic AMP, whose production is in turn stimulated by the neurotransmitter serotonin released from neurons responsive to the unpleasant stimulus.

Much of this work on learning and memory in *Aplysia* has been carried out under the direction of E. R. Kandel at the Harvard Medical School. The results obtained by Kandel's team stand in sharp contrast to the vague speculations available concerning learning mechanisms in higher organisms. What lies ahead is the difficult task of generalizing the findings in small neural systems to the incredibly complex neural circuitry of the human brain.

Chapter Twelve

Simulation of Neural Networks

The modelling of neural networks has a long history, dating at least back to the early 1940s when McCulloch and Pitts designed formal "neurons" as a contribution to automata theory. In more recent decades, there have been several books devoted to mathematical models of the nervous system. A comprehensive survey of this work is beyond both the scope and purpose of this chapter. Several selection criteria have led to the specific systems and models discussed below.

First, the restriction to models studied by simulation remains in effect. Second, discussion of models involving neural input-output systems is deferred until Chapter 14. Third, an attempt has been made to coordinate the present material with the neural networks treated most extensively in the previous chapter (i.e., cerebellar cortex, cell assemblies). Finally, there is continued attention to simulation *systems*, which facilitate study of broad classes of models (see Sections 2.2.3 and 5.1).

12.1 NETWORK SIMULATION SYSTEMS

Each of the three simulation systems considered in this section permits study of a considerable range of neural network models. The *real system* is always a sort of composite "classical" vertebrate CNS, with chemical synapses of the well-studied kind and no explicit representation of sensory or motor functions. Even within these general boundaries, each system must further limit the range of *experimental frames* available to its users.

Because there is an endless list of parameters that could be varied, the system designer must set most of them to the default values he hopes will be acceptable to his community of users. In this way, a neural network simulation system inevitably embodies a high level *lumped model*, representing at least a general theory about how neurons interact. For example, none of the authors even mentions the presence of glial cells in the *base model*; this seemingly automatic simplification rules out study of models in which neurons and glia interchange information. Nor does any of the systems permit representation of axons which do not transmit action potentials, or of dendrites which do.

The three studies make quite different uses of the *computer*. The simulation methods span the continuous, discrete event, and discrete time paradigms. Another major variation among the systems is the degree of interactive participation in the simulation process that is afforded the user.

12.1.1 PABLO

Developed in the early 1970s by Perkel (1976) and his colleagues at Stanford, PABLO is a well-organized and versatile network simulation system. It is not, however, an interactive system, since the user specifies a model on a sequence of rigidly formatted control cards and then loses all contact with the simulation process. The user may, however, include control cards which alter the model in prespecified ways at predetermined interrupt times. In addition to the network entities and their parameters and interconnections, control cards are used to indicate when and from what components output is to be produced. The output facilities are not very extensive, being simple line-printer displays of impulses (action potentials) over time. Finally, PABLO control cards can be used to save (at any interrupt point) and later retrieve a network.

In sum, PABLO seems to offer the nonprogrammer nearly every good feature of an interactive simulation system that can be preserved in a batch environment. A more flexible and natural command language could be introduced, however. And more elaborate output facilities would be nice.

Because it runs in a batch environment and is written in FORTRAN, PABLO is much more portable than most interactive simulation systems. Virtually any installation that can cope with PABLO's space requirements (3500 line source program, 276 kilobyte object module) should be able to implement it with little difficulty.

Models to be studied using PABLO are built from three types of components: sources, fibers, and neurons. A *source* is an external input, representing the influence either of a neuron which is not part of the model, of a sensory input, or of an experimental stimulation. Two kinds (*genera*) of sources were described in 1976, with the addition of others planned. A source of genus 1 delivers impulses stochastically in time, the interval between impulses being governed by some user selected probability distribution. More than 20 distributions and ways of combining distributions are available for this and other stochastic features of models. A source of genus 2 delivers regularly spaced impulses, according to a user specified period.

Individual sources belonging to the same genus may have the same or different parameters. If they have the same parameters, they are considered to belong to a common *species*, for which the parameters are stored only once in the data structure, as described below. Thus a model might have several cyclic inputs of period 5 (one species) and several others of period 10 (another species of the same genus). This organizational scheme of species within genera applies also to fiber and neuron components, where the savings in storage when, say, 50 neurons all have common parameters can be considerable.

Fibers are axons or portions of axons, some of which terminate in synapses. Fibers of genus 1 terminate either as free endings (external outputs from the net) or as branch points where groups of other fibers can attach. The latter option allows explicit representation of axonal ramification. Fibers of genus 2 terminate in synapses which influence the postsynaptic neuron in a fashion unrelated to the previous activity in the fiber. The amount of transmitter released is either constant or drawn from a specified probability distribution. Fibers of genus 3 terminate in synapses whose influence is a function of past activity. Appropriate parameter settings can induce facilitatory behavior, in which the amount of transmitter tends to increase with fiber activity, or antifacilitatory behavior. Each synapse is designated either excitatory or inhibitory, regardless of fiber genus.

Each fiber has an associated transmission lag parameter, allowing the modeller to incorporate axon length as a factor in the structure of his model. Also, any fiber may be assigned a nonzero probability of failing to transmit an impulse. Fibers, or chains of fibers, connect sources to neurons, neurons to other neurons, and neurons to external outputs. Fibers are understood to conduct impulses only in the forward direction.

Neurons, like sources, initiate impulses on the fibers attached to them. An impulse is generated whenever the neuron's membrane potential equals or exceeds its threshold, provided the neuron is not in an absolute refractory period following the previous impulse. The ARP may be constant or stochastic. The threshold may be constant (genus 1 neurons) or variable (genera 2 and 3); in the latter case the threshold undergoes exponential convergence to its asymptotic value at a rate governed by a time constant.

The membrane potential undergoes a similar convergence process in all neurons, but is perturbed by synaptic activity. Upon impulse arrival, a synapse emits some fraction of its maximum possible amount of transmitter release. The postsynaptic membrane potential is displaced toward a value which is that same fraction of its present distance from the IPSP or EPSP equilibrium potential. Impulses reaching synapses have effects that combine over time. The target amount of membrane displacement at a particular synapse is seldom exactly achieved, because of additive or cancelling effects attendant upon other synaptic events. Integration of postsynaptic potentials cannot, however, be modulated by spatial summation, since neurons are treated as points, with no natural way to represent the distance of a synapse from the axon hillock.

Both membrane potentials and (except in genus 1 neurons) thresholds are assigned post-ARP values which are linearly weighted sums of their pre-impulse values and their "reset" levels. It is thus possible for either variable to emerge from the ARP with its previous value, and continue convergence from there, or with some designated reset level, or with any weighted combination of the two values.

A genus 2 neuron has all the characteristics mentioned above. Genus 1 neurons are simpler in that their thresholds do not change with time. Such neurons provide a basis for modelling cells with rhythmic spontaneous outputs ("pacemakers"), since the threshold can be set somewhat below the asymptotic value of the membrane potential. A genus 3 neuron is like one of genus 2 except for the addition of a variable for modelling postinhibitory rebound. This exponentially decaying quantity is augmented whenever an IPSP reaches the cell. The cell's excitability is enhanced in proportion to the magnitude of this rebound variable, through an increased membrane potential and/or a decreased threshold.

The major data structures in PABLO are two large one dimensional arrays and an associated two dimensional array of subscript pointers. Each one dimensional array contains information about all the sources, fibers, and neurons in a model, ordered sequentially by species within genus. One array contains the parametric information common to species. A genus 2 fiber species, for example, would be specified by its species number, lag time, failure probability, a distribution number for determining stochastic amounts of transmitter release, and logical variables indicating whether the synapse is excitatory or inhibitory.

The second one dimensional array contains the current values of the "state variables" of the individual members of each species. Not all of these quantities (e.g., the number of fibers attached to a genus 1 fiber) are actually state *variables*; the term is generalized to include any static features that may vary from one member of a species to another. A genus 2 fiber individual would be represented by the identifying number of the neuron to which it connects, the number of the component from which it originates, and a logical variable indicating whether its impulses are to be part of the output record.

The indexing array has two rows and one column for each model component, all of which have unique identifying numbers. The value in the first row of column N is the subscript of the start of data concerning component N in the parameter array; the second row specifies the corresponding point in the state variable array. The ease, or even possibility, of altering these data structures, by addition of components, for example, is necessarily limited by the sequential and nonextensible nature of FORTRAN arrays.

PABLO's simulation methodology is an interesting blend of all three major paradigms. The fundamental strategy is a standard next event scheduling technique. But, unlike many discrete event implementations, the time base in PABLO has been discretized to a resolution significantly less than that of the computer. All time values in the data structure are stored as integers, in units corresponding to 100 microseconds of real system time, roughly a tenth of a typical ARP. Although no explanation for this arrangement is given, it would seem to confer a nice combination of temporal flexibility and computational economy. Asynchronous networks, for example, are routine in PABLO (and presumably in real nervous systems) but often awkward or impossible in standard discrete time formulations. PABLO also makes use of continuous simulation techniques, as described below.

The event scheduling algorithm maintains an unordered list of "interesting times" for all network components. An active source has a next event at the time of its next impulse emission. Processing this event adds new entries to the event list, for the arrival of the impulse at the ends of all fibers attached to the source, and for the new next firing time of the source itself. The next event for a fiber (which has not failed to transmit an impulse) is the arrival of an impulse that has been in transit. At a branch point, new events are scheduled for the emanating fibers. At a synapse, a rather complicated computation determines the effect on the membrane potential of the postsynaptic neuron. The threshold is compared to the new potential to see if the neuron should fire immediately.

The next event for a neuron that has just fired is its emergence from the absolute refractory period, when the state variables must be updated and the time of next firing scheduled (unless it is immediate). Computation of the next firing time requires finding the next expected intersection of threshold and membrane potential curves. This is where continuous simulation enters the algorithm. A reasonably accurate and rather efficient root finding method is employed. Even so, this is by far the most computationally costly event to be scheduled. The asserted economy of PABLO's discrete event treatment of neural networks hinges on the cost of this numerical method and how often it must be employed in the simulation of a given model. The savings achieved by not simply comparing potential and threshold in each neuron with each advance of a discrete time clock could easily be swamped by the costs of repeated root finding. This issue of comparative simulation methodologies has not been directly addressed.

This synopsis of the PABLO neural network simulation system has omitted three details of the implementation. (1) there is a sort of "divide and conquer" method to accelerate search of the event list (another unaddressed issue is whether it would be more efficient to maintain an ordered list). (2) The random number generation techniques involve an innovative "recursive" aspect, in which the parameters of a distribution can themselves vary stochastically.

(3) The possibility of more than one impulse traversing a fiber at the same time generates a considerable amount of bookkeeping; such multiple impulses are considered unlikely in real neural networks, however.

12.1.2 NET

NET is an interactive simulation system that employs the discrete time paradigm and has most of the desiderata outlined in Section 2.2.3. The system evolved from the modelling and simulation of a particular single neuron model by Sampson and Peddicord (1974). Although NET has been described by Covington, Sampson, and Peddicord (1978), the present account incorporates features added to the system since that publication.

The current implementation of NET is in the C programming language under the UNIX operating system. Since UNIX installations are becoming more common, NET is somewhat portable, but obviously much less so than PABLO. The compensation for not being bound by FORTRAN occurs in the ease of organization and modification of NET's data structures; C has both mixed type structures and a link (pointer) data type. NET is also suited to smaller installations than PABLO since its object code requires only about one tenth of the storage space of Perkel's system. One factor in this space reduction, however, is a lesser range of modelling capabilities, as will become apparent in the description to follow.

A time step in a model implemented in NET corresponds to about one millisecond of real time. Each neuron in a network is a point source of output impulses, which are faithfully transmitted with a one time step delay to the synapse(s) at the end of the axon. Like PABLO, then, NET cannot represent nonspike propagation along axons or the effects of relative placement of synapses on a neuron. A further limitation in NET is the inability to vary either the length or transmission fidelity of axons.

Output of an impulse from a NET neuron may occur as often as once per time step and is determined by comparison of the membrane potential at the axon hillock with the current threshold. The hillock potential is computed as a running exponentially decayed average of values derived from an equivalent circuit model of the neuron cell body. The circuit contains parallel components representing excitatory and inhibitory synaptic input resistance, as well as the capacitance, resistance, and voltage source supplied by the passive membrane. After the one millisecond ARP, the threshold decays exponentially to an asymptotic level. Neuron parameters include both this level, the threshold decay time constant, and the asymptotic value of the hillock potential. The membrane time constant for convergence to the asymptotic hillock potential is presently fixed at one millisecond, but could be converted to a parameter with little difficulty.

All of this is much the same as in PABLO. Also similar are the ways in which the two systems handle external outputs (axon branches that do not lead to synapses) and inputs. The latter type of component in NET may be (1) cyclic, (2) stochastic with uniform, normal, or exponential distribution, or (3) governed by a user specified pattern. Although there are fewer distributions available than in PABLO, NET offers a more flexible method of specifying the intervals between impulses. A pattern like 3 7 2(1 2 3), for example, would cause continuous repetition of this sequence of intervals: 3 7 1 2 3 1 2 3.

Perhaps the most innovative physiological assumption in NET occurs in its treatment of the synapse, which is characterized by two parameters. The *gain* G of a synapse is the number of transmitter molecules released into the synaptic cleft. PABLO has an equivalent parameter. But there is no counterpart in the Perkel system to the *persistence* P of a synapse in NET. This parameter is the inverse of the rate constant governing disappearance of transmitter molecules. Concentration of transmitter molecules in the cleft is expressed as a linear differential equation in terms of G and P. For appropriate boundary conditions, Peddicord and Sampson (1974) derived the following discrete analog of the solution to that differential equation:

$$x(t) = [x(t_0) + G] \exp[(t_0 - t)/P], \tag{12.1}$$

where $x(t)$ is the transmitter concentration at time t and t_0 is the time of the previous firing of the axon.

NET has been designed to make it easy for a user unfamiliar with computers to specify, modify, and experiment with any neural network model that falls within the domain of the physiological assumptions outlined above. There is no limit other than the pragmatics of running time on the size or complexity of model networks. To specify and exercise a network the user deals with a simple, concise command language. An attempt has been made to provide useful error messages in response to incorrect commands. Recovery of normal operating status is possible after any error except an overflow of available memory.

Commands occur in three environments: (1) the main control and simulation environment, which the user enters initially; (2) the editor environment, invoked from environment 1 by the command *edit*; and (3) the output formatter or trace environment, invoked from environment 1 by the command *trace*. The multipurpose *quit* command causes a return to environment 1 from environment 2 or 3 and termination of the simulation session when invoked from environment 1.

In addition to those already mentioned, there are six other main control and simulation commands. (1) *read* [net], where the optional "net" is the name of a file containing a previously stored network, in a form ready for simulation; if "net" is omitted, the system prompts the user for definition of a model, asking for all parameters of each neuron and its associated synapses, then for the external inputs and their connections. As in PABLO, every model component has an identifying number and also acquires a status with respect to output (tracing) when the component is defined. (2) *write* [net] may be used to create a file suitable for *read*ing; omitting "net" causes the description to be printed at the terminal, but not in the more readable form that can be generated from the editing environment.

The remaining four commands are all related to the control of simulation runs. (3) *go* initializes the current network, prompts the user for the number of time steps of simulation, and then carries out the simulation for that period. (4) *again* restarts the simulation from the point of the most recent *go* command, but without reinitializing the network; so, instead of an exact repeat, a replication with different stochastic elements is produced. (5) *continue* resumes simulation from the most recent stopping point, possibly after the network has been modified by use of the editor. Both *again* and *continue* generate prompts for new run durations. (6) *test* will be explained in conjunction with its associated trace mode command.

Most of the commands in the editor environment occur in natural triples for dealing with external inputs, neurons, or the synapses associated with the neuron most recently edited. The prefix "x,y" specifies a range of component numbers to be edited. There are four groups of editor commands. (1) *ae*(*n*)(*s*) adds another input (neuron) (synapse) to the network, prompting the user for the relevant parameters and trace options. (2) x,y*ce*(*n*)(*s*) causes the user to be prompted for any changes to inputs (neurons) (synapses) having numbers x through y. (3) x,y*de*(*n*)(*s*) deletes inputs (neurons) (synapses) x through y. (4) x,y*pe*(*n*)(*s*) prints out the current characteristics of inputs (neurons) (synapses) x through y.

In the trace environment, the command x,y*stats* generates tabular data for the variables (e.g., number of outputs, average output interval, maximum hillock potential) that were specified for trace when neurons x through y were defined. The remaining trace commands produce graphical output, the appearance of which is greatly enhanced on a graphics terminal; but there are commands to indicate that graphs will be appearing on either hardcopy or other types of CRT terminals.

Two types of histogram summary of accumulated data are available in the NET trace environment. An *interval* histogram shows the distribution of intervals between impulses for any given neuron whose firings have been traced. In a *stimulus-response* histogram, the contributing intervals are between the firing of one neuron (or external input) and some other neuron to which it is connected. Both kinds of histograms have proven useful in the study of models where relevant laboratory data are presented in the same format.

There are three varieties of the "*plot* x,y variable" command, all of which superimpose (if the output device permits) displays of the specified variable for neurons x through y during the previous simulation run(s). The variable may be "fire" (the neurons' times of impulse generation are displayed), "hillock" (the axon hillock membrane potential is graphed against time), or "test." This last option is a special feature for studying frequency response characteristics (see below) of networks. In the main control environment, the *test* command is used to reset a counter subsequently incremented by every *go* command. The "*plot* x,y *test*" command then summarizes the maximum obtained hillock potentials for neurons x through y for an entire series of experiments in which (typically) an input frequency has been systematically varied.

The data structures employed in NET have been designed to provide maximum flexibility for network modification. Linked allocation of storage is used throughout. There is a list of external inputs, in which each node represents one input and identifies its type and parameters. Nodes here and elsewhere take advantage of the mixed type records allowed in C; typically, node elements include integers, real variables, character strings, and pointers.

The data structure for neurons is a doubly linked circular list. Each node represents a neuron and contains information about its resting potential, threshold, threshold decay time constant, firing history, variables to be traced for output, and a pointer to a list of the synapses which influence it. In such a synapse list, each node specifies the type (excitatory or inhibitory), source (neuron or external input), and parameter values (gain and persistence) of the synapse. Thus, when any particular component of a given network is to be deleted, all that needs to be done is to reassign a few pointer values. Addition of a component requires instantiation of its data fields as well. Such operations would appear to be significantly easier and more flexible than those required to modify a PABLO network.

NET has been used to study a variety of simple neural networks. Of first concern was the frequency response paradigm of Peddicord and Sampson (1974). That study showed that a single neuron driven by one excitatory and one inhibitory synapse, both tied to a single external input, could exhibit band pass filter characteristics with proper settings of the synapse gains and persistences. It was proposed that the rather poor drop off on the low frequency side seen in those experiments could be remedied by construction of a 3-neuron filter. Two such filters have been studied using NET and have shown the desired improvements in behavior. Several pacemaker networks and a possible basis for certain laboratory recordings from cerebellar cortex have also been studied using NET. Details can be found in Covington, Sampson, and Peddicord (1978).

12.1.3 BOSS

PABLO and NET are intended for study of relatively small neural network models, usually well below 100 neurons in size. Despite some superficial similarities to PABLO (FORTRAN implementation, event scheduling approach), Wittie's (1978a) "Brain Organization Simulation System" (BOSS) is intended for much larger models, having thousands of neurons and tens of thousands of synapses. The treatment of all neurons as distinct components becomes impossible. Rather each of a few dozen neuron classes, similar to species in PABLO, has a single set of parameters.

These parameters govern the stochastic placement of the neurons in a three dimensional lattice, the shape of the axonal and dendritic fields, and the firing behavior of the neurons. Once all neurons have been positioned, other parameters determine how many and what types of synapses should appear in any given region of space where an axonal field intersects a dendritic one. Synapses in BOSS can be modifiable, according to a cell assembly type learning rule. The other major physiological departure from PABLO and NET is the provision of internal neuron geometry, with inputs on smaller dendrites propagating more slowly and attenuating faster.

The output of a BOSS simulation run consists of pictorial maps showing which neurons fired during each "quantum" of simulated time (see below). The average firing rates of all neuron classes are also printed periodically. Finally, the modeller can designate certain neurons for which he wishes to see a more detailed report, including the individual firing frequency and some aspects of the internal state.

The BOSS data structures consist of three major arrays. SOMA has one 13-integer entry for each individual neuron; SYN has one 11-integer entry for each individual synapse. These arrays are analogous to the "state variable" array in PABLO. The potentially very large number of components requires special attention to compact packing of data in these arrays. The third array is VNT, the event "list." By use of subscript pointers, VNT is partitioned into 257 sublists, one of which contains free locations. Each of the remaining 256 lists contains events that are to be regarded as occurring simultaneously in a single "quantum" of time, interpreted as corresponding to about half a millisecond. Events to be scheduled are placed in the VNT sublist that will be processed at the appropriate future time.

The simulation algorithm advances the clock in uniform quantum steps, cycling through the event sublists and scheduling all consequent future events. This arrangement represents an interesting blend of the discrete time simulation paradigm with the event scheduling strategy from the discrete event paradigm.

12.2 MODELS OF CEREBELLAR CORTEX

12.2.1 Wittie

Wittie (1978b) has employed his BOSS simulation system to study a proposed learning mechanism in cerebellar cortex. The model contains 22,000 synapses connecting 1800 neurons of 12 different classes; some 500 parameters have to be specified. The *real system* for this work is an arbitrary square of cortex, approximately 3 millimeters on a side. Since such an area actually contains around 30 million neurons, a major simplifying assumption in deriving the *lumped model* is a considerable reduction in neuron density. Impulse effects are adjusted to compensate for this simplification. Model components include both climbing and mossy-parallel fiber inputs as well as three kinds of cerebellar neurons: Purkinje, Golgi, and granule cells. Omission of stellate and basket cells is another major simplification.

In the principal reported simulation study, synapses from parallel fibers to Purkinje cells were strengthened during common excitation of the Purkinje cell by both input systems. Training periods were alternated with test periods in which the mossy fiber input fired alone. Training and test periods overlapped asynchronously in two different regions of the model. Mossy fiber inputs in both regions, however, affected all Purkinje cells studied. The point was to see if Purkinje cells would become relatively more sensitized to the stronger, earlier influence of mossy fibers in the same region.

The cost of computer time made it necessary to terminate the simulation experiment after about 3 seconds of simulated time, which involved executing roughly a billion instructions. This limitation curtailed the number of test periods to a total of three in each experimental region.

At the outset, no Purkinje cells fired during test periods; climbing fiber input was essential. By the end of the experiment, the test periods contained moderately selective responses by Purkinje cells to the type of mossy fiber input that had been associated during training with climbing fiber input. If climbing fibers could actually thus "educate" Purkinje cells concerning relevant patterns of mossy fiber input, an important speculation about the function of the dual input system would be confirmed. Unfortunately, as Wittie himself observes, there is not yet sufficient data about the real system, nor sufficient interaction between modellers and laboratory experimenters, even to begin work on validation of this model.

12.2.2 Mortimer

After extensive study of anatomical and physiological data, mostly derived from cat cerebellum, Mortimer (1970, 1974) chose to model a square of cerebellar cortex approximately 6 millimeters on a side. In the lumped model, the region is divided into 400 smaller squares, about 300 micrometers on a side. Mortimer's principal simplifying assumption is to treat these regions as identical primitive model components, with the same state space and regular pattern of interconnection to neighboring components. The result is a *cellular automaton* model (Sampson 1984). Development of the model along these lines was undoubtedly influenced by Mortimer's access to an interactive simulation system for study of models specified as cellular automata, discussed below.

Other simplifying assumptions include omission of climbing fiber input and lumping of basket and stellate cells into a single population of neurons. There remains a component whose output state can be specified in terms of the mean firing frequencies of 5 neuronal populations: mossy fibers and granule, Golgi, Purkinje, and basket/stellate cells. The input state consists of the average firing frequencies of 6 axonal plexuses entering each component: mossy and parallel fibers, Golgi and basket/stellate axons, and two classes of Purkinje recurrent collaterals. The internal state of a given neuron type is defined as the temporally summated amplitudes of each synaptic input impinging on neurons of that type. Another major simplification is to assume that the time course of postsynaptic activity is determined only by its presynaptic origin, not by the nature of the postsynaptic cell. When this is the case, the internal state of a component can be defined over the set union of the internal states of all 5 neuron types.

Of course not all inputs actually affect all neuron types. The circuit diagram of the model shows 9 types of synapses. The 5 excitatory connections run from mossy fibers to granule and Golgi cells and from granule cells (via parallel fibers) to Golgi, Purkinje, and basket/stellate cells. The 4 inhibitory connections are basket/stellate to Purkinje, Golgi to granule, and (via collaterals) from Purkinje to Golgi and basket/stellate. These connections constitute a large subset of those in the more complete circuit diagram of Figure 11.4 (Section 11.3.1).

After consideration of the distributions of axonal plexuses and dendritic fields of each neuron type, Mortimer could define the *neighborhood* (group of components supplying input) of a component with respect to each input type. Parallel fiber input, for example, is assumed to come from a linear array of 10 components, 5 on either side of the component receiving the input. Weights are assigned to neighbors, reflecting the lesser influences of the more remote ones. The largest neighborhood is for Purkinje recurrent collateral influence on Golgi cells, a 3 by 7 array centered on, and including, the receiving component. The composite neighborhood of a component is the set of neighborhoods of its 5 neuron types.

The transition function of a component is a system of 6 difference equations of the form

$$Sn(t+1) = AnSn(t) + In(t), \qquad (12.2)$$

where $Sn(t)$ is the internal state of the nth axonal plexus at time t, In is the input state of the nth plexus, and An is a time constant representing the rate of synaptic decay in the synaptic terminations.

The output function implicitly specifies the circuitry and contains most of the critical model parameters. All 5 equations of the output function are shown below. The indexing of outputs (On) is as follows: (1) mossy fiber, (2) granule, (3) Golgi, (4) basket/stellate, and (5) Purkinje. The indexing of the internal state components (Sn) is: (1) mossy fiber, (2) parallel fiber, (3) Golgi, (4) basket/stellate, (5) Purkinje collateral to Golgi, and (6) Purkinje collateral to basket/stellate.

$$O1(t) = I1(t)$$
$$O2(t) = R1*S1(t) - R6*S3(t) - T2$$

$$O3(t) = R2*S1(t) + R3*S2(t) - R8*S5(t) - T3$$
$$O4(t) = R5*S2(t) - R7*S6(t) - T4$$
$$O5(t) = R4*S2(t) - R9*S4(t) - T5 \qquad (12.3)$$

The 9 parameters Rn in Equation set 12.3 are the gains of the synapse types in the model circuitry. These parameters incorporate both the strength of the individual synapses and the relative number of synapses of that type. $T2$ through $T5$ are the thresholds of the respective neuron types. When viewed over the space of interconnected components, the above transition and output functions induce a transition function for the cellular automaton as a whole. Simulation of the model amounts to a pseudoparallel updating of this transition function at each successive time step.

As already mentioned, Mortimer was able to carry out his simulation experiments using an interactive simulation system for models specified as cellular automata (Brender 1970), implemented on a local and somewhat unusual assembly of hardware. The command language for this system is powerful and flexible, permitting storage and retrieval of model states, definition of experimental sequences via macros, control of time advance (and even single-step backing up), and control of graphical output displays. Aspects of how this system handles input and output deserve further comment, since these features have not been seen in any interactive simulation system previously discussed in this book.

Experimenter input to a cerebellar cortex simulation, whether it represents natural mossy fiber activity or the counterpart of stimulating electrodes in the laboratory, is specified in two steps. First, the model components that are to receive the input are indicated (via light pen picks), thereby defining an *input map.* A particular component may be part of several different input maps. Each map is subsequently assigned an *input stream*, some function of time which specifies the arrival intervals of inputs for the components belonging to the map. This arrangement confers a nice flexibility on the system, since either the map or the stream can be changed without affecting the other element.

The system also makes *output maps* available. The experimenter can assign different combinations of the variables making up a component's state to different maps. Since the cerebellar model produces 2000 data points (400 components times 5 neuron types) each time step, some such selectivity is evidently needed. Mortimer constructed output maps that allowed him to display the output states of any given cell type. Whenever the simulation has been interrupted, it is possible for the user to look at the current state display under control of more than one output map.

Mortimer carried out five series of simulation experiments with his model. Few details are given in this brief summary; they may be found in Mortimer (1970). The first series of experiments was exploratory in nature. A working value for the time step was established, corresponding to about 3 milliseconds of real time. The effects of variations in the synaptic gain parameters were studied to determine what regions of the parameter space permitted stable, steady state behavior of the model. Although no single parameter exerted a controlling influence, it was found that stability was most easily maintained when the ratio of inhibitory to excitatory input was below 1/2 for all cell types, a situation which is probably found in the real system. The second series of experiments used data from published reports of laboratory research in an attempt to discover parameter settings that would replicate the behavior of cerebellar cortex in the anesthetized cat. There was little difficulty in duplicating the time course of inhibition of Purkinje cell background activity following the application of a stimulus through an electrode on the surface of the cortex.

Using the parameters established in the second series, the third tested the range of experimental phenomena with which the model could cope. In addition, some predictions for behavior in as yet untested experimental paradigms were generated. The fourth series of experiments returned to an analysis of effects of parameter variation. One rather surprising finding was the relatively strong influence of the two Purkinje recurrent collateral paths, despite the small magnitudes of the gains of the synapses involved. Also investigated were the kinds of parameter changes required to alter the model's responses to those seen in the awake cat. One consequence of these experiments was a new hypothesis concerning the mechanism of operation of anesthetics.

The last series of experiments investigated the still largely mysterious transfer function of mammalian cerebellar cortex. The time course of Purkinje cell excitability was examined for a variety of mossy fiber input patterns (impulse, square wave, sinusoid) under conditions corresponding to anesthetized, unanesthetized, and an intermediate ("decerebrate") state of general cortical excitability. Among other conclusions, Mortimer suggested that the cerebellar cortex of the anesthetized preparation produces some responses not characteristic of normal function.

12.2.3 Pellionisz

With several co-investigators, Pellionisz has been involved in simulation of cerebellar cortex models since the late 1960s. This discussion is based solely on a pair of more recent papers (Pellionisz and Llinas 1977, Pellionisz, Llinas, and Perkel 1977), both concerned with modelling the cerebellar cortex of the frog. The first paper reports results of simulation, by numerical integration, of a continuous model of the frog Purkinje cell. Using the same system of dendritic ramification employed in the network model (see below), the investigators divided the neuron into 62 compartments and then studied activity of the model cell in response to simulated activation by parallel fibers, by climbing fibers, and by artificial stimulation propagated along the axon toward the cell. Results largely validated the model and its parameters.

The network model (Pellionisz, Llinas, and Perkel 1977) was designed to study, at the level of individual neurons, events anywhere in the cerebellar cortex of the frog. This objective is not quite so ambitious as it might seem, since the frog cerebellum is considerably smaller and simpler than the mammalian one described in Section 11.3.1. Stretched into a flat sheet, frog cerebellar cortex measures about 4 by 1.5 millimeters, an area comparable to those modelled by Wittie and Mortimer. Further, there are no counterparts in the frog to Golgi and basket cells; and, although stellate cells exist, the Pellionisz model ignores them. The model need thus contend only with mossy and climbing fibers and with Purkinje and granule cells. Yet these components number nearly two million in total, substantially in excess of the number that can practically be represented and processed in a computer.

The solution is to represent the entire cortical network *implicitly*, in the form of regular or stochastic distributions of components and connections. In a particular simulation experiment, only those elements actually encountered in tracing the course of a stimulus are explicitly "grown" according to the implicit representation rules. There is a clear analogy to the way in which a chess playing program makes explicit for local search a part of the impossibly large game tree that can be fully represented only implicitly (by the piece movement rules). The Pellionisz model can thus explore the behavior of only a small fraction of the cortex in any one experiment. Nevertheless, it is nice to be able to implement a model with the potential to track events throughout a counterpart of the entire real system.

The rectangular slab of cortex (approximately .5 millimeters in depth) is discretized so that model components occur at regular intervals of 5 micrometers (or multiples thereof) in each dimension. All of the roughly 30,000 lattice points so engendered along the shorter vertical faces are potential entry points for mossy fibers. Granule cells can occupy every second point, yielding a total of about 1.7 million possible cells. Each Purkinje cell occupies a vertical region of 2 by 14 units, yielding a potential population of about 8000. A single climbing fiber is associated with each Purkinje cell.

Extensive anatomical data provides the basis for growing mossy fiber terminal arborizations. The length of the trunk fiber, from point of entry to the cortex until first division, and the overall length of the planar tree of terminals are computed according to established distributions. Other parameters affecting the shape of the tree were varied in the early experiments, but proved to have little effect on the pattern of granule cells activated. Granule cell activation required simultaneous activity in some small number (usually 3-5) of the mossy fiber ramifications terminating at the cell's position.

Other simulation experiments focusing on the mossy-granule interface included studies of the effect of clustering versus dispersal of mossy fiber inputs, and of the interaction of two remote patches of such inputs. The frequency with which relatively compact regions of granule cell response were encountered in these experiments seemed to conflict with the general belief that the interface serves to scatter mossy fiber input widely among Purkinje cells.

The Purkinje dendritic arborizations, providing sites of input for the parallel fiber axons of the granule cells, were grown as needed on a 29 by 29 lattice with 10 micrometer spacing, in a plane above the cell. Some 15 branch points were designated in accord with simple distributions for branch length and deviation angle. Synaptic sites were spaced randomly along both the Purkinje dendrites and the intersecting parallel fibers. Spatial coincidence of one site of each type was required for establishment of a synaptic contact. The last reported experiment involved simulation of input to the cerebellar cortex from the frog vestibular system and an analysis of the effects of that input in terms of the spatial distribution of Purkinje cell activation.

To conclude this review of selected cerebellar cortex simulation studies, it is instructive to compare the aggregation and simplification techniques used in the three projects to cope with a real system having millions or billions of components. Wittie and Mortimer both began by selecting presumably representative pieces of the large mammalian cortex, while the Pellionisz group opted for the comparably sized complete frog cortex. At this point, all three investigations still faced millions of neurons and fibers.

Three quite different approaches were developed. Wittie reduced the actual density of all model components to a point where BOSS, which was designed to deal with large numbers of neurons, could handle the regular arrangement; to compensate for the reduced density, impulse effects were adjusted in an unspecified manner, a step that is probably the weakest link in Wittie's derivation of a lumped model. Also open to criticism is Mortimer's decision to treat small cortical regions as identical to each other and as containing homogeneous populations of each neuron type. Pellionisz elected to represent explicitly only that portion of a very large implicit lumped model which was involved in the current simulation experiment; yet this approach seems to have curtailed the detailed representation of neuron (or neuron class) behavior and synaptic parameters found in the other two studies. One interesting similarity between the BOSS and Pellionisz implementations is the discovery of points of synaptic interaction by virtue of the intersection of appropriate elements.

Dropping of model components was also employed in deriving the lumped model in all three cases. Wittie did not consider stellate and basket cells, while Mortimer left out climbing fiber input. The Pellionisz investigation, beginning with fewer real system component types, nevertheless neglected stellate cells. The diversity of these three approaches, all of which can make some claim to have produced interesting results, only begins to suggest the vast repertoire of simplification and aggregation techniques that can, and must, be brought to bear on the natural complexity of most real biological systems.

12.3 CELL ASSEMBLY MODELS

Attendant on the development of cell assembly theory by Hebb and his followers (Section 11.3.2) have been a number of efforts to validate and employ that construct via simulation. Many of these projects were undertaken by John Holland's students at the University of Michigan. Holland himself participated in the first attempt to simulate the formation of cell assemblies, a 1950s effort that was not fully successful, partly because of the embryonic nature of the theory and partly because of the limited computing power of the machines then available.

As investigators have sought to explain increasingly complex behavioral phenomena by simulation of cell assembly models, the primitive model components have necessarily become more abstracted from the single neuron level. It would appear that only in the work of Finley (Section 12.3.1), which followed directly on that of Holland and his colleagues, has there been intensive investigation of how cell assemblies could arise in neural networks. Sampson (Section 12.3.2) used subassemblies as atomic units. More recent work has employed the cell assembly itself as a primitive element, in order to model aspects of perception and cognition; some of these projects are considered in Section 14.2.

12.3.1 Formation and Development of Cell Assemblies

The class of neural network models studied by Finley (1967) involves square arrays of regularly spaced neurons, usually 400 in number. The orderly arrangement is not intended to mirror a similar pattern in the real system, as was the case in the cerebellar cortex models of the previous section. In some of Finley's models, inputs to a given neuron can come from anywhere else in the network, a so-called uniform random connection scheme. In models with distance biased connections, inputs can arise only from neurons randomly selected within a surrounding circular region of specified radius. In addition to the network interconnections, external inputs can be directed to any neuron. To avoid the complications of edge effects, Finley wraps the array around in both dimensions, effectively mapping the model space onto the surface of a torus.

As in the PABLO and NET systems, neurons are point sources of impulses. In a strict discrete time model and associated simulation paradigm, typical parameter values allow minimum interspike intervals of a few time steps. Neurons are never spontaneously active, requiring a sufficient excess of excitatory over inhibitory input to fire at all. The firing decision is based on a comparison of the sum of external inputs and (signed and weighted) synaptic inputs with the neuron's current effective threshold. The latter quantity is the sum of threshold and fatigue components. Synaptic weights change according to a Hebb type rule.

For speed of computation and maximum flexibility, a table lookup method determines the major components of the transition function. The threshold function is tabulated in terms of the neuron's recovery state, defined as the number of time steps since it last fired. The first few values in this table are usually impossibly large thresholds, providing an absolute refractory period. The remaining values normally decrease to some resting level. Finley could, however, employ a threshold function of any desired shape, simply by inserting a new table. This flexibility may be contrasted with the options in a NET model, where the user controls only the time constant of the exponential decay process. There is an inevitable extra burden associated with Finley's scheme, since part of the assessment of any model for its physiological realism must be an estimation of the naturalness of the threshold and other functions.

The fatigue component of a neuron's effective threshold is computed from a tabulated function of its fatigue state. Within established bounds, this state is adjusted up or down each time the neuron does or does not fire. The fatigue function is normally arranged so that a neuron with a higher rate of recent activity has a higher effective threshold. This factor can be eliminated from a model altogether, however, simply by specifying a constant function.

The value of a synaptic weight, which corresponds to gain in a NET synapse, is determined in Finley's models from a function of the synapse state. This state is adjusted whenever the presynaptic neuron has fired in the previous time step, upward if the postsynaptic neuron fired in the current time step, downward if it did not. These alterations in synapse state can be given a stochastic character, with independent probabilities of the upward and downward adjustments actually occurring when the proper conditions obtain. And, as for fatigue and threshold, the synapse state can vary only within a bounded range. When the synapse weight is a monotonically increasing function of synapse state, a reasonable interpretation of Hebb's postulate for synapse modification is in effect.

Having to specify three functions and a host of individual parameters, Finley faced an enormous number of possible models. As is generally the case in simulation of large neural networks, he encountered great difficulty in merely obtaining reasonable steady state behavior, even in the absence of experimental inputs. By far the most common results of simulating arbitrary networks are cessation of all activity after a short period of time or wild oscillation of activity, neither of which can be considered characteristic of the resting nervous system. During his early parameter tuning efforts, Finley also frequently ran into another common problem, a network that begins with reasonable steady state characteristics but soon lapses into rigidly periodic behavior.

Through a laborious combination of mathematical analysis and parameter tuning via simulation, Finley was able to delimit parameter ranges for various types of networks that would yield acceptable steady state behavior and thus serve as a basis for the study of cell assembly formation. The best such candidates turned out to be networks with both excitatory and inhibitory synapses (as opposed to excitatory only), with distance biased connections, and with fatigue operational. In addition to this intuitively satisfying combination of characteristics, such networks could operate with reasonable threshold functions. When simulation was initiated with a subset of some 20 of the 400 neurons designated to fire at time 0, the best of these networks maintained stable, aperiodic, background activity for many hundreds of time steps.

To see if a subset of neurons resembling a cell assembly could form in such a network, Finley elected to provide periodic external input to a selected group of neurons. Such input could be considered the counterpart of a repeated perceptual experience of the sort discussed by Hebb. After an initial "running in" period of several hundred time steps, the external input was applied every seventh time step during alternate 100 time step periods. The spatial pattern of neurons receiving this input turned out to be surprisingly critical. The optimal arrangement was a group of 9 cells comprising every second cell in a 5 by 5 region of the network. Under these conditions, and after several thousand time steps, a group of 8 neurons, involving only one of the input neurons, had formed a closed cycle based on excitatory interconnections of noticeably increasing weight. The ability of this group to respond to the input stimulus and then maintain a short period of coherent self-sustained activity (reverberation) led Finley to regard it as an embryonic cell assembly.

To study the potential for cell assembly interactions, Finley continued the experiment, introducing a second periodic stimulus, with slightly different parameters (every eighth time step in alternating 150 time step periods). This second input pattern was applied to a different block of 9 neurons, arranged in the same way as the first input group but located a considerable distance away. After another 1000 time steps, the original cell assembly was still firmly established, while another embryonic cycle of neurons with increasing interfacilitatory connections was emerging. Of perhaps greatest interest, there was a suggestion of the development of inhibitory connections between the two groups. This phenomenon was interpreted as supporting the inhibitory priming mechanism proposed by Holland and others.

Finley was regrettably unable to continue this experiment or to undertake others that could have bolstered his results, because his budget for computer funds was exhausted. Nevertheless, his research does strongly suggest that a Hebb type of synapse modification rule, embodied in a large and suitably tuned neural network, can lead to the formation and development of subnetworks with some of the characteristics of cell assemblies.

12.3.2 Subassemblies

Sampson (1969, 1971) began with the major assumption that Finley's results sanctioned the use of small, strongly interconnected groups of neurons as primitive model elements in the further study of "neo-Hebbian neural dynamics." Although he adpoted Good's term *subassembly*, Sampson revised and extended the concept in a number of important ways. The transition from neurons to subassemblies is a very clear example of lumped model simplification by component aggregation. In many ways the subassembly is a kind of superneuron, in which the coordinated activity of many cells leads to a smoothing of behavior and to somewhat more sophisticated information processing functions than are usually attributed to individual neurons.

Sampson's subassembly (SA) unit is a finite automaton with a single output that branches to carry information both back to the unit of origin and to every other SA unit in the network. In an N-unit network, then, each unit has N-1 inputs, which are initially routed through gating elements that determine the (possibly zero) magnitude and sign of the connection. Gating elements represent the net excess of excitatory over inhibitory connections between subassemblies. Gates among SA units belonging to a common cell assembly have large positive values; those between units in different assemblies can have negative values. All of the synapses thus implicitly represented are assumed to be modifiable by a Hebb type rule. The composite effect of all such modifications is applied to each gate output by an "external priming" element which correlates past activity on the input line and in the unit.

A running, time-decayed average of the resulting signals is gathered into a single transmission line by a spatial and temporal summation element. This information is then passed to an internal priming element, which can amplify positive input in proportion to recent activity in the unit. The idea is that an active subassembly will be more responsive to excitatory inputs than a resting one. After processing by all these switching circuits, the collected and modified input to an SA unit reaches the state transition device, where it can modify the excitation and fatigue levels of the unit. These two state variables are considered to correspond to the composite activity and exhaustion of a unit's component neurons. Although output is determined exclusively by excitation level, higher fatigue values proportionately reduce the maximum excitation attainable.

Simulation of the above model (Sampson 1971) preserved the discrete time approach and was carried out in three phases, using two programming languages. In the first two phases, the ease of program development and modification in APL were exploited. Phase I explored the behavior of a single SA unit and was intended primarily to test the coherency of the model and gain an intuitive feel for the effects of the various parameters. Phase II built on the algorithms from the first phase, in order to simulate small networks, having fewer than 20 SA units. An explanation (Sampson 1969) for superior recall of visual as opposed to verbal stimuli was tested in this implementation and found reasonable.

The larger networks of phase III dictated conversion to a more economical programming language; the choice was PL/I, with the most frequently executed iteration as an assembler subroutine. A 50 unit network was used to simulate a free recall experiment (Sampson 1969) and demonstrate how the model could account for improved memory for the first few and last few items in a list (phenomena known to psychologists as "primacy" and "recency").

Chapter Thirteen

Neural Input-Output Systems

The elaborate information processing apparatus in the central nervous system of a higher animal interfaces with a complex array of channels by which the organism communicates with its environment. These input-output systems are conventionally divided into *sensory* and *motor* systems, according to whether the primary function is acquisition of information or control of behavior. As will be seen in the ensuing descriptions, however, most neural input-output systems have information flowing both to and from central processing stations. These stations can be very simple, such as the single synapse that connects the sensory and motor neurons of one type of spinal reflex arc (Section 13.1), or can involve widespread regions of the brain, as in vision (Section 13.3) or the coordination of complex movement.

The information processing mechanisms of input-output circuits, or at least of their more peripheral parts, have come to be understood rather well, in contrast to most brain circuits. This understanding has arisen from the relative ease of experimental access, measurement, and control of sensory and (to a lesser extent) motor systems. Some of the basic work on spinal reflexes, for example, was done in the late nineteenth century by the noted British physiologist Sir Charles Sherrington. Despite the lack of modern investigative technology and the consequent understanding of signal propagation and synaptic transmission, Sherrington was able to learn a lot about neural information processing by clever manipulation and measurement of the external elements of reflex systems in experimental animals.

13.1 REFLEX AND MOTOR SYSTEMS

The term *reflex* is loosely applied to any sensory-motor channel containing chains on the order of a few neurons in length and normally operating with little or no intervention from those higher brain centers that involve consciousness and volition. Hundreds of kinds of reflex "arcs" have been identified in higher vertebrates. The autonomic nervous system operates largely in reflex mode. The principles of reflex function are illustrated here by instances from two groups of mainly somatic reflexes, those mediated at the level of the spinal cord and those concerned with the mechanics of vertebrate vision. A preliminary discussion of the interface between nervous system and skeletal muscle sets the stage both for consideration of these reflexes and for the brief discussion of motor systems which follows.

13.1.1 Innervation of Skeletal Muscle

The designation *skeletal* is applied to all mammalian muscles of the type arrayed in parallel with bones. But the category also includes muscles involved in activities like respiration and eye movement. Not included are the *smooth* muscles of the digestive tract and other viscera. A skeletal muscle is composed of many parallel fibers, each of which will contract in response to release of ACh at the neuromuscular junctions (see Section 10.5.2) distributed along its length. The contraction mechanism is not fully understood but seems to involve a sliding of protein filaments, in a process mobilized by enzymes that respond to an increase in calcium ion concentration.

A group of muscle fibers innervated by the axonal ramifications of a single motor neuron is called a *motor unit*. Motor units are classified according to the characteristics of the fibers that comprise them. "Fast" units contain on the order of 1000 fibers, all of which respond to stimulation quickly (within 10 to 50 milliseconds), reach relatively large peak tension values, and fatigue easily. The 10 to 200 fibers in a "slow" motor unit are more resistant to fatigue, but take longer to reach their lower peak tension values. The mix of fast and slow units in a particular muscle is related to the response speed and load bearing functions it must provide.

Some structural components and neural connections of a skeletal muscle are illustrated in Figure 13.1. There are two types of efferent neurons, both having their cell bodies in the ventral horn of the spinal cord. The *alpha motor neurons* (also often called "motoneurons," as in the figure) send the primary signals for muscle contraction along their large diameter axons to the extrafusal muscle fibers. The smaller diameter axons of the *gamma motor neurons*, comprising about one third of the efferent lines, carry signals which can cause contraction in the intrafusal fibers that are located in the elaborate sensory-motor organs called *muscle spindles* (described below). Spindles are also the source of two classes of afferent fibers, known as Group I and Group II. A third afferent channel originates in the *tendon organs* that are situated near the junction of muscle and tendon.

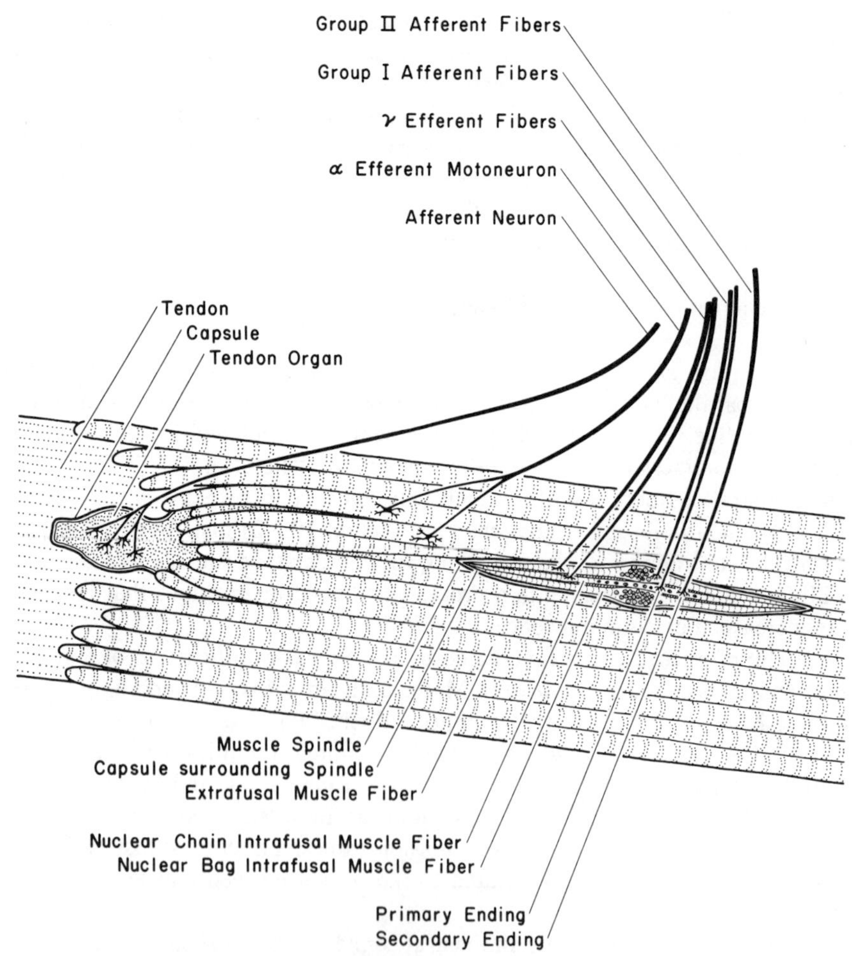

Figure 13.1 Innervation of mammalian skeletal muscle.

The tendon organs, numbering one or two dozen per muscle, are encapsulated structures about half a millimeter in length. Each tendon organ is connected in series with up to 20 extrafusal muscle fibers, which may belong to a number of different motor units. The tendon organ is a sensory element that is primarily responsive to the total amount of tension in the muscle fibers to which it connects. This information is relayed to the spinal cord, encoded as action potential frequency in the afferent axon associated with the organ.

Usually two or three times as numerous as tendon organs, muscle spindles are bundles of 10 or so intrafusal fibers, each around two millimeters in length. These fibers are connected in parallel with the larger extrafusal fibers in the region of the spindle. There are two kinds of intrafusal fibers, called "nuclear chain" and "nuclear bag" because of their shape and the distribution of their cell nuclei. Both kinds of fibers can be caused to contract by activity in the gamma motor neurons that form neuromuscular junctions with them. Two groups of spindle afferents both carry information from both kinds of fibers. But the chain fibers seem to have the preponderance of *secondary endings* (from Group II axons), while the *primary endings* (from Group I axons) are distributed more or less equally on both kinds of fibers.

The information conveyed to the spinal cord from both primary and secondary endings relates to the amount of stretch, or change in length, of the muscle. But there is a difference in the information reported by the two kinds of endings. Whereas the rate of action potentials arising from a secondary ending is mainly a measure of current muscle length, the primary endings generate additional information about the rate of change of stretch. The situation is further complicated by the fact that the stretch monitored at an ending can arise from two different sources. First, the extrafusal fibers, and hence the embedded spindle, may be stretched, in response to movement (of the limb or other associated structure) normally caused by contraction of other muscles. Second, the contractile ends of the intrafusal bag fibers may shorten in response to gamma motor neuron signals, leading to stretch at endings located centrally in the fibers.

The second consideration means that signals from the brain directing contraction of a muscle can actually follow two pathways. Direct excitation of the alpha motor neuron causes faster muscle response. Slower but more refined control can be exerted indirectly. When gamma motor neuron activity causes spindle stretch, a reflex pathway (see below) leading to alpha motor neuron excitation is activated. Recent theories propose that a combination of these mechanisms is operative in the initiation of many movements, in the form of a cooperative *alpha-gamma coactivation* system.

13.1.2 Some Spinal Reflexes

A considerable amount of nonconscious reaction to stimuli and coordination of movement are mediated primarily through the circuitry of the spinal cord. Included in such behavior is the reflexive withdrawal of a limb from a noxious stimulus and the systematic alternation of limb activity required in locomotion.

Superimposed on the more elaborate spinal reflex circuits is some direction and control from higher CNS motor-system centers. The present discussion is limited to three particularly simple spinal reflexes, where there is nearly complete understanding of the underlying circuitry. A diagram showing two of these circuits is presented in Figure 13.2.

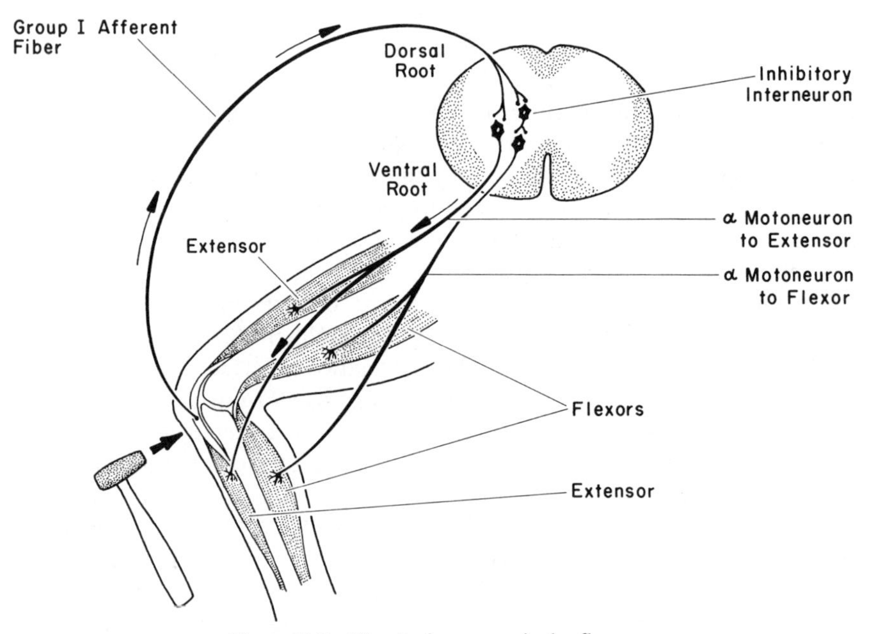

Figure 13.2 Circuits for some spinal reflexes.

Possibly the simplest neural circuit that can be directly correlated with behavior in higher mammals is the two-neuron arc of the *stretch reflex*, most commonly seen in the "knee jerk" test. A sharp rap on the tendon near the knee causes a brief stretch in the muscle along the front of the calf. Responding to this stretch, the primary spindle endings send a volley of action potentials along the Group I afferents. In the ventral horn of the spinal cord these afferents make excitatory synaptic connections with the alpha motor neurons that influence the same muscle. There is thus increased firing of the motor neurons, leading to contraction of the muscle and a forward kick of the leg.

The stretch reflex thus depends on a pathway of just two neurons and one synapse (see Figure 13.2). The phenomenon can be demonstrated in virtually all skeletal muscles, although the strength of the response varies considerably. In addition to its value in diagnosing pathologies of the spinal cord, the stretch reflex has an evident normal role as a length control system. The feedback loop insures that larger amounts of stretch on the muscle will be met by comparable amounts of contraction, to promote constant muscle length.

A second prevalent feedback control reflex (not shown in Figure 13.2) employs a three neuron arc to promote constant tension in a muscle. The tendon organ afferents, which transmit information about tension, have excitatory synapses onto short-axon neurons within the spinal cord. These cells in turn inhibit the alpha motor neurons responsible for contraction in the same muscle (as well as those controlling "cooperative" or synergistic muscles). This circuit thus provides for an alteration in muscle length that can maintain constant tension in the face of variable load.

Since the objectives of this tension control system are incompatible with those of the previously described length control system, some higher level of control is evidently needed in order to choose the appropriate mode of feedback regulation in any given situation. In general, the more neurons in a reflex pathway that are interposed between the primary sensory and motor links, the greater the opportunity for modulation of the circuit by information flowing down from higher centers.

The last spinal reflex system considered here is one in which the sensory information from one muscle is used to inhibit activity in another muscle with an opposing function (an antagonist). The most common examples of antagonistic pairs of muscles are found in the flexors and extensors designed to move a limb in opposing directions. In the case where the extensor system is attempting to extend the limb, the consequent stretch on flexor muscles would give rise to a stretch reflex opposing the extension. Such a conflict of interest is avoided through the mechanism of *reciprocal inhibition.*

As show in Figure 13.2, the afferent signal from the primary (and possibly secondary) spindle receptors runs via an axon collateral to a neuron that has an inhibitory connection to the alpha motor neuron of the antagonist. This circuitry is actually symmetric, so that whichever act has been "ordered" from above, there will not be a reflexive counter argument from the opposing system. This use of reflex systems to control the effects of other reflex systems only begins to suggest the complexity of neural information processing in the spinal cord.

13.1.3 Visual Reflexes

With the exception of the kinds of limb reflexes just considered, some of the most apparent human reflexive actions are associated with the visual system. Less is known about the particular neuronal pathways involved, however, since the pivotal synapses are mostly within the brain stem.

As background to this discussion of visual reflexes, and to the later treatment of visual information processing, Figure 13.3 presents the major structural features of the human eyeball. In order to reach the visual receptors at the back of the retina, light must successively pass through the following structures: (1) the transparent, protective cornea, (2) the "aqueous humor" that fills the anterior chamber, (3) the crystalline lens which focuses the incident light, inverting the image both vertically and horizontally, (4) the gel-like "vitreous humor", and (5) the neural tissue of the inner layers of the retina. A variety of tissue specializations have evolved which maximize transparency in this pathway. Most notable is the absence of blood vessels in many of the structures. Oxygen and other metabolites reach the cornea and lens primarily by diffusion through the aqueous humor.

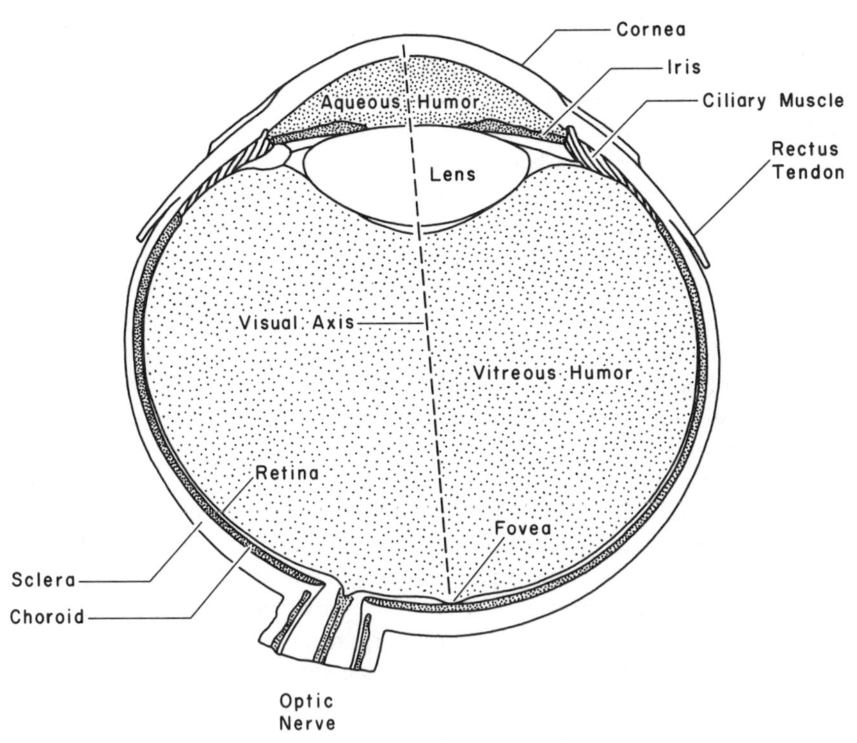

Figure 13.3 The human eyeball.

Of course virtually no light reaches the retina when the eyelid is closed. One purpose of blinking is protection against excessively bright light or rapidly approaching objects. Some of the information supplied by the retina about such events is shunted to midbrain nuclei. The reflex path is completed by neurons which send information to the eyelid muscles by way of cranial nerves III and V (see Section 10.1.2). Another reflex pathway for blinking arises from the eyelashes, the follicles of which contain nerve endings responsive to bending of the lashes, as would occur if an object were about to strike the cornea.

The movement of the eyeball in its socket is controlled by three pairs of skeletal muscles. The motor units of these *extraocular* muscles are richly innervated and quickly responsive to neural signals, which reach them by way of cranial nerves III, IV, and VI. Some eye movements are mainly reflexive, such as the compensatory rotation that keeps the eyes level when a tilting of the head is detected in the vestibular sensory apparatus near the inner ear. A multineuron pathway serving this reflex runs through a couple of brain stem nuclei and out along the oculomotor nerve.

More complicated eye movements, such as occur in reading or in tracking a moving object, seem to have both voluntary and reflex components. The information that guides these movements must come from elements of the visual scene, as processed at the retinal and higher levels of the visual system. It is likely that at least a dozen neural links are involved in the chains that mediate even the reflexive aspects of most eye movements.

Other visual reflexes involve the autonomic nervous system. Parasympathetic outflow along cranial nerve III reaches the ciliary muscle that can modify the shape of the lens to aid in focusing. The stimulus for this response is a blurred retinal image. The sensory side of the reflex pathway has not been fully identified. Finally, there is the reflexive response of pupil diameter to level of illumination. The iris contains two muscles, one subject to parasympathetic control and the other to sympathetic. Within about 0.2 seconds of the onset of increased illumination, signals from the retina traverse a pathway of some six neurons, returning via cranial nerve III to effect pupil constriction.

13.1.4 Aspects of Higher Motor Systems

The preceding accounts of muscle innervation and spinal reflexes have already encompassed the relatively well understood portions of mammalian motor systems. And the circuitry of a major motor center, the cerebellum, was treated in Section 11.3.1. It remains to identify and characterize, insofar as possible, those brainstem and cerebral regions which seem to function primarily in the control of movement.

With one important exception, the descending spinal tracts that converge on the alpha-gamma motor neuron system come from brainstem nuclei. These nuclei are more than simple relay stations, as evidenced by the ability of animals deprived of any higher brain regions to walk and perform other functions involving coordination of the entire body musculature, tasks not executed effectively by the spinal cord alone. Yet these brainstem motor centers do relay instructions from higher centers (basal ganglia, cerebral cortex, cerebellum). The role of the basal ganglia is regrettably most noticeable in certain patterns of aberrant motor behavior associated with injuries to those structures.

The disorders just mentioned are usually classified by neurologists as *extrapyramidal*, in recognition of a long standing dichotomy between those central motor pathways that traverse the brainstem without synapsing (passing through the "pyramids" of the medulla) and all the other pathways. As this characterization suggests, the *pyramidal motor system* is rather easily delimited. It seems probable that both kinds of systems have input from several regions of the cerebral cortex. Since the phylogenetically older extrapyramidal brain pathways have proven less accommodating to experimental investigation, the remainder of this discussion is devoted to some known features of the pyramidal system.

Most fully developed in higher primates, the pyramidal motor system seems to be an evolutionary response to a need for fine, voluntary control of the peripheral musculature. The source of the majority of pyramidal signals is in the neurons of primary motor cortex, which runs along the precentral gyrus (see Section 10.3.3). As in the primary cortical areas for the sensory systems discussed later in this chapter, motor cortex shows an orderly representation of the associated environmental field, in this case the body's musculature. Within this map, body areas having more fine muscular control, such as the lips and fingers, are allocated correspondingly inflated cortical space.

After passing through the medulla, the pyramidal system axons descend in the spinal cord as the corticospinal tracts (Section 10.2.1). Relatively few of the axon terminals connect directly to motor neurons, however. Most end on short axon neurons within the cord, which suggests an indirect and intercoordinated form of control. There is also a good deal of evidence for modulation of the ascending sensory systems by some motor fibers. Clearly the input from somatic receptors, as well as that from sensory organs in the muscles themselves, is essential to the proper functioning of motor systems.

13.2 SENSORY SYSTEMS

Most neural arrangements specialized to convey environmental stimuli to the brains of higher animals are susceptible to the following general characterization: a *sensory system* is an information channel that originates in a collection of *receptors*, sensitive to a particular form of energy, and terminates in a localized region of the cerebrum. The study of such systems, however, is providing increasing evidence for a number of important and broadly applicable corollaries to this characterization.

First, the channel is seldom a passive pipeline. Multiple intermediate data processing stations, the first of which may be right at the receptor level, transform the sensory information before sending it further. The signals that leave the retina, for example, are already a sophisticated summary of the patterns of activity in the individual photoreceptors. Perhaps the simplest and most pervasive transformations executed within sensory channels are based on *convergence*, in which a number of units at one level project information to a single unit at the next higher level, and on *divergence*, in which a single unit feeds many higher level ones.

Convergence is the basis for *receptive fields*, the fundamental building blocks in many sensory systems. The receptive field of a neuron at any level is defined as the group of receptors which can influence that neuron's activity. Frequently found in parallel with convergence, divergence allows receptive fields to overlap, one receptor contributing to many fields at any given level. Such organizational principles have been extensively studied in the vertebrate visual system, as discussed later in this chapter.

A second corollary is that sensory channels often transmit information in both directions. Higher level centers can send signals to lower level ones, signals that may influence how much information is passed on, and in what form. These "gating" kinds of influences from above are commonly called *centrifugal control*. Instances of this phenomenon will be encountered in the ensuing discussions of somatosensory and auditory systems.

A third general consideration in the characterization of sensory systems is that many of them would not function properly in the absence of cooperating motor systems. (The converse of this proposition, that motor systems depend on sensory input, was acknowledged above.) Vision offers an obvious example. Not only are eye movements an intrinsic component of visual activity, but it has also been discovered that an image which is artificially locked onto one group of retinal receptors will lose some of its formerly perceptible features. Thus, in addition to reading, scanning, and tracking, sight itself seems to depend on motion. A similar critical involvement of motor systems has been found in the process of identification of objects by touch, an ability that degrades severely when motion of the hand and fingers is not involved.

A final observation about sensory systems is that recent research strongly suggests a variety of neural coding mechanisms, going well beyond the simple pulse frequency modulation scheme described in Section 11.2.3. One widely accepted notion is that some neural systems code information in the statistical parameters of the distribution of intervals in trains of pulses. Another idea receiving growing support is that coordinated responses in populations of neurons may be a fundamental mode of representation for sensory information.

13.2.1 Receptors and Generator Potentials

The collections of receptors from which sensory signals arise have remarkable structural variety. The simplest receptors are just the peripheral terminals of afferent neurons. Such "free nerve endings" are extensively distributed in the skin and various internal organs. The skin also contains several types of specialized nerve endings, axonal fibers embedded in various tissue structures. Some examples of this second type of receptor are considered in the treatment of somatosensory systems in the next section. A third major class of receptors involves specialized non-neural structures, such as muscle spindles, the cochlear hair cells subserving audition (Section 13.2.3), and the photoreceptors in the retina.

Regardless of their structural complexity, all receptors have in common the primary objective of transducing a particular form of energy into neural signals. Different kinds of mammalian receptors are particularly sensitized to the following energy forms: mechanical energy, as in the stretching of spindles, pressure on the skin, or the bending of hair cells; thermal energy, as in the effects of heat and cold on certain skin receptors; chemical energy, as in the reaction of taste receptors to substances in solution; and electromagnetic energy, in the effect of visible light on photoreceptors.

Since most cells respond in some way to most forms of energy, receptors cannot be exclusively tuned to the appropriate form. Experimental stimulation of receptors, for example, is frequently done by application of electrical energy. An important point is that the resulting sensation is independent of the form of energy applied. Thus electrical stimulation of the retina causes visual experiences. Clearly, the cerebral destination of the sensory channel, not the energy form, determines how the neural signals are interpreted.

The various mechanisms by which receptors transduce energy are in most cases not well understood. A unifying concept, however, is that of the *generator potential*, a change in membrane polarity at the nerve terminal which gives rise to neural signals that can propagate down the channel. Generator potentials resemble postsynaptic potentials in that they can be either depolarizing or hyperpolarizing (but only one or the other in a given sensory system) and can interact by summation. Generator potentials can arise in at least two distinct ways: (1) by direct response to incident energy, as in the depolarizing response of a free nerve ending to mechanical deformation; and (2) by consequence of activity in specialized receptor cells, as in the hyperpolarization of retinal neurons via synapse-like connections from photoreceptors.

The consequences of a generator potential depend on the nature of the first-order afferent nerve fiber. If this fiber does not conduct action potentials, as in the retina, the generator potential invades the sensory channel by passive propagation. In the more common case, the cumulative effect of generator potentials is continuously compared against the fiber's threshold, thereby determining the frequency with which action potentials are initiated.

This frequency is seldom a linear function of stimulus intensity. In many sensory systems there is a roughly logarithmic transfer function between stimulus and neural signal. Such a relation has physiological utility in that it allows sensitivity over a wide range of stimulus intensity, yet retains discrimination of smaller differences at lower levels of stimulation. It is not known whether the nonlinear transformation is generally introduced in the conversion of stimulus intensity to generator potential or in the conversion of generator potential to firing frequency; but the first alternative is considered more likely.

A final important variable in the response of sensory systems is their degree of *adaptation* to stimuli. Since "news" from the environment is normally associated with changes in stimulus parameters, most sensory channels are less responsive to unvarying stimuli than to sudden increases and/or decreases in stimulus intensity. But the extent to which an afferent fiber shows diminishing response to a constant stimulus varies considerably, not only among sensory systems but also among components of individual systems.

Adaptation is a consequence of one or both of two processes. First, non-neural parts of the receptor structure may have physical and/or chemical characteristics which lead to adjustment or fatigue not long after stimulus onset. Two commonly cited examples are the cushioning tissue layers surrounding some mechanoreceptors in the skin (a specific case is described in the next section) and the loss of critical chemicals in photoreceptors. The other possible factor in sensory adaptation is the responsiveness of the nerve fiber itself. In a phenomenon known as *accommodation*, some axons evince decreasing action potential frequency in response to sustained depolarizing stimulation.

There can be yet a third factor in lessening awareness of a repetitious stimulus, such as regular background noise. Higher information processing centers can in some cases exert centrifugal control in such a way as to reduce the amount of input reaching the brain through a particular channel. Additionally or alternatively, the neural mechanisms of attention, whatever they may be, can probably divert sensory information from conscious processing. When an experimental subject shows diminished response to a repeated stimulus, the behavioral phenomenon is called *habituation*; an example was described in Section 11.3.3.

13.2.2 Somatosensory Systems

The several kinds of sensations that humans detect on the body's surface originate in a variety of receptors distributed in and below the skin. Free nerve endings report pain, heat and cold, or aspects of touch and pressure. Specialized nerve endings, embedded in various kinds of tissue capsules, are sensitive to vibration, or touch, or touch and pressure. Some of the variation in receptor structure seems to have arisen as an alternative, in hairless skin, to the widespread hair follicle receptors found elsewhere. Other differences may relate to the simultaneous provision of both rapidly and slowly adapting receptors for a particular sensory modality.

Because of its size and accessibility the *pacinian corpuscle* is among the more thoroughly studied somesthetic receptors. A rapidly adapting vibration receptor found in the skin and various internal structures, the corpuscle is a nerve ending wrapped in many layers of tissue, usually compared to the layers of an onion. The resilience of this capsule is at least part of the basis for the very fast loss of response by this receptor to a sustained stimulus. When the surrounding tissue is removed, the nerve ending shows less adaptation.

The axons of the skin receptors enter the spinal cord as part of the dorsal root cable and then assort into two ascending systems that are differentiated by both location and function. The two pathways are diagrammed in Figure 13.4. The *dorsal column* system (Figure 13.4a) consists of collaterals of the original

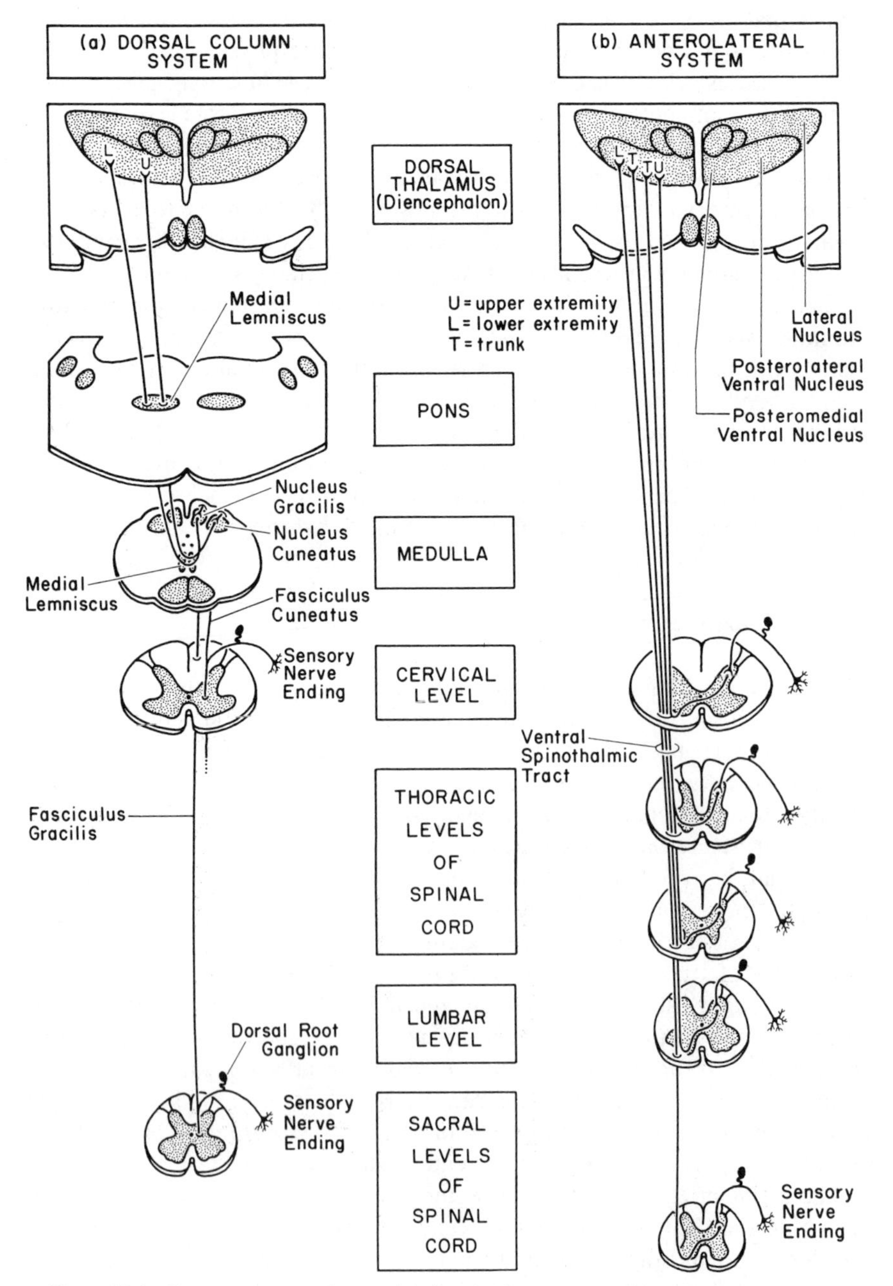

Figure 13.4 Somatosensory pathways: (a) dorsal column system, (b) anterolateral system.

axons, which ascend through the fasiculi gracilis and cuneatus (Section 10.2.1). The first synapse in this pathway is in the nuclei gracilis and cuneatus at the level of the medulla. The axons of the second stage neurons then cross the brainstem and continue upward via a tract called the *medial lemniscus* to the next point of synaptic interface, in the posterior ventral nuclei of the dorsal thalamus.

The *anterolateral* somatosensory system (Figure 13.4b) follows a different route to the same thalamic destination. Collaterals of axons from the receptors synapse with second stage neurons in the dorsal gray matter of the spinal cord. The second order axons cross the cord immediately, ascend in the lateral and ventral spinothalamic tracts, and frequently reach the thalamus without further synaptic interruption. Thus both systems project to the contralateral thalamus via a two-neuron path, but with differing locations for synapsing and for crossing to the contralateral side of the nervous system.

The two pathways also carry contrasting kinds of information. The dorsal column system rapidly and reliably transmits data concerning fine, highly localized tactile discriminations, as well as precise information about limb position. There is good preservation of somatotopic relations and mode specificity in the axonal cables. The system evidently fills the need for detailed, conscious somatosensory information. The anterolateral system has a more generalized role, one which includes primary responsibility for pain and temperature information as well as crude, poorly localized touch. There is more mixing of modalities and less precision in somatotopic mapping in this presumably more primitive somatosensory system.

From the thalamus, where there is considerable opportunity for mixing and resorting of information, both systems project to a number of cerebral centers. The most specific and best understood of these is the primary somesthetic cortex that occupies the postcentral gyrus (Section 10.3.3). The precise somatotopic map found in this region suggests it is primarily a destination for the dorsal column system. Within the map, a finer level of organization divides the cortex into small regions, orthogonal to the cortical plane. Each of these columns contains neurons responsive only to one particular sensory modality (touch, pressure, vibration). This columnar arrangement of primary sensory cortex will also be encountered in subsequent discussions of the auditory and visual systems; it may be a basic organizing principle throughout the cortex.

From cortical neurons back down to primary afferent fibers, it is possible to identify receptive fields at most stages in somatosensory systems. That region of the body's surface where stimulation can influence a particular neuron seems mainly to increase in size (through convergence) as signals ascend through the somatosensory processing stations. Representative values for channels originating in hair follicle receptors might be .5 square centimeters in the primary afferent, 20 square centimeters in a thalamic neuron, and 80 square centimeters in the cortex. At all but the first level, these receptive fields seem to be internally organized in much the same way as the retinal fields that are thoroughly discussed later.

Finally, it should be noted that somatosensory systems are probably subject to considerable centrifugal control. Projections from both sensory and motor cortical regions enter the thalamic and brain stem relay stations. It is thought that these pathways serve mainly to sharpen contrast discriminations in the afferent signals.

13.2.3 The Auditory System

The apprehension of audible environmental stimuli begins with the transformation of sound waves to mechanical action by the elaborate multistage channel of the ear. The ultimate sensory receptors transmit information via the acoustic nerve (cranial nerve VIII) into an array of ascending pathways.

The external ear focuses local variations in air pressure on the tympanic membrane (eardrum), which vibrates in response. In the middle ear, the tympanic membrane is linked mechanically to an orifice of the helically coiled *cochlea*, which is the actual sensory organ for audition. The linkage, executed by the articulation of three small bones, provides an essential mechanical advantage that effects an impedance matching between tympanic membrane and cochlea; only about a twentieth of the proportion of input energy that actually reaches the cochlea could do so in the absence of middle ear function.

A cross section of the cochlea is diagrammed in Figure 13.5. The fluid filled chambers above (scala vestibuli) and below (scala tympani) the triangular *cochlear partition* are connected by a small hole at the apex of the cochlea. Motion in the inner ear gives rise to surging of the fluid, which in turn induces vibrations in the *basilar membrane* at the bottom of the partition. This vibration leads to local bending or shearing of hair-like fibers whose ends are embedded in the relatively stable *tectorial membrane*. The fibers arise from the *hair cells*, which respond to hair deformation with a membrane change that somehow causes a generator potential in the apposed neural terminals. These terminals come from the bipolar neurons in the *spiral ganglion* adjacent to the cochlea. The axons of these neurons are bundled together to form the cochlear division of the acoustic nerve.

The response of cochlear nerve fibers to controlled acoustic stimuli can be assessed through microelectrode recording of action potentials. Each fiber responds only to a fairly narrow range of sound frequencies, the threshold decreasing to a minimum near the middle of this *tuning curve*. There is an orderly sequence of maximal frequency sensitivities, correlated with position along the length of the basilar membrane. The traveling waves induced in the membrane have peaks for higher tones at the end near the middle ear and for lower tones at the far end of the membrane. In addition to the nerve fiber tuning curves just mentioned, it is possible to plot "mechanical tuning curves"

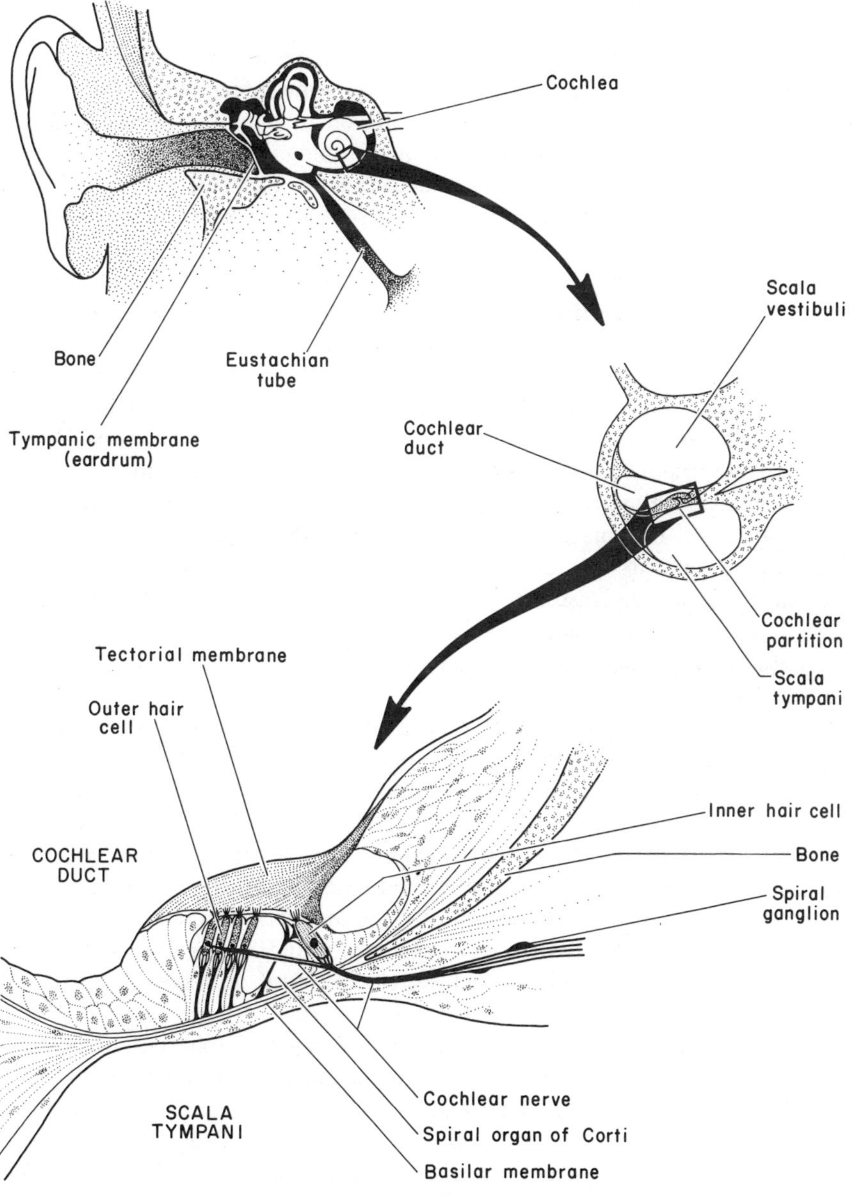

Figure 13.5 Cross section of the cochlea.

that show the increasing attrition of membrane deformation at greater distances from a point of mechanical stimulation.

If the cochlear nerve fibers were reporting merely a passive read-out of membrane behavior, then the neural and mechanical tuning curves centered on the same frequency would be quite similar. Actually, comparisons of this sort show that the neural tuning curve is generally narrower and sharper than the mechanical one. Thus there has already been some information processing by the time the neural representation of audition reaches the primary afferent axons. One explanation for the sharpening of tuning curves could be in terms of inhibitory interactions at the receptor-neuron interface. This kind of phenomenon will be examined in the retina (Section 13.3.2), where the mechanism is better understood.

The ascending pathways which carry auditory information from the eighth nerve to the primary cortical receiving area in the temporal lobe involve a number of major processing stations at several levels. As shown in Figure 13.6, the axons of the spiral ganglion cells all terminate in the cochlear nuclei at the level of the medulla. The output of these nuclei is distributed, some crossing to the opposite side, to other nuclei at the same level (e.g., the superior olivary nucleus) and also upward by way of a tract called the *lateral lemniscus*. In the nucleus midway along this tract, most auditory pathways encounter another synaptic relay. The major auditory center of the midbrain is the inferior colliculus, where information can again pass between the two sides. From the colliculi and, in some cases, directly from the brain stem, the auditory paths flow to the medial geniculate nucleus of the thalamus. From this last major relay station, the *auditory radiations* project directly to auditory cortex in the temporal lobe.

The precise tonotopic map that originates in the orderly arrangement of eighth nerve fibers is preserved at most of the higher levels of the auditory system. In some cases the map is fully replicated, three copies being found in the cochlear nuclei and two in the inferior colliculus. The tuning curves of the individual neurons comprising these maps become progressively sharper at higher levels. There is also increasing interaction with collateral systems, such as the motor systems involved in auditory and audiovisual reflexes like those for orienting the body and sight to an unexpected sound. Other collateral systems mediate centrifugal control, through an elaborate set of descending pathways (suggested on the left in Figure 13.6).

In primary auditory cortex, the tonotopic map is still apparent, organized in the typical cortical arrangement of vertical columns. Within a column, neural tuning curves are centered over a narrow range of frequencies. The tuning curves of cortical neurons vary widely in sharpness and often have multiple peaks of sensitivity. These characteristics are consistent with the tasks of complex sound recognition that must be executed at the cortical level.

The binaural character of auditory input imposes some special information processing demands and affords some corresponding advantages. In the

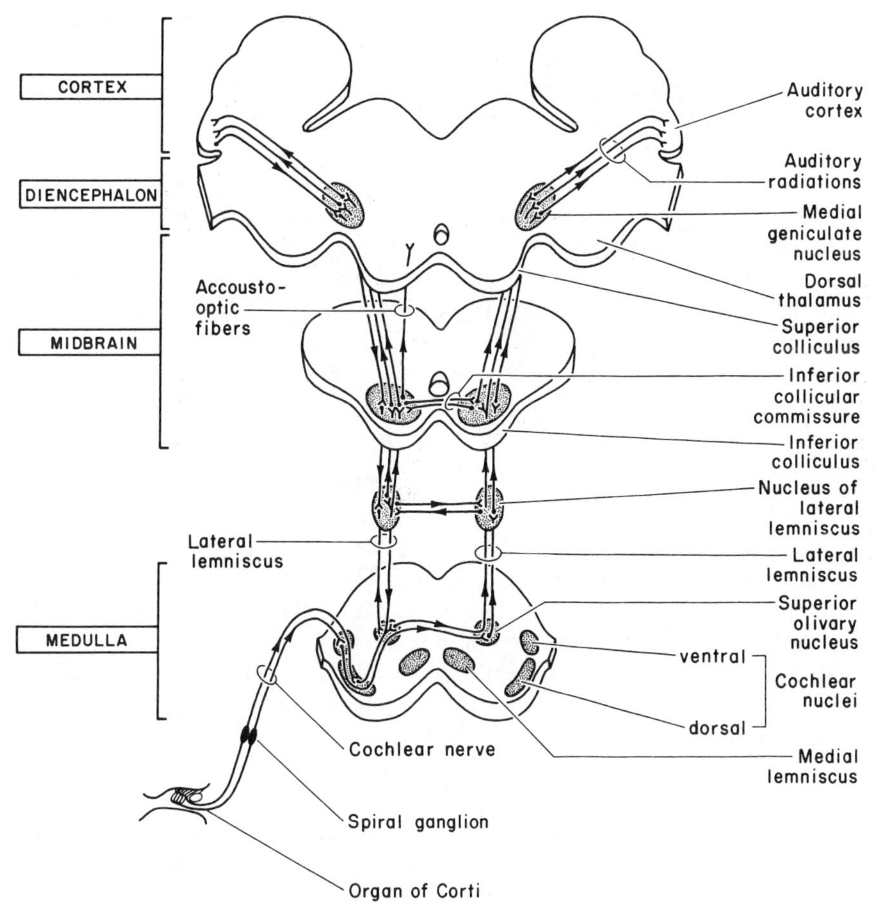

Figure 13.6 Auditory pathways.

auditory cortex, most neurons respond to stimuli originating at either ear, but usually with greater sensitivity to inputs from the opposite side (in line with the general principle of contralateral control). Differences in both amplitude and phase information, as well as differential time delays in the arrival of binaural signals at higher centers, are used by auditory circuits responsible for localization of sounds.

13.3 THE VISUAL SYSTEM

It is not difficult to extend the rather superficial characterization of particular sensory systems begun in the last section. There are variable amounts of data concerning receptors and pathways for the senses of taste (gustatory system), smell (olfactory system), and balance (vestibular system) in higher vertebrates.

And the catalog of sensory systems can be expanded to include many specialized input channels found elsewhere in the animal kingdom.

It is more appropriate, however, to examine in greater detail one sensory system that has been the object of especially extensive study. The visual system is probably best understood among all those even distantly rivaling it in complexity. As has been the case throughout these chapters on neural information processing, much of the detail in what follows is oriented toward the systems found in mammals, particularly man. The more general considerations, however, apply to most vertebrates. On the other hand, there is no attempt to encompass quite different visual information processing mechanisms, such as those found in the compound eyes of insects.

13.3.1 Visual Pathways

The structures of the eye involved in the transmission of light to the retina were discussed in Section 13.1.3. The visual receptors, near the back of the retina, are the origin of the first true information bearing signals in the visual system proper. There are two classes of receptors. The 100 million *rods* in a human retina are more extensively distributed away from the center of the retina. The 5 million *cones* are most numerous in the central region, or *fovea*.

Rods and cones serve different aspects of vision. Much more sensitive to light, the rod system is responsible for vision at night and under other conditions of low illumination; color differences are not perceived through this system. The cone system requires high levels of illumination to carry out its functions of fine detail and color vision. The high acuity of foveal vision results from the close spacing of cones that have a high degree of information flow into retinal neurons, as discussed below. Color vision is a consequence of three populations of cones, maximally responsive to three different wavelengths of visible light.

Rods and cones are structurally similar. The *outer segments* contain highly convoluted membranes and are the location of the photosensitive substances whose chemical conversion by light is the basis for vision. The chemical change leads to alteration in membrane polarity which propagates to the receptor's *inner segment*. Since this portion of the receptor strongly resembles a neuron, some investigators argue that rods and cones should be considered highly specialized nerve cells. The end of the inner segment enters into multiple specialized synaptic arrangements (see Section 10.5.3) with retinal neurons. A rather recent and surprising discovery is that the generator potential in a receptor's inner segment is hyperpolarizing, opposite in sign to most known generator potentials. The reaction of a visual receptor to stimulation by light is actually a reduction in the so-called dark current that flows from the inner segment in the absence of light.

The organization of neural pathways in the primate retina is depicted in Figure 13.7. Receptors project in both convergent and divergent relations onto the *bipolar cells*, which stand in the same relations to the *ganglion cells*. The axons of the ganglion cells are gathered together and exit the rear of the eyeball as the *optic nerve*. The amount of convergence and/or divergence in these pathways varies with requirements for visual precision; cones in the center of the fovea have nearly private lines of communication into the postretinal stages of the visual system. By contrast, hundreds of rods in the periphery may be served by only a single bipolar cell, which is in turn one of many served by a single ganglion cell.

Figure 13.7 also shows two other kinds of neurons, ones that interconnect retinal elements in a lateral fashion. The *horizontal cells* participate in triadic synaptic relations with receptors and bipolar cells. *Amacrine cells* have a similar relation to bipolar and ganglion cells. Like the receptors, horizontal cells

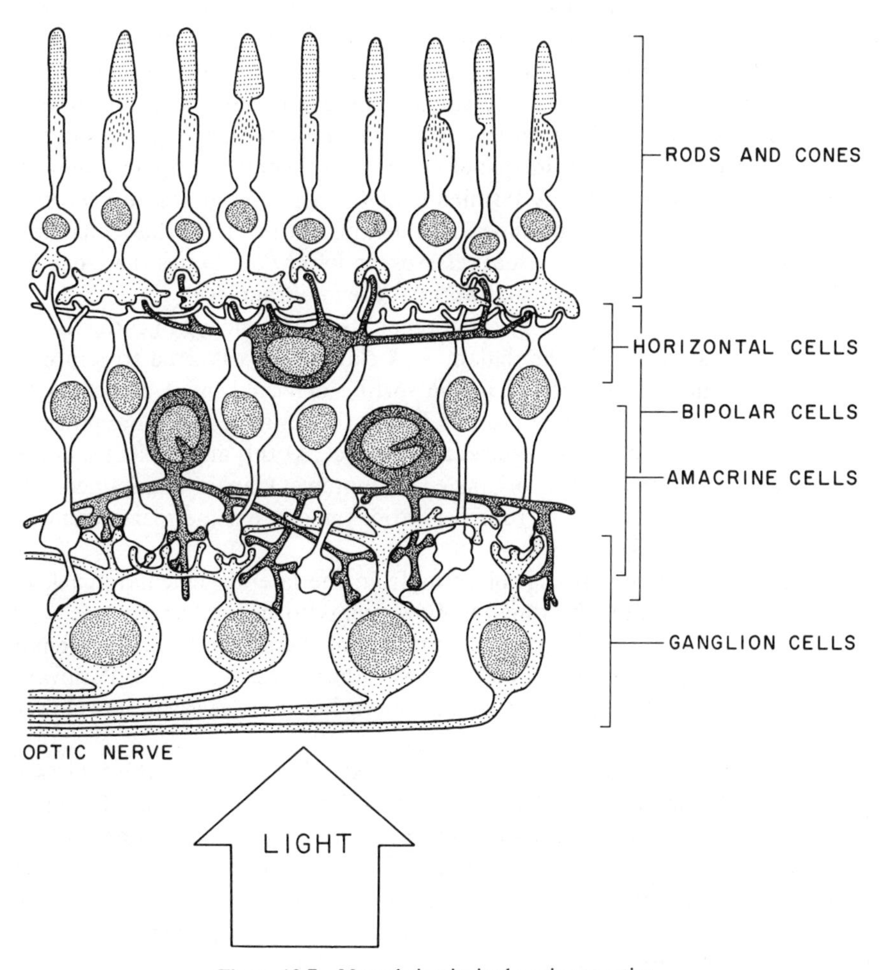

Figure 13.7 Neural circuits in the primate retina.

usually respond to the effects of light with membrane hyperpolarization. (A depolarizing response may also be involved in a proposed role for these cells in the sorting of color information.) Bipolar cells show both hyperpolarizing and depolarizing responses, according to the cell and the circumstances.

The responses of amacrine cells consist of depolarization and occasional action potentials, the first point in the system where this impulse form of encoding is encountered. Ganglion cells appear to communicate primarily through action potentials, which is not surprising given the distance over which their signals must be sent. It may be that the relatively compact retinal layers have permitted evolution of sophisticated nonspike signalling mechanisms, employing specialized synaptic arrangements, for the first stages of visual information processing.

The visual pathways from the optic nerve to visual cortex are shown in Figure 13.8. The first reshuffling of information occurs at the *optic chiasm*, where (in man) the fibers from the interior or "nasal" half of each retina cross to the opposite side of the nervous system. The result is that the *optic tract*, as the channel is known from this point, carries on the left (right) side information from the two left (right) hemiretinas. Since this information originates in the right (left) visual field, the principle of contralateral control is again followed. Some of the optic tract fibers terminate in (or send collaterals to) various midbrain nuclei, especially the superior colliculus. In primates these channels serve as the afferent stages of visual reflexes. In lower vertebrates there can be considerable processing of the visual world at the midbrain level.

The major staging point for information on its way from eye to cortex is the lateral geniculate nucleus of the thalamus. A elaborate six-layered structure in primates, this nucleus is involved in the sorting of visual input according to retina of origin and centrality of position within the retina. Some more processing of color information also transpires at this level. At this and all stages in the visual ascending pathways an organized retinotopic map is maintained, although the relative positions of retinal locations undergo some rearrangement (shown at the left in Figure 13.8).

There are no centrifugal control paths from the lateral geniculate back to the retinal level. In fact, information flow seems to be exclusively one way throughout most of the primate visual system, an interesting departure from a general organizational feature found in most other sensory systems. Pathways mediating centrifugal control of vision have been found, however, in some lower vertebrates.

The last link in the route to the cortex consists of the *visual radiations* (see Figure 13.8). These fibers are the axons of lateral geniculate neurons. They terminate exclusively in primary visual cortex, mostly synapsing with neurons of the fourth cortical layer. Once again, the cortical map is organized in a columnar fashion, with a hierarchy of functional levels that is described in Section 13.3.3. An approximately equal amount of cortical space is given to the information originating in each ganglion cell. The much more extensive

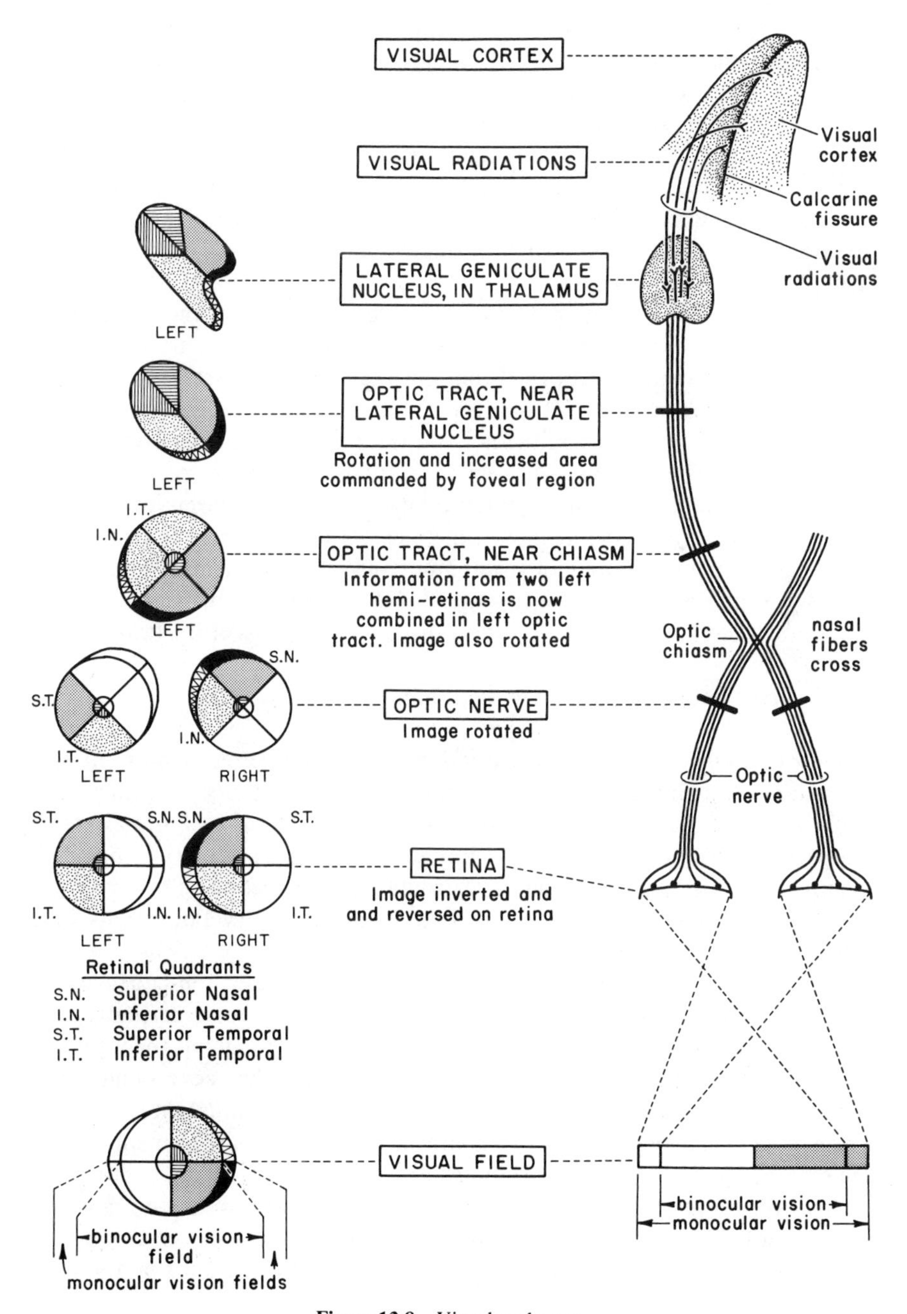

Figure 13.8 Visual pathways.

representation of foveal regions (in both cortex and geniculate) is thus based on differences in convergence ratios within the retina itself.

The area of the occipital lobe that is mainly involved in vision consists of three distinguishable areas. Primary visual cortex, the most posterior region, is also called *striate cortex* because of a characteristic striped pattern created in the middle layers by the arriving visual radiation fibers. Primary visual cortex is area 17 in the most widely used numerical reference system. The immediately anterior areas 18 and 19 are together known as prestriate cortex. They correspond approximately to two functional regions that receive most of the information flowing out of primary visual cortex. Henceforth the more functionally oriented terminology, Visual I (primary or striate cortex), Visual II, and Visual III, is used here.

Some knowledge is beginning to accumulate concerning the postoccipital destinations of visual information. Visual II and III project to a temporal lobe area called inferotemporal cortex. From this region and all three occipital areas there are connections to identifiable portions of frontal lobe cortex. Understanding of the functional significance of most cortical areas involved in vision is incomplete. Visual II and III seem to participate in object recognition and simpler kinds of visual learning. Inferotemporal cortex has been implicated in more complex learning and memory functions. At least one role of some frontal visual fields is the integration of visual input and eye movements. In any case, it is hardly surprising that a sensory system of such singular importance (at least in primates) interacts with many widely distributed regions of the brain.

13.3.2 Information Processing in the Retina

The intricate circuity of the primate retina provides numerous opportunities for coding and recoding information supplied by the receptors. A fundamental mechanism, also found in other sensory systems, is *lateral inhibition.* Activity in a particular receptor-bipolar unit may tend to suppress activity in its nearby neighboring units, possibly through horizontal cell connections. Presumably the strength of the inhibitory influence is proportional to the level of activity in the unit.

This kind of suppression can have a number of significant consequences. The first is an enhancement of the signal-to-noise ratio. If all units normally manifest a low background level of activity, as seems true throughout the visual system, a localized stimulus will stand out more clearly if the background "noise" in nearby units is suppressed to some extent. This kind of situation is actually a special case of the more general phenomenon of contrast enhancement. Along a line separating dim light from brighter light, for example, the units on the dim side of the edge will be the most strongly suppressed among all those responding to the dim light, as a consequence of the more powerful inhibition emanating from the nearby brightly lit units. Similarly, the units on the bright side of the edge will be the least strongly suppressed among all those responding to the bright light, because only they have neighbors with lesser activity.

These effects are not sharply demarcated but rather decline smoothly with distance from the edge in both directions. The net result is that the difference in light intensity reported in the neural signals is greater than that in the visual field. Since edges and other kinds of contrasts are usually more critical to visual information processing than uniform fields, the enhancement has evident value. Similar processes may be at work in the sharpening of tuning curves in the auditory system, as well as in the detection of edges by somatosensory systems.

Another critical kind of visual stimulus involves motion. If lateral inhibition is biased so that it operates more effectively in some particular direction, then a stimulus moving across the field in the opposite direction will elicit a greater overall response than one moving in the direction of the bias. There will be lesser or greater inhibitory resistance to be overcome as the moving stimulus invades new territory. Since the higher level visual receptive fields, discussed in the next section, are especially sensitive to edges and motion, it seems likely that the lateral suppression mechanism is not confined to the retina, but is rather a general principle of interaction among units at all levels.

As further evidence of the importance of stimulus change in visual information, most retinal elements are not particularly responsive to sustained illumination. Bipolar and amacrine cells tend to show transient membrane polarity changes when light appears and/or disappears in their receptive fields. And ganglion cell responses can be grouped according to whether the cell emits a burst of impulses upon stimulus appearance (an *on* response), disappearance (an *off* response), or both (an *on-off* response). The physiological basis for these types of responses is not fully understood. *On* responses could result merely from rapidly adapting receptors; but more elaborate multilevel processing is probably involved. *Off* responses would seem to require some sort of signal inversion, perhaps in the form of postinhibitory rebound enhancement of background activity.

The receptive fields of ganglion cells have been studied in experimental mammals for some three decades. The typical modern technique is to insert a microelectrode in an optic nerve fiber (a ganglion cell axon) and monitor its firing behavior while a small spot of light is moved around on the immobilized retina. Virtually all ganglion receptive fields have been found to be approximately round in shape and to occupy a retinal region with a diameter on the order of one millimeter. Such a field usually has three discernable regions, in which spots of light produce different effects. In one type of organization, the circular center region generates *on* responses, the annular surround generates *off* responses, and a narrow ring in between generates *on-off* responses. Complementary to these *on center* fields are *off center* ones in which the responses are just the opposite. Uniform illumination of the entire field generally has little effect.

The responses just described are actually characteristic of ganglion receptive fields only in light adapted retinas, where the rod system is relatively inactive. In the dark adapted state, the center region becomes larger and the surround loses some or all of its influence on the cell's behavior. Another receptive field variable relates to position on the retina. In accord with visual acuity requirements, the field centers become smaller nearer the center of the retina.

Mammalian retinal receptive fields are actually among the simplest found in vertebrates. This is because more sophisticated visual information processing has been deferred to well developed higher centers, where greater flexibility is possible. In animals with relatively little cortical processing capacity, a considerable amount of pattern detection and classification can occur right in the retina, which has a correspondingly more complex neural structure.

A classic study of the frog retina was carried out in the late 1950s by Maturana, Lettvin, McCulloch, and Pitts. They found five different classes of optic fiber receptive fields, some responsive mainly to pattern information (e.g., boundary detectors), others mainly to illumination (e.g., dimming detectors), and still others to both of these stimulus parameters. Of special note was a type of field maximally responsive to small, convex, dark, moving objects; this could be the frog's "bug detector." The investigators also proposed a convincing correlation between receptive field properties and five classes of frog ganglion cells distinguished by the nature and distribution of their dendritic ramifications.

13.3.3 Information Processing at Higher Levels

Neurons of the lateral geniculate nucleus generally have receptive fields much like those of ganglion cells. In primates, however, some of the cells are sensitive to color information, receiving different kinds of inputs from different classes of cones. One example is a *red on-center, green off-surround* field, which responds maximally to appearance of red light in the center or disappearance of green light in the peripheral part of the field. Different combinations of colors and center-surround relations give rise to eight kinds of fields of this general type. There is also a type of field with no center-surround organization, which will give an *on* response to one color and an *off* response to its "opponent."

Much of the present understanding of the receptive fields of visual cortex neurons can be attributed to more than two decades of research by Harvard Medical School investigators David Hubel and Torsten Wiesel. At first in cats, later in monkeys, Hubel and Wiesel laboriously cataloged the responses of hundreds of cells in dozens of locations. Their work has also contributed significantly to knowledge of the mechanisms of binocular vision (discussed at the end of this section) and visual maturation (discussed in Section 13.3.4), as well as leading to an improved understanding of the organization of visual cortex.

In Visual I, Hubel and Wiesel have discovered three kinds of receptive fields. In cortical layer IV, where the geniculate input enters, there are some cells with fields much like those of geniculate neurons. The other two kinds of receptive field organization in Visual I are specific to the cortex and have been named *simple* and *complex*. As at lower levels, a simple field consists of both excitatory (*on* response) and inhibitory (*off* response) regions. But instead of a circular arrangement, the regions are positioned along side one another in an elongated field. The two areas can be of varying relative size; and either can be flanked on both sides by the other. Some examples of simple (and other) cortical receptive fields are diagrammed in Figure 13.9.

The maximally effective (stationary) stimulus for a simple field is a bar of light that exactly covers its excitatory region. Even slight deviation of the bar's orientation from the field's axis leads to a dramatically reduced response. At 90 degrees from optimal there is virtually no reaction, as there is none to diffuse lighting of the entire field. An especially strong response can often be obtained when the bar is moved across the field in a direction orthogonal to its axis. Presumably, as the bar is both leaving the *off* region and entering the *on* region, nearly all of the field's sensitivities are activated at once.

A cortical cell with a complex receptive field retains the simple field properties of optimal orientation axis and enhanced response to orthogonal movement. But complex fields do not have excitatory and inhibitory areas. Any properly oriented edge evokes a response (see Figure 13.9). Light and dark bars are both generally as effective as simple edges.

Visual II and Visual III do not have neurons with simple receptive fields. In addition to the prevalent complex fields, a new type, called *hypercomplex*, is found in these cortical regions. Still sensitive to orientation and motion, hypercomplex fields generally fail to respond if even a properly oriented stimulus invades antagonistic flanking areas that may be found at one or both ends of the axis (see Figure 13.9). Thus these fields seem to be oriented toward detection of "stopped edges," as would occur at corners. Hypercomplex fields are generally larger than simple or complex fields. They can also show a preference for one of the two directions of motion orthogonal to the axis.

Hubel and Wiesel have offered some hypotheses about how higher level receptive fields might be synthesized by convergence of lower level ones. A simple field with a central *on* strip, for example, could result from the overlapping of several *on* center geniculate fields arrayed along the simple field's axis. And perhaps a complex field is the result of convergence of many adjacent simple fields having a common axis. These kinds of arguments can become a bit strained, however, and seldom account for all the known field characteristics. But even if the wiring is more complicated, it is satisfying to comprehend a rather natural progression of increasingly more sophisticated pattern detection mechanisms at successively higher levels of the visual system.

In addition to differential pattern sensitivity, visual cortex neurons vary in their responsiveness to inputs from the two eyes. Because of the segregation of left and right retinal inputs into different layers of the lateral geniculate, the cortex offers the first opportunity for binocular interaction. Most cortical

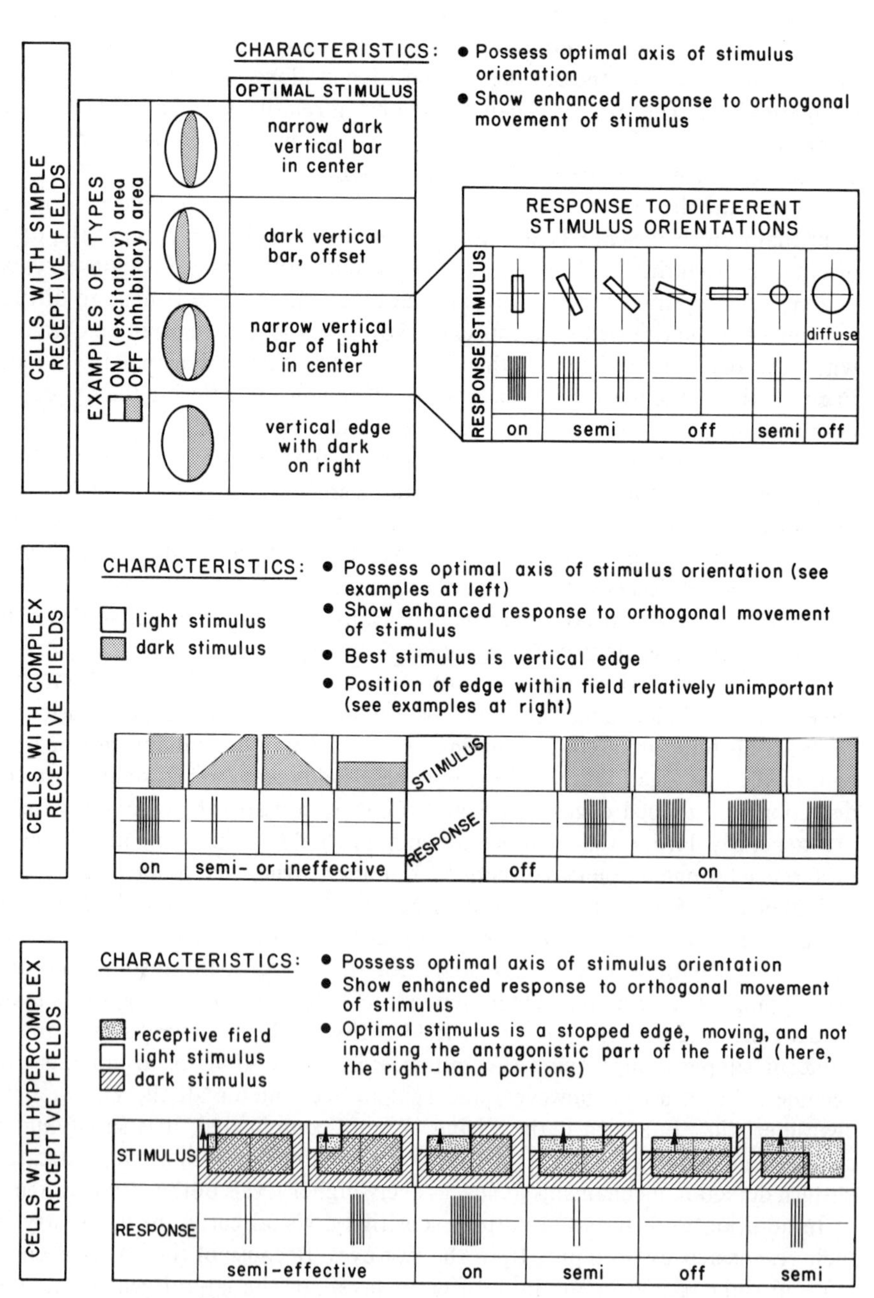

Figure 13.9 Some receptive fields in the visual cortex.

neurons possess closely or exactly corresponding receptive fields in each retina. But the relative influence of, say, the contralateral field has been found to range from very little to nearly complete control of the cell's behavior. Thus cortical neurons can be classified along an *ocular dominance* spectrum. Further, some neurons have receptive fields in the two retinas that are not perfectly in register. This disparity has been proposed as a neurophysiological mechanism for depth perception.

A neuron in the visual cortex can now be seen to possess at least three independent attributes: (1) the position of its receptive field on the two dimensional retina; (2) the amount of ocular dominance; and (3) the critical axis of orientation in its receptive field. In the columnar organization of visual cortex (at least in Visual I), these three attributes are the basis for a hierarchical organization. Within a square millimeter or so of cortical surface, receptive field position does not change much, mainly because of the extensive overlapping of fields.

Within this basic cortical block are large columns that sort the neurons according to ocular dominance. And within a column having virtually constant ocular dominance are many smaller columns, in each of which all the neurons have the same preferred stimulus orientation in their receptive fields (whether simple or complex in type). There may also be intermediate levels of columnar organization, for preferred direction of movement and aspects of color processing. This multilevel organizational plan of visual cortex helps to explain how the system copes with so many variables at once, and why it needs so many neurons to do so.

13.3.4 Effects of Early Visual Experience

The organization of the adult nervous system is a joint product of genetically specified developmental programs, like those seen in Chapter 8, and environmental experience. The relative roles and interactions of these influences have long been objects of considerable but largely frustrated curiosity. In recent decades, however, studies of the mammalian visual system have yielded a number of important clues. Most of the experiments have been conducted on young cats and monkeys by Hubel and Wiesel and several other research groups.

Questions about the extent of genetically programmed visual system organization and about the role of early visual experience in maintaining and developing that organization can be approached through a variety of experimental techniques. The characteristics of receptive fields can be examined in newborn and maturing animals. The effects on normal visual development of unusual early environments can be studied. And vision itself can be modified, by depriving the animal of light and/or pattern input to one or both eyes for selected periods of time, or by interfering with normal binocular interactions.

Three kinds of measures can be employed to assess the effects of such experiments. Anatomical studies can reveal changes in tissue structure, such as might be associated with alterations in neuron density, size, and pattern of interconnection. Physiological studies can identify changes in neural response to controlled stimulation. And behavioral tests can be employed to assess the impact, if any, of anatomical or physiological abnormalities on the animal's ability to function in its visual world.

Most studies of newborn monkeys, and of kittens before the eyes open at about 10 days after birth, suggest that the organization and properties of retinal, geniculate, and cortical receptive fields are in large part genetically specified. Nearly all the features described in the preceding two sections are present, although responses are generally weaker (even absent in some cells). If an animal is kept in total darkness for some time after birth, it may (depending on the species and the time period) show permanent behavioral blindness when normal vision is permitted. Even then, however, receptive fields are largely normal, the most notable difference being a large reduction in cortical neurons responsive to inputs from both eyes.

Exposure to diffuse light does not counteract the above behavioral and physiological consequences, since they are also seen in animals reared wearing opaque goggles. Similarly, humans born with cataracts that do not receive sufficiently early attention can be blind for life. Yet none of these problems arise from even very long term interference with sight or pattern vision in older animals. Adults routinely recover full vision after surgical removal of cataracts acquired years before.

There is clear evidence for a species-dependent *critical period* in which absence of normal vision can have profound and permanent consequences. Some studies with kittens have led to such changes after interference for as little as a few hours a day during the fourth or fifth week of life. There is no susceptibility after about three months. In primates, the critical period has been less accurately mapped, but may last for up to a few years and manifest maximum susceptibility to abnormal experience around the second or third month.

In contrast to these results of binocular deprivation, the physiological and anatomical consequences of early prevention of patterned input to just one eye are much more noticeable. In the lateral geniculate layers serving the deprived eye, the nerve cell bodies are much smaller than normal, although receptive field properties do not seem to be much affected. In the cortex, the distribution of ocular dominance is biased almost exclusively to the eye that was open. There is a corresponding shift in the width of the ocular dominance columns, those for the normal eye becoming much wider at the expense of cortical space for the deprived retina.

A widely accepted interpretation of these results is in terms of a competition theory. The amount of normal activity in an ascending pathway is seen as determining its success in maintaining synaptic influence at the cortical level. During monocular deprivation, pathways from the open eye are virtually unopposed in their attempts to expand their spheres of influence.

Not even pattern vision in both eyes is sufficient to maintain the large initial proportion of cortical neurons responsive to binocular inputs. The eyes must work together during the critical period of visual experience. If monocular deprivation is alternated, or binocular coordination prevented by cutting an eye muscle to produce a squint, nearly all aspects of visual system organization remain normal. The important exception is that most cortical receptive fields respond to input from only one retina.

The last type of abnormal early visual experience considered here allows the animal normal vision in a modified environment. In one widely reported experiment, kittens were raised in darkness except for a few hours each day that were spent in a cylindrical container. The interior of the cylinder was painted with stripes, horizontal for one group of kittens, vertical for another. After a few months in these unusual "worlds," the kittens' cortical receptive fields were found to have critical axes for stimulus orientation that were mostly very close to the angle of the stripes to which they had been exposed. The theory of competition for synaptic influence might therefore be extended to include orientation preference as well as eye preference.

At least in the mammalian visual system, then, a rather clear picture of the cooperative influences of heredity and environment is emerging. Genetically encoded developmental programs provide broad organizational patterns having the same general features as those that normally become immutable in adults. During a critical period of early experience, however, the maintenance, strengthening, and fine tuning of these patterns seems to depend almost entirely on exposure to a normal visual world. The fine tuning is not genetically encoded, probably for two reasons. First, detailed information about so many synaptic strengths and placements might overload even the prodigous storage capacity of DNA. Second, the arrangement may have adaptive value whenever an organism is born into an environment having new kinds of important visual features.

Chapter Fourteen

Simulation of Vision and Cognition

The first modelling and simulation projects considered in this final chapter relate directly to aspects of the mammalian visual system intensively examined in Chapter 13. The work on vision leads rather naturally into some models of greater scope, dealing with such higher mental functions as perception and cognition. Although the real systems serving such functions have not yet been identified, a unifying theme in the three models described in Section 14.2 is the use of Hebb-like cell assemblies (Sections 11.3.2, 12.3) as building blocks.

14.1 MODELS OF VISUAL INFORMATION PROCESSING

14.1.1 Kabrisky's Model

Kabrisky (1966) developed a model of human visual cortex in his 1964 doctoral thesis at the University of Illinois. The *real system* is thus the circuitry of the millions of neurons in Visual I. In deriving his *lumped model* Kabrisky employed two major simplifying assumptions.

First, the primitive model components are columnar cortical regions about half a millimeter in diameter and containing on the order of 300 neurons. The neurophysiological evidence for columnar organization of cortex, much of which became available after Kabrisky's work, supports his choice of such a component as his *basic computational element*, or BCE. The second assumption is to characterize the input and output of a BCE in terms of single numerical quantities, proportional to the average firing rates in fibers entering and leaving the column. This assumption has less physiological realism, given current ideas about the possibilities for encoding information as patterns of activity in arrays of sensory neurons.

Kabrisky's basic model is of the discrete time, cellular automaton variety. The BCEs are regularly arrayed in a plane, each communicating with its four nearest neighbors. Kabrisky also offers some appealing speculations about the kinds of transformations of two dimensional patterns that could be effected by convergent and divergent connections between planes of BCEs. These ideas do suggest mechanisms for scale and position invariance in pattern perception, but were not tested by simulation.

The state transition and output functions executed by an individual BCE are characterized in Figure 14.1 and Equation set 14.1. Some simplifications and other minor alterations of Kabrisky's notation have been introduced. The numeric weighting factors shown in the diagram are the values of those model parameters that were actually employed in the simulation studies, after some preliminary tuning experiments. Not shown is a model feature that allowed for individual variation in the connection weights from different neighbors, an option not employed in the simulation work.

The output Q of a BCE at time t is a function F of its input P and of two memory parameters: SS (for *S*low *S*torage) is a function G of recent values of

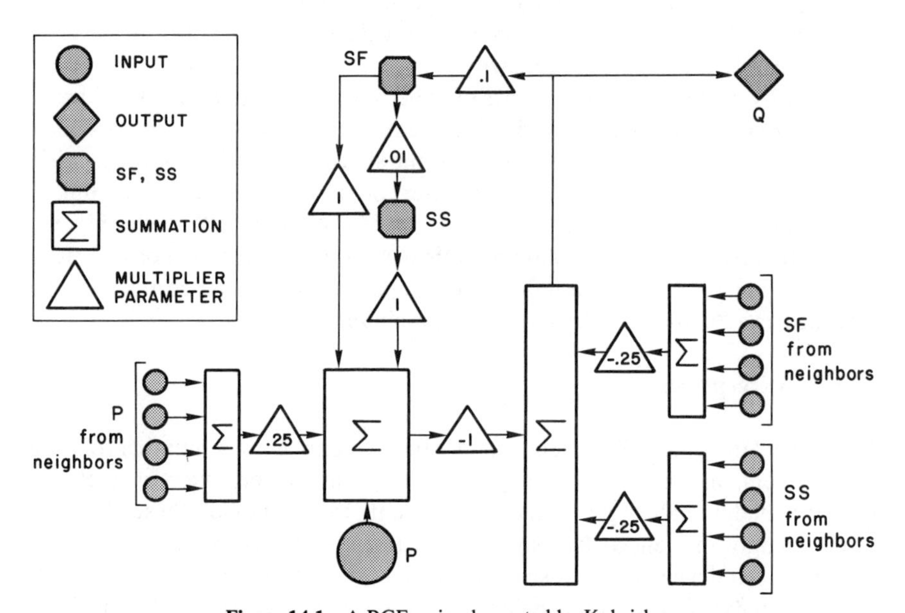

Figure 14.1 A BCE as implemented by Kabrisky.

SF (for *F*ast *S*torage); *SF* is in turn a function *H* of past values of itself, *SS*, and *P*. The number of past values that influence the current computation is a consequence of the extent to which some weights in Figure 14.1 are less than 1.

$$
\begin{aligned}
Q(t) &= F\,(P(t), SF(t), SS(t)), \text{ where} \\
SS(t) &= G\,(SF(t), SF(t\text{-}1), SF(t\text{-}2),\ldots) \text{ and} \\
SF(t+1) &= H\,(SF(t), SF(t\text{-}1),\ldots, SS(t), \\
&\qquad SS(t\text{-}1),\ldots, P(t), P(t\text{-}1),\ldots)
\end{aligned}
\tag{14.1}
$$

The coupling of a BCE to its neighbors adds arguments not shown in the above equations. As can be seen in Figure 14.1, the determination of *SF* is based partly on values of *P*, *SS*, and *SF* from the four neighbors.

Hardware limitations (an IBM 1620 with 20,000 characters of memory) restricted Kabrisky's simulations to a 10 by 10 array of BCEs in a single plane. The FORTRAN II program contained about 120 statements. Even in this extremely limited experimental context, the model revealed some modest abilities for classifying groups of simple patterns along certain dimensions of similarity.

When a subset of the BCEs, such as the 30 comprising a horizontal bar, are presented with a sustained nonzero input (say, $P = 50$), the *SF* values decline from 0 to -*P* and then return to 0, while the *SS* values more slowly approach -*P* and stabilize there. Meanwhile the output *Q* exhibits a mildly oscillating convergence on 0, from its initial value of -*P*. The pattern may be considered "recognized" when the array sum of the *Q* values remains within some small magnitude threshold. After being so "trained," the array can be presented with patterns having various amounts of similarity to the original one. The initial magnitude of the array sum of the *Q* values is found to be inversely proportional to the degree of similarity.

Kabrisky conducted several series of experiments, in which the model proved able to discriminate degrees of displacement, rotation, and change in scale of various simple patterns. There was also some success in preferential recognition of distorted versions of a training letter over other letters. Although its capabilities as a pattern recognition device do not compare favorably with efforts directed primarily to that end, this partial implementation of Kabrisky's model may have helped shed light on some operations of visual cortex, which he regarded as a "two-dimensional-pattern manipulator."

14.1.2 Perceptrons

Since the late 1950s Rosenblatt of Cornell, along with a number of other investigators, has been developing and analyzing the behavior of a class of network models called *perceptrons*. Although the design of these devices is often strongly influenced by aspects of neural systems for vision, perceptrons have not been developed exclusively as models of such systems. Other work has involved study of perceptron behavior, by analysis and/or simulation, in the context of artificial pattern recognition tasks. This discussion begins with a brief characterization of perceptrons and then focuses on one project where the objective was clearly to model and simulate mammalian visual information processing.

A perceptron is composed of three kinds of units. S (sensory) units are regularly arrayed on an input layer (plane) that is often referred to as the "retina." S-unit outputs are typically binary, reflecting the presence or absence of part of a stimulus pattern in the grid square of the retina. Outputs that report an intensity value are also possible.

Signals from the retina converge on a layer of A (association) units. The connections may be excitatory or inhibitory. The A units are typically threshold logic devices, emitting 1 whenever the total excitatory input exceeds the total inhibitory input by an amount equal to or greater than some predetermined parameter (threshold). There may be more than one layer of A units, with convergent and divergent connections between layers. In cross-coupled perceptrons, the A units in any layer may also be interconnected.

The final layer of A units sends signals to a layer of one or more R (response) units. These are also decision elements, but ones that are sometimes not limited to binary output. The connections among A units and from A units to R units have weights. Usually at least one level of interlayer connections has modifiable weights. In a typical learning experiment, a series of training patterns is presented to the retina and some reinforcement rule is used to adjust weights according to whether the R units report correct classification of stimuli.

In the simplest reinforcement method, weights of connections between A units and R units are adjusted, by small amounts, only when there has been a wrong classification to which the A unit contributed. The sign of the adjustment is such as to move R-unit output in the direction of the proper classification. A major and somewhat surprising theoretical result is that if there exists an assignment of weights which solves a given classification problem, an elementary perceptron (one with a single A-unit layer) will converge on that assignment when the above reinforcement method is applied over a sufficiently long series of training stimuli.

In practice, however, difficult classification problems, especially if presented on high resolution retinas, may require impractically large numbers of A units and/or impractically long training sequences. For these reasons, there has been a considerable amount of attention given to more sophisticated designs, including multiple cross-coupled A-unit levels and convergence pathways organized to exploit critical features of the stimulus class. The latter idea was used by Rosenblatt (1966) in an attempt to model the kinds of receptive fields discovered by Hubel and Wiesel. The visual cortex of the cat is thus the *real system*.

The 5-layer perceptron used as a *lumped model* in this study is diagrammed in Figure 14.2. The 32 by 32 cell retina is wrapped around in both directions, creating a torus in which there is no need for special processing at edges, a technique seen previously in Finley's neural networks (Section 12.3.1). Each of the roughly 18,000 elements of the first A-unit layer (A^1) receives inputs from a rectangular region of the retina. The 40 or so contributing S units supply excitatory and inhibitory connections in patterns mostly resembling simple cortical receptive fields maximally sensitive to narrow bars of various critical orientations and lengths ("line detectors").

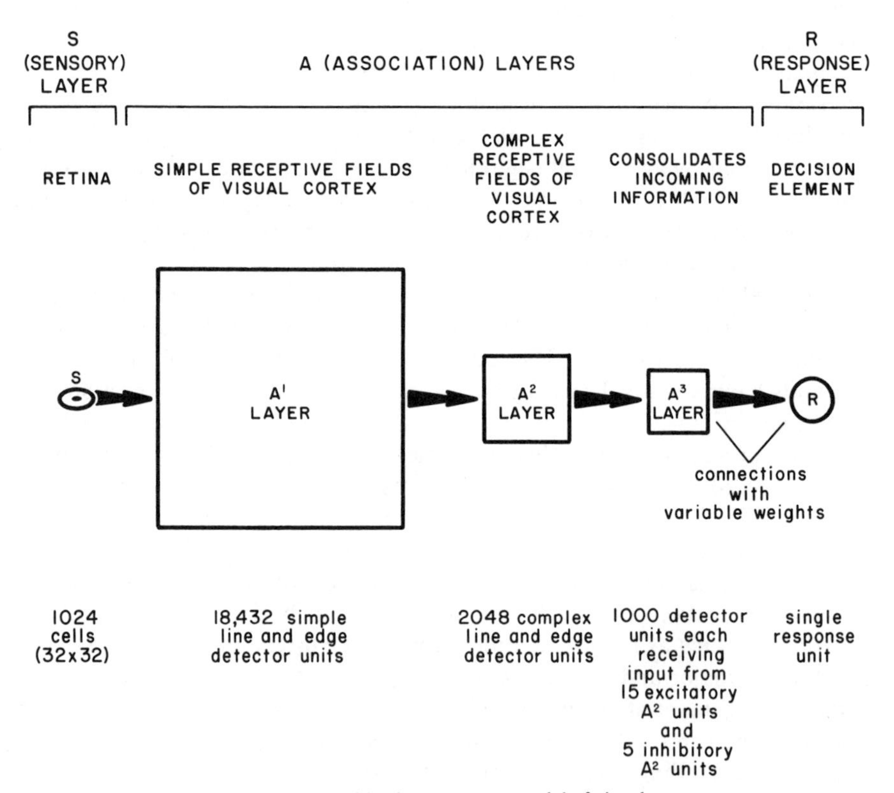

Figure 14.2 Rosenblatt's perceptron model of visual cortex.

The roughly 2000 A^2 units are each responsive to activity in any of a block of 9 A^1 units, all having the same line angle. Adjacent A^1 blocks are displaced from one another in a regular overlapping fashion. The A^2 units thus represent the complex receptive fields of visual cortex. Each of the 1000 A^3 units receives inputs from 20 randomly chosen A^2 units, 15 of the connections being excitatory and 5 inhibitory. The connections from the A^3 layer to the single R unit have modifiable weights.

In some initial simulation experiments, the perceptron model of Figure 14.2 proved far superior to an elementary perceptron in tasks requiring classification of horizontal and vertical lines. But the significance of this result is open to question on two counts. First, the "cortical" model had not only specially tailored connections but also an overwhelming advantage in terms of number of A units; the competing elementary perceptron had only a single layer of 1000 units. Second, the difference in performance of the two systems was greatly diminished in a task requiring classification of letter shapes ("E" and "F"-like patterns).

Of much greater relevance in the modelling context, however, is Rosenblatt's (1966) comparison of the perceptron's discriminatory powers with those of young children. Psychological data were available concerning the tendency of 4-year olds to confuse all possible pairs of upper case letters. Two versions of the perceptron model, differing only in the stochastic realization of the same parameters, were each tested in separate experiments on all 325 letter pairs, a procedure requiring a substantial amount of computing time. The training stimulus sequence involved presenting the letters in alternation at random retinal positions, 60 times each. The confusion rate was then measured for further samples of 50 letters. These scores could be compared between the two model realizations and with the psychological data.

Rosenblatt first noted that the performances of the two perceptrons had a correlation coefficient of only about 0.6. This surprising amount of variability between two identically specified models would have made further simulation desirable; but the computational cost was prohibitive. The average of the two models' performance records correlated to the human data with a coefficient of about 0.3. Some further analysis of the particular letter pairs that were easy for the children and hard for the models, or vice versa, suggested certain kinds of pattern features to which the two systems were differentially sensitive. Among those features not well discriminated by the perceptrons were some that seemed closely related to hypercomplex receptive field sensitivities (e.g., stopped lines). Such results were not published by Hubel and Wiesel until shortly after the perceptron simulation studies had been carried out.

14.1.3 The Models of Baron

In a series of major contributions, Baron (1970a, 1970b, 1974) has described complex and interrelated models for memory, vision, and language. Some aspects of these models have been studied by simulation. Baron (1970b) notes that his vision model embodies two significant extensions to prior approaches like those of Kabrisky and Rosenblatt. First, Baron makes overt functional distinctions among a variety of visual processes, notably search, selection, storage, correlation, recognition, and recall. Second, the Baron models employ explicit *control networks* to regulate the flow of information, visual and otherwise.

To appreciate the vision model Baron studied by simulation, it is useful to have some understanding of his basic approach to cortical memory (Baron

1970a). In a first major simplification of the real system, he defines a *mathematical neuron* as a composite of four 3-dimensional components: cell body, dendrite field, axon, and axon field. Coupling fields with positive (excitatory) or negative (inhibitory) coefficients arise when the axon field of one neuron intersects the dendrite field of another. A neuron's behavior (defined as its firing frequency) is determined by the behavior of the neurons which couple to it and is modulated by the state of the background non-neuronal medium.

A *memory block* is a mutually intercoupled group of neurons together with the background medium in which they reside. The block is partitioned into many identical *memory cells*, each containing one neuron of each of six different types: (1) input and (2) output neurons, respectively, carry elements of received and recalled patterns of information to and from the location and hence, as groups, to and from the block as a whole; the combined state of the block's array of (3) memory neurons constitutes the currently stored pattern(s); (4) memory-effector and (5) recall neurons, respectively, cause input patterns to be stored and stored patterns to be retrieved; finally, (6) recognition neurons measure the similarity of input patterns to those previously stored.

A memory timing mechanism encodes temporal firing patterns as spatial neuronal changes. During storage or retrieval a wave of activity in the non-neuronal medium sweeps smoothly across the array of memory locations, "activating" rows of them in turn. Although there is no established physiological counterpart of this timing mechanism, its effect in the model is storage of neuronal information in a fashion analogous to the way a Fourier hologram stores optical information.

Through a detailed derivation and analysis of interneuronal coupling coefficients, both within and between memory locations, Baron shows how his memory block can mediate the functions of pattern storage, recognition, and recovery, and thus learn to provide proper responses to given inputs. Forgetting is seen as a consequence not of simple decay of stored information but rather of an elevated "threshold of recallability" resulting from disuse of an association, or as a consequence of the block having been overwritten by new information with stronger associations.

A memory store like that just described is an integral component in Baron's "model for the elementary visual networks of the human brain" (1970b). The *real system* and hence the *base model* for this study encompass the retina, visual pathways to the brain, primary visual cortex, and, presumably, a number of additional associated cortical areas. Baron's *experimental frame* limits the scope of the study to six networks that will be described shortly. Two major simplifying assumptions are used in deriving the *lumped model*: (1) the organization and activity of individual neurons are ignored because such details are not adequately understood and, in any case, are not relevant to the behavioral characteristics of the system; and, (2) for similar reasons, the ways in which *external networks* (see Figure 14.3) govern operation of the primary networks are not considered. Analysis of many model processes by *computer* simulation was undertaken, with results to be described at the end of this section.

Baron's three information processing networks are the retina, the visual selection network(s), and the permanent visual memory store. His three control networks are a selection control network, a memory effector system, and a memory coordinating system. The flow of information and control among these six components is sketched in Figure 14.3.

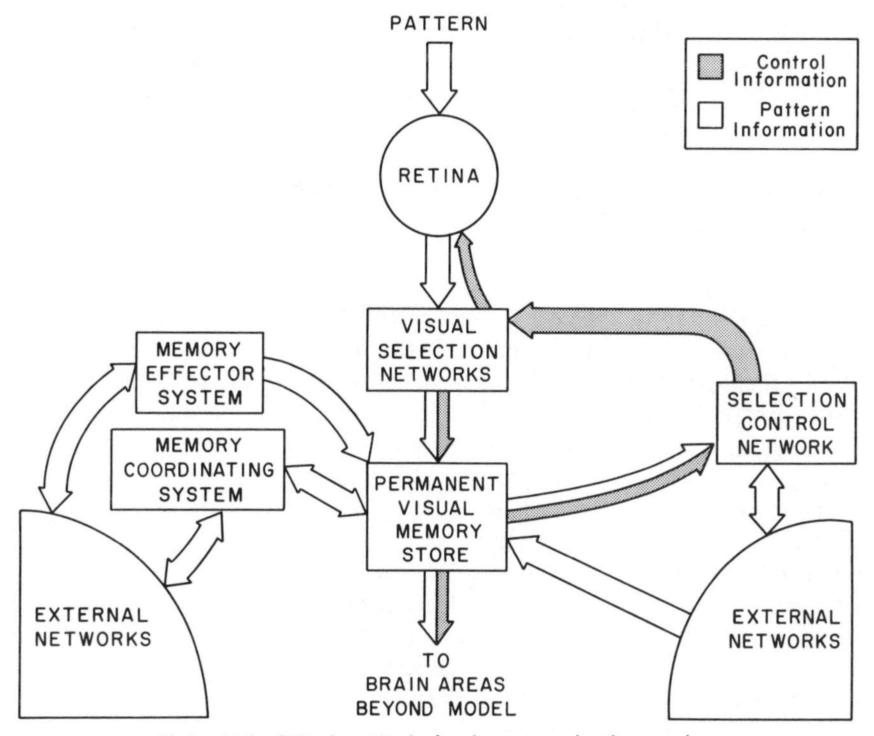

Figure 14.3 Baron's networks for elementary visual processing.

The *retina* receives input from the external world, in the form of a time-varying pattern of light, and (indirectly) from the selection control network. The retinal output, which is the set of firing rates in all ganglion cell axons, is a pattern that has been preprocessed by neural networks in the retina itself (see Section 13.3.2) and perhaps also influenced by the selection systems. (It should be recalled that there is no good anatomical evidence for pathways in primate visual systems that could mediate the latter form of control.)

The *visual selection networks* process the retinal output ("primary visual information") and send the result ("attended visual information") to the permanent memory store. Visual selection transformations are directed by the *selection control network*. As shown in Figure 14.3, information from previously stored patterns is available to the selection control mechanisms. In addition to choosing particular regions of the visual field for further processing, the visual selection networks can enlarge, reduce, rotate, reflect, and otherwise alter perspective of the selected region. Although visual information may well be permanently lost by such operations, the memory store does receive, as part of its input, the nature of the selection operations that have been performed.

The *permanent visual memory store* is a memory block of the type developed in Baron's earlier paper. Its memory effector neurons are governed by input from the *memory effector system*, which directs storage of input patterns into appropriate memory locations. The *memory coordinating system* receives correlation information from recognition neurons in the memory store and, when the threshold of recall is exceeded, directs the recall neurons to deliver output information. This output, which contains both pattern and control components, is sent back to the selection control network and forward to brain areas beyond the model.

Baron describes the operation of his model in four elementary visual scanning tasks. (1) In the general, undirected scan, regions of the visual field involving motion, many small details, or bright and/or contrasting colors are selected for further attention. If the memory system is able to locate an object in such a region, identification (task 2) is attempted, and the entire process repeated until a map that allows immediate location of any object in the field is built up in memory. (2) In the scan for object identification, the selection control networks attempt transformations until the memory store responds with a sufficiently strong recognition signal. (3) In the test for exact equality of an object with one previously experienced (such as recognition of a particular letter during reading), matching of all transformations is required. And (4) in the location of a specific object, all of the previous scanning strategies are employed, in conjunction with verbal processing networks not included in this model (but see Baron 1974) to direct the visual search process.

In a discussion that will not be summarized here, Baron shows how half a dozen classes of clinical disturbances in the visual system can be systematically related to malfunctions in various elements of his model.

Simulation of portions of the model was done using FORTRAN programs on a medium sized computer. Retinal input was encoded as integral brightness levels on a 60 by 60 grid. The only retinal preprocessing simulated was lateral inhibition for contrast enhancement, which transformed an input into a pattern of contours with a reduced range of intensity levels. Both the original and transformed patterns were made available to the selection networks, where further processing was limited to translation and size reduction for a set of 10 selected square regions of varying sizes. Regions were selected either by recall of coordinates for a region in a previously stored pattern or by correlation against a test pattern contained in the memory store. The numerical constraints indicated above were primarily a consequence of limited computational resources and did not, in Baron's opinion, significantly affect interpretation of the simulation results as indicators of model performance.

To simulate recognition, each memory cell had an activity level proportional to the largest of the 100 correlations of its 10 stored images with the 10 current input images. This simulation of the general scan terminated when a particular memory location responded with a high recognition signal. Simulation of equality search was then effected by using control information to direct the selection network to send forward all regions of the visual field corresponding to the remaining stored images in the test pattern. A sufficiently high overall correlation constituted recognition. Recall of information from the memory store was simulated by reconstruction of a 60 by 60 pattern from the 10 stored images.

Baron summarized the results of his simulation experiments as follows: (1) specific regions chosen by the selection networks depended on the memory location used by those networks; (2) regions scanned in fine detail contributed significantly to the fidelity of a recalled image, whereas regions scanned only briefly led to loss of fine detail; and, most importantly, (3) the system was always able to achieve correct identification in equality search, even for input images which were translated or perturbed by random noise.

More extensive and equally encouraging simulation results are reported by Baron in connection with his (1974) elaboration of the vision model into a theory of the neural basis of language. But that work is beyond the scope of this section.

14.2 COGNITIVE MODELLING

The three modelling and simulation projects considered in this final section deal with diverse and sophisticated aspects of human (and animal) behavior. Although the *real system* in all cases is understood to be most or all of the central nervous apparatus and its associated sensory and motor systems, the ties to known mechanisms of biological information processing are tenuous at best. Because so little is known in detail about the global neural basis for perception and cognition, the investigators (with one exception) begin with *lumped models*.

One feature shared by all three models is the assumption that they may be based on neural structures resembling Hebb's cell assemblies. But since this assumption arises more from the investigator's preference for one type of neurophysiological speculation than from any essential role for Hebb's constructs, the results of these studies should not be taken primarily as evidence for (or against) the existence of cell assemblies.

14.2.1 Whitehead's Model

Although it is chronologically last of the three projects, Whitehead's "Neural Network Model of Human Pattern Recognition" (1978) is considered first because its behavioral emphasis relates to vision and pattern classification tasks that have some elements in common with the models of the previous section. Also, as suggested above, Whitehead's is the only project in this section that does contain some analysis of the neuronal mechanics underlying the cell assemblies in his model. That mathematical development (which Whitehead did not simulate) will not be examined here, however.

In Whitehead's discrete time lumped model, each component "netlet" (cell assembly) has three state variables: the proportion of neurons in netlet *i* that are firing at time *t* may be denoted ACTIVITY(*i,t*); RECOVERY(*i,t*) represents lumped threshold information, the mean "readiness to fire" of the neurons in netlet *i* at time *t*; and COVARIANCE(*i,t*) represents the inhomogeneity in recovery state arising from the fact that inputs from different sources arrive at different places in the netlet. Interconnection between netlets is specified by arrays of temporally fixed parameters, CONNECT(*i,j*), representing the combined total strength of excitatory and inhibitory synapses from neurons in netlet *i* to neurons in netlet *j*, and CONNECTSQ(*i,j*), representing the variance of CONNECT(*i,j*). For unconnected netlets, these parameters will have values of zero. Connections internal to a netlet are indicated when *i* equals *j*. By deriving appropriate transition functions for the state variables, in terms of themselves and the connection parameters, Whitehead shows exactly how ACTIVITY(*i,t*+1) may be computed as the sum, over all *j*, of the products of ACTIVITY(*j,t*) and CONNECT(*j,i*).

In pattern recognition, the unit of storage is the abstract representation of an object (an equivalence class of stimuli) in the environment. The features used internally by this representation consist of whatever have been determined to be the perceptual similarities among the stimuli in the equivalence class. In Whitehead's model, the unit of storage is implemented as a network of connected feature detectors, each of which is a netlet that both represents and receives input from one sensory feature. CONNECT(*i,j*) is a monotonic function of the degree to which the feature that netlet *i* represents tends to predict the presence of the feature that netlet *j* represents. If the network representations of two different objects are likely to be activated by the same features, then there are inhibitory connections between the netlets belonging to one representation and those belonging to the other, connections which prevent simultaneous activation of the two networks.

The feature decomposition of a perceived object is presented in parallel to some number of Whitehead's networks. In a particular network some of the feature detectors (netlets) may become active, while others receive insufficient stimulation to respond. Each active detector will facilitate other, inactive, ones in proportion to the validity with which the detected feature tends reliably to predict the presence of the as yet undetected feature. Netlets representing features that are strongly predicted by aggregates of those already active will have their thresholds substantially lowered and may join the group of active detectors. Since all of this is going on in many networks at once, the "winner" will be the one that succeeds in remaining most active despite the competition of other candidates via their inhibitory connections. In such a fashion is the perceived object identified.

It should be noted that this approach, in which a feature contributes to a classification decision partly on the basis of temporal context (according to which other features have been *previously* activated) differs significantly from many traditional approaches to pattern classification, whether they be biological models or purely pragmatic systems. In these other schemes, features make strictly parallel contributions to the global decision. Such is the case, for example, in perceptrons, where an A-level unit's response is independent of any order of activation of its contributing lower level inputs. Baron's approach, on the other hand, is more like Whitehead's, in that new regions may be selected during a scan partly on the basis of "what has been noticed so far."

For purposes of simulation, Whitehead developed three versions of his model. The first two were based on pattern classification tasks drawn from the psychological literature. The results of simulating these studies helped determine the parameter settings for simulation of the third model version, which was based on a rather challenging discrimination task designed by Whitehead himself. He wanted to test the validity of sequential, conditional contribution of features, the fundamental hypothesis implicit in his network model. The binary classification problem involves a relatively large number of features each of which alone contributes very little information for correct pattern recognition. On the basis of using this task both in experiments with human subjects and as a domain for computer simulation, Whitehead was satisfied that the hypothesis embodied in his model had been confirmed.

The test stimuli are line drawings of houses, like those shown in Figure 14.4a. The task clearly challenges human pattern classification skills, since casual or even careful inspection yields few clues concerning why half of these houses belong to the "Hatfield" pattern and the other half to the "McCoy" one. The 9 features, and the assignment of ranges of their values to the two stimulus categories, are shown in Figure 14.4b. On arbitrary scales, normalized to 0-100, Hatfield houses vary uniformly over a 10-70 range on each feature, whereas the range for McCoy houses is 30-90. The figure also shows the grouping of features into three different sets of three features each. Grouped features (e.g., Hatfield height, roof flatness, and overhang, or McCoy height, window difference, and window tallness) have values with mean interfeature correlations of 0.9 in the experimental stimuli. Nongrouped features (e.g., Hatfield height and windows, or McCoy height and roof) are uncorrelated.

As shown in Figure 14.5, Whitehead's model for this categorization process contains 18 netlets, 9 for each type of house. The netlets are numbered in the

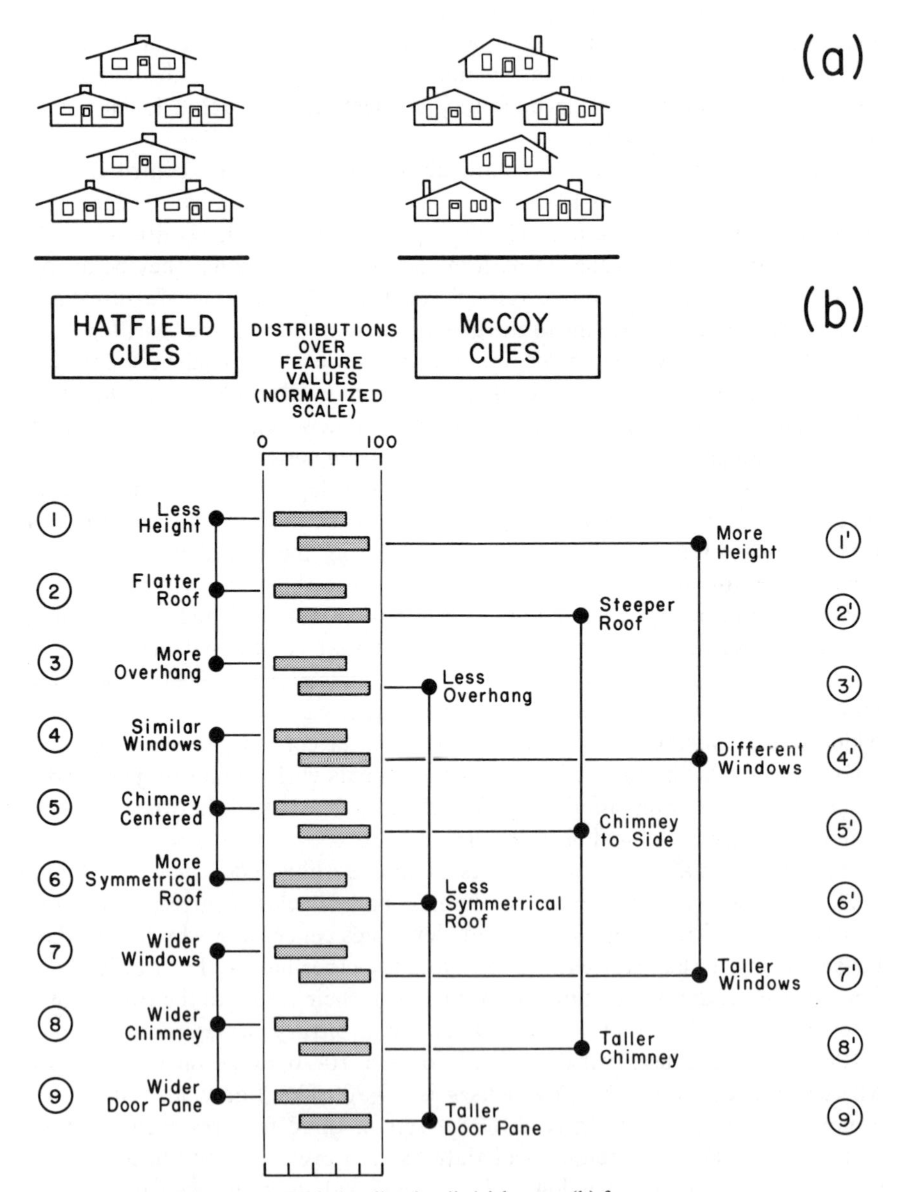

Figure 14.4 Whitehead's stimuli: (a) houses, (b) feature ranges.

figure in the same way the cues are numbered in Figure 14.4b. Thus there are 2 netlets for any feature, each exclusively excited by one extreme of the feature value range, but both equally excited by intermediate values. Other excitatory and inhibitory connections implement the grouping of correlated features and the separation of the two parts of the network. When Figure 14.5 is viewed as an encoding of a "knowledge base," each component of the neural structure stands for a cue (an individual netlet) or cue correlation (the lumped connections between the netlets).

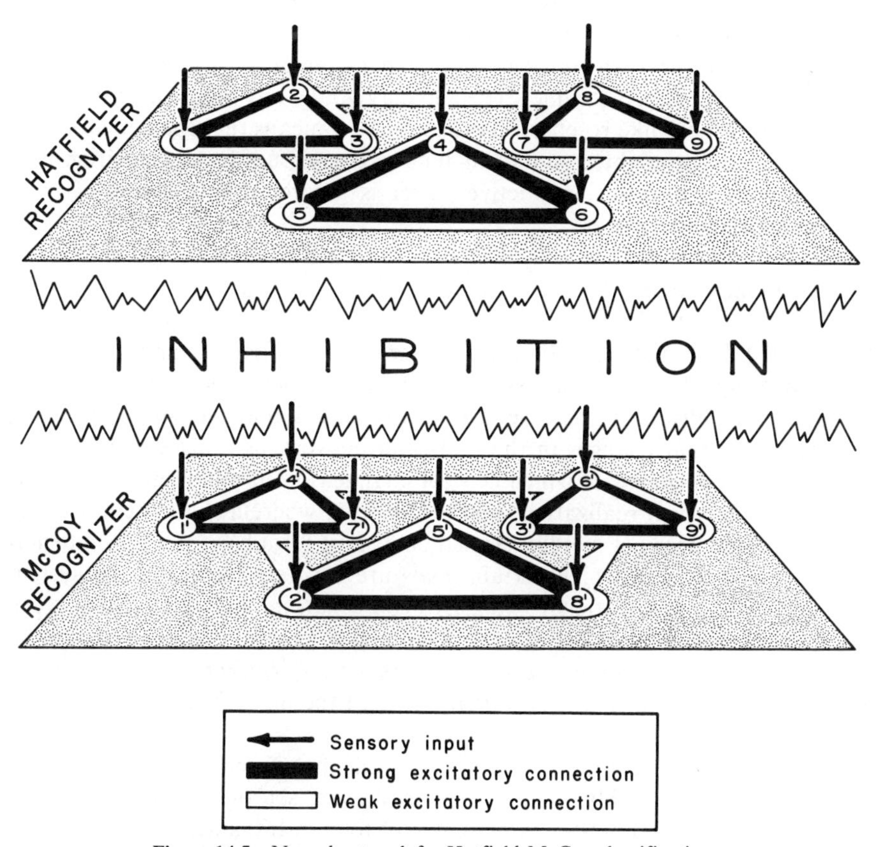

Figure 14.5 Neural network for Hatfield-McCoy classifications.

To see Whitehead's sequential feature prediction hypothesis in action, consider a house with identical very tall windows, a very tall centered chimney, and intermediate values on the remaining five cues. Initially, this stimulus generates maximum activity in netlets 4, 5, 7′, and 8′, and no activity in the other netlets. With just two Hatfield and two McCoy cues all maximally present, a model based on feature independence could make no classification. But the Whitehead correlations will cause netlets 4 and 5 to reinforce each other and their converging output to "predict" cue 6, providing enough additional input, when added to the modest generalized input already present, to bring the netlet to full activity. The influence of netlets 7′ and 8′, on the other hand, is dispersed among 1′, 2′, 4′, and 5′, which gives none of them a sufficient boost to achieve activity (the latter two did not even have initial sensory input).

As the total activity in the Hatfield subnetwork grows, it begins more strongly to inhibit the McCoy subnetwork than to be inhibited by it. So the model rapidly converges on the Hatfield decision, partly because a more symmetric roof (even though this one is not) tends to "go with" identical windows and a centered chimney. (People may not be any more objective.) The nonlinear threshold-like response curve of a netlet means that the network does not simply take a weighted sum of its inputs. If two cues fit a correlation pattern, then their combined predictive power is greater than the sum of their individual predictive powers.

14.2.2 Plum's Model

In his 1972 University of Michigan doctoral dissertation, "Simulations of a Cell-Assembly Model," Plum asserts the following desirable features of such models: *parallelism*, in that several units may be active simultaneously; *adaptation*, in that changes in the units and network take place according to localized rules; *memory*, of which new (or changed) units and their connections constitute a physical realization; *semi-autonomy*, wherein the system actively seeks goals and does not merely respond on a short term basis to environmental inputs; *motivation*, in which certain goals are innately fixed and the system maintains this organizational integrity; *coordination*, in which parts work together in a dynamic pattern of control without a "central executive" assigning them roles; *extensive interconnection* of units; *ease of interpretation* of units in terms of neurons; and finally *specificity*, in that the model should be explicitly implementable.

In Plum's discrete time lumped model the basic component is a single aggregated cell assembly, which he views as a set of neurons and their interconnections. Each unit may be in one of four states, among which there is a natural progression from "ready" through "competing," "active," and finally "successful." These steps correspond to increasingly strong levels of firing. Some units are *sensory*, their internal dynamics being controlled by ongoing interaction with the environment. *Action* units directly affect the environment. And *intermediate* units are governed by certain internal processing rules. The whole system is an aggregate of units of the various types, connected by *evidence* and *priming* relations (see below).

The network is organized into *competition classes*. Several units can respond to or control the same subset of an organism's resources; the units in such a group are mutually competing. When two units control the same resources, negative connections between them insure that they cannot both be very active at the same time. (Note the similarity between a competition class and Whitehead's inhibitory relation between networks that would be activated by the same environmental features.) A unit is the winner in a competition class if it has the largest amount of *expected utility*. That unit is then put into the active state and controls the resources of the competition class.

There are four aspects to the internal dynamics of Plum's model. First there is the current evidence value, a function of evidence inputs and their associated weights. The evidence inputs may come from other units or, in the case of sensory units, from the environment. An evidence input can alter any state of the unit.

The second aspect has to do with the current priming value. This is a measure of the utility of a unit in the organism's current situation, the goal-relatedness of the unit. To arrive at a priming value for the unit as whole, Plum sums (1) basic priming, which resides in a unit intrinsically and is determined by the prior motivational relevance of the unit, (2) temporary basic priming, which is awarded on a short term basis by routines modelling subcortical motivational structures, (3) prediction priming, from an active unit that can supply evidence signals which aid the unit in question in becoming active, and (4) priming that arrives along connections from other units in the so-called evaluation list.

The third aspect of internal dynamics is a probability estimating evidence/priming function. Overall, a unit that can contribute evidence to another is in either an active or a successful state. Thus it is necessary to determine the success of a unit based on states of other units. To arrive at this evaluation, Plum derives a function of the a priori probability of a unit's success, given that it advances to the active state in the first place. The total priming (utility) of a unit is the linear sum of all its sources of priming. The probability of success--given-activation in this situation, multiplied by the utility, is the expected utility, which is the total lumped strength of a unit.

The fourth and final aspect relates to the activity states, denoted REDY (ready), COMP (competing), ACTV (active), and SUCC (successful). When a unit is in REDY, it is awaiting evidence that exceeds a ready threshold; it has no effect on other units. In COMP, a unit is competing with others in its competition class. If its firing is strongest, it enters the ACTV state, controls the resources of the class, and sends prediction-priming to its subgoals, the units on its evidence list. Whenever the given threshold function of those subgoals' signals exceeds the active state threshold, the unit advances to SUCC. A unit in SUCC sends evidence to units waiting for it to have become successful. In general, signals sent from a unit consist of its activity state, its utility value, and its expected utility value.

Using the APL programming language, Plum did two distinct types of simulation studies of his model. The first was a "model of mental arithmetic," which was designed to give preliminary information about such matters as how to represent general rules or processes (as opposed to specific actions) within cell-assembly theory, how to represent variables within units, how a parallel network can perform a coordinated sequence of actions without a supervisory central executive, and how to realize hierarchies in a network. The crucial issue, for purposes of simulation, is how to represent variables. Plum's solution was to create a temporary new unit which linked the variable with attributes by which it could be accessed.

Plum's major simulation was of a model of simple animal learning behavior. The animal, known as "Rat," lives on a runway with a nest at one end and an object at the other. The object is of fixed size; from the nest end it looks small, from close up, large. The end with the object sometimes delivers food which Rat can eat if it is close enough. But this end can also deliver a painful electrical shock. There are red and green color signals in the environment. Rat's possible actions are FLEE, APPRO(ach the food), and EAT. The situation categories are DANGER, OBJECT, FOOD, and SAFETY. With respect to Rat's interaction with its environment, there are three units, LARGE, SMALL, and ODOR, which concern Rat's position relative to the object and the food. The color units are RED and GRN. The payoff units are PAIN, TASTE, and NEST. These last three classes are environmentally driven sensory units; so the model must include a function to set the relevant values for these units based on the organism-environment interaction. In the simulation, this function is simply supplied by "outside forces," namely Plum himself. The motivational system is based on two simple parts. The emptier an internal nourishment reservoir gets, the more desirable are APPRO and EAT. PAIN, on the other hand, makes FLEE very desirable.

The entire simulation consists of three interrelated parts: the environment, the interface systems, and the network of units. In the environment are the location of Rat, the location of food, the current color, and the presence of pain. These factors are initially set and can later be altered by the investigator; they are also altered by the behavior of Rat. The factors affect Rat's perception of color, of food size/location, of payoff, and the degree of "hunger." The first three of these are units of Rat, whereas the last is one of the two interface systems. The other interface system, the motivational system, is directly affected by the degree of hunger and in turn sends goal-setting priming to one of the action class units in order to indicate what action is called for by the current situation. The action competition class formed by FLEE, APPRO, and EAT sends information directly to the nourishment reservoir and to the environment, reporting what action was performed and how to update the reservoir or environment appropriately. The motivational system also receives direct information from the payoff competition class.

In addition to the direct effects of one unit on others, and to the goal-setting priming of the motivational system, there are also connections from one competition class to another, where the first sends evidence signals to units in the second (and the second sends prediction priming to the units of the first). The units in payoff, color, and size/location send evidence signals to each of action and situation, which in turn send such signals to each other.

Each of the units was encoded as a separate APL function, designed to contain and produce the following sorts of information. A unit was placed in a competition class and given a list of priming sources, including the basic priming parameter value and a list of other units which might supply further priming to that unit. There then followed a series of computations, each related to the unit's being in one of the states (REDY, COMP, ACTV, SUCC), iterated a number of times, during which a unit tested whether it could advance to the next higher state. For example, a unit competing with other units in its class was usually allowed 100 attempts to surpass the succeeds threshold before being demoted to REDY. Within each iteration, the unit received evidence input from other units, indicating whether those units were in ACTV or SUCC states. These data were multiplied by evidence weights and added to the initial evidence level to produce an overall value for comparison against the threshold level. Within the FLEE unit, for example, while competition was continuing for up to 100 time steps, if PAIN was in SUCC its input to FLEE was multiplied by 2, whereas if DANGER was in ACTV, its input was multiplied by 3. Such numbers were added to the initial support to give the current FLEE value.

As can be seen, the behavior of Rat is largely determined by the modeller, who supplies information about which other units are relevant to the evaluation of a given unit, about how much emphasis the unit is to place on that input, and also about how long any unit can wait for more input. Nevertheless, with a network of this sort, many learning processes are possible. Any of the connections between units could be altered by making some of the aforementioned constants into variables that change with the current state of the entire network. But Plum incorporated only one learning mechanism, the continual alteration of evidence connections. In this "cue learning" process, the initial support and the weights assigned to the various evidence connections were subject to conditional modification, by methods not described here.

Plum reports a rather lengthy simulation of Rat's behavior. This experiment provided a training regimen for the animal in which food was available only when the green light was on, and shocks were administered only when the red light was on. At random times (every tenth simulation step on the average), the light changed. Two steps later, PAIN switched from its prior value and food could appear or disappear. Initially, Rat had no evidence connections from RED or GRN, so its behavior was controlled entirely by its motivational state. It repeatedly approached the food when hungry and fled upon encountering pain. It often attempted to eat in foodless locations. When fully hungry, it approached despite pain, but when full it sat in the NEST. By time step 330, Rat had learned enough about cues to danger so that it could avoid pain when RED was on.

One obvious bit of learning Rat needed was to obtain cues about situations in which eating could succeed. Some good negative cues were learned quickly. When size/location unit SMALL won in its class, Rat could not be in a proper location for food. Because of early failures, EAT soon acquired a negative connection from SMALL; and Rat never again tried to eat unless SMALL was off. In addition, EAT quickly developed a strong negative evidence connection from RED.

The positive cues for EAT were more tricky. Success of ODOR, GRN, and TASTE, and activity in SAFETY all appeared as positive cues. TASTE was the most conservative and reliable cue; but ODOR was also reasonable. SAFETY was tested only once, and happened to precede a successful EAT. GRN was necessary but not sufficient. Because of the small number of cues, Rat never fully learned where to eat. LARGE, SMALL, and ODOR, of which only ODOR could lead to successful eating, were all in the same competition class. But SMALL and RED had become negative evidence for EAT; and Plum allowed only two negative evidence weights. Thus there were many cases in which Rat approached the food and, as soon as size/location sensing switched from SMALL to LARGE, stopped and tried to eat while still just short of the food. With no facility for creating new units, this was the best Rat could do.

Toward the end of the simulation run, between time steps 300 and 400, SMALL ceased to be a negative cue for DANGER. During most of this time Rat was very hungry because of unsuccessful attempts to eat. It therefore approached the food even in the presence of PAIN. Many perceptions of DANGER took place in regions where SMALL was active, causing Rat to learn that SMALL did not negate DANGER.

14.2.3 The Work of Cunningham and Gray

In his book, *Intelligence: Its Organization and Development* (1972), Cunningham proposes a synthesis of the "building block" theory of cell assemblies and the well known Piagetian "black box" theory of mental operations in early human development. Piaget's elements are *schemata*, integrated patterns of behavior capable of participating in circular reactions and of entering into assimilation-accommodation relations with environmental objects and other schemata. More will be said about Piaget's ideas shortly.

There are four common characteristics to which Cunningham attributes the power of Hebb's and Piaget's theories in describing intelligent processes: (1) complexes at each level are built out of components from the next lower level, with the same rules of operation in effect at all levels; (2) the developmental and functional processes are independent of the "meaning" of the sensory input and of the motor output; (3) the components can affect each other's activity as well as interact with environmental objects; and (4) the same elements encode experience and act as units of the system's internal functioning, so that the unit of information is the same as that of operation.

There are two basic aspects to Cunningham's model. The first is the *data structure*, which consists of Hebb-like building blocks called *elements* and their interconnections. The data structure encodes accumulated experience as structural developments and represents the transitory state of the model in terms of activity levels in the elements. These elements are distinguished by type (input, output, or memory) and by order (degree of removal from the input or output elements). Interconnections are strictly facilitatory and pass activity along *links* to the receiving element. To avoid overwhelming complexity, it is often convenient to consider a battery of interconnected elements as a single unit at some higher level of organization. The second aspect of Cunningham's model is the *operating system*, a set of rules or constraints defining how the data structure functions and interacts with the environment. The operating system is considered to be a largely genetically determined collection of Piagetian black boxes. Cunningham develops his model by presenting postulates or rules concerning the data structure and the operating system. These are given here, mostly in Cunningham's own words.

The 10 data structure rules are as follows. (1) The system is initially endowed with a set of input elements and a set of output elements, with links called reflexes from the former to the latter. (2) Each element has a transitory parameter called its activity. (3) Activity of the input elements depends upon external conditions of the environment. (4) Activity of the output elements may alter the external conditions of the environment. (5) Each input must have a link pointing to at least one output; and each output must be similarly connected to at least one input. (6) Each input and output element has its own maximum capacity above which its activity cannot rise; this capacity may increase or mature with time and regardless of experience. (7) The relation between inputs, outputs, reflex links, and the objects of the environment must be such that the objects activate and exercise these reflexes and thereby set up "reflexive circular reactions" with them. (8) When two or more links have been functioning at the same time, then (if such has not already been done) a new element is formed with links from the elements on the tail end of the said functioning links, and with links to elements pointed to by the functioning links; the new element is a memory element. (9) Memory elements are created once and for all, though subsequently links to and from them may be added in accordance with rule 8. (10) Memory elements and links are not removed, though subsequent development may make it impossible to bring an element into reverberation (compare Baron's treatment of "forgetting").

Governing the above data structure are the 8 principles of the operating system. (1) Under certain conditions (described below) an element with a high level of activity may, as a function of its own activity, increase the activity of each of the other elements to which its links point; such an element is said to reverberate. (2) Only a subset of elements, those having greatest activity, reverberate; their number is called the *attention span* and the subset itself, the *attention.* (3) An element will continue reverberating for some short period of time, even if the elements facilitating it have stopped and stimulation from the environment has ceased. (4) Aside from the input elements, the activity of an element is the sum of the activity passed to it by the reverberating elements of the attention span. (5) Whenever the reverberation ratio (the sum of the activities of the reverberating elements divided by the sum of the activities of all the elements) becomes smaller than a certain fraction, the size of the attention span will increase until that fraction is reached; the amount of activity passed along a link will decrease. (6) Whenever the reverberation ratio becomes larger than a certain fraction, the size of the attention span will decrease until that fraction is reached; the amount of activity passed along a link will increase. (7) With time and increasing complexity of the data structure, the size of the attention span is increased. (8) Memory elements are treated the same as the primitive input and output elements, except for those characteristics that refer explicitly to input and output elements.

In Piaget's theory of human development, learning during the so-called sensorimotor period (roughly the first 18 months of life) proceeds through six stages, some of which have several substages. In Cunningham's "element" terminology, Piaget's major sensorimotor stages may be approximately described as follows.

Stage I: reflex exercise. This is the presumed initial state of a newborn. The elements interact with the environment in a purely reflexive fashion, without benefit of training or prior experience.

Stage II: primary circular reactions. In this stage, the environment sometimes causes several of the reflex systems of Stage I to reverberate at the same time, which leads to the formation of new elements that coordinate the simultaneously reverberating reflex systems. This coordination manifests itself in two different ways. First, if the new element coordinates roughly parallel systems, then it appears that the infant has developed an enhanced sensitivity to certain input patterns and/or has become more proficient at producing certain output patterns. Second, if the new element coordinates different systems, then it appears as if the infant has learned to connect the two systems.

Stage III: secondary circular reactions. This stage is marked by the formation of new elements that coordinate, in a hierarchical fashion, the elements that produced Stage II behavior.

Stage IV: familiar procedures in new situations. As more and more learned behavior patterns are linked together in secondary circular reaction, chains of second order elements eventually arise. If reverberation on this chain allows some first level element to reverberate, and this reverberation then allows the original element to reverberate, the infant has discovered a means to an end. For example, suppose that two second level elements have a first order procedure (behavior pattern) in common, all activated by some sensory input. Suppose the response is "reaching" but it fails because the object is too far away. If Cunningham's operating system rule 5 is invoked, the attention span will increase and may invoke the reverberation of some other first level element. If by chance this changes things so that reaching is now appropriate, then these complex chained paths are said to have allowed the development of *signs*, sets of sensory inputs which signal remote procedures.

Stage V: active experimentation. In Stage IV the infant is groping about for something to try; in stage V he is persistently trying something out, acquiring true independence from environmental stimulation. Some elements become linked together in tight little mutually interconnected knots (as opposed to the long chains of Stage III). Once such a knot of interlinked elements is large enough, it can maintain its own reverberations. This represents short term memory, providing the ability to coordinate actions with deferred results by allowing the infant to keep a goal in mind while trying out different procedures, one after another, to see how they relate to the goal.

Stage VI: mental recombination. Further increase in the attention span and continued formation of memory links eventually allow a child to hold two different short term structures in the attention span without direct sensory facilitation. This in turn enables "experiments" without recourse to the environment (as required in Stage V). The eventual actions thus appear insightful.

In his book, Cunningham argued informally that his data structure and operating system postulates contained sufficient power and flexibility to carry out the sort of organization called for in the six Piagetian stages. He described some proposed computational aspects in considerable detail. In a later study, Cunningham and Gray (1974) devised assembler language programs for simulation studies.

The simulated behavior was that of an infant learning to produce speech sounds. There were a number of parameters describing lip and tongue positions, nasalization, and vocal cord activity. Represented as elements, these parameters could be combined in a variety of ways to produce a particular type of "sound" (in the form of printed output). There were also sensory elements responsive to various properties of these sounds, including six different frequency ranges, a noisy sound, and an abrupt sound shift. Finally, there were 30 other sensory elements for such inputs as coldness and olfactory sensations. The sensory apparatus responded not only to input from the environment (typed in by the user) but also to the system's own output of sounds.

The Piagetian Stage I was "prewired" into the program, so that reflexive production of various sounds occurred in apparently random alternations of letter strings and blanks (silence). If learning by formation of new elements was not permitted, and the only changes allowed were in the size of the attention span, the amount of activity passed by an element to its successors, and the relative strength of auditory input elements, then sequences that could be interpreted as the result of maturation occurred.

As soon as learning was introduced, however, the simulation failed. Once the model learned to coordinate some sound with the shape of the mouth that produced that sound, it tended to make that sound endlessly. All new learning contributed only to strengthening of that particular behavior pattern. More and more new elements were created which did essentially the same thing. To resolve this endless growth problem, Cunningham and Gray decided not to allow the creation of a new element if there was already an old element whose successors included more than a certain fraction of the would-be successors of the new element, or whose predecessors included more than that fraction of the would-be predecessors of the new element. Thus a minimum amount of "newness" was required for an element to be created. As became apparent after a number of runs, the greater the amount of newness required, the more likely the system was to continue its behavior without learning anything, to assimilate the new situation to established modes of behavior. After much experimentation, Cunningham and Gray found a useful value for the "newness threshold" fraction.

The model was then in a position to move on to the Piagetian Stage II, typified by primary circular reaction. Sounds should have become coordinated with the positions of the mouth that produced them. These sounds should also have been sustained or repeated for longer periods of time because they would act as positive sensory feedback to the movement that produced them. Sounds typed in by the user should also have provided positive feedback, but only if the simulation had just recently been making the sound itself. As a sound pattern developed a number of coordinations with the different mouth positions that could produce it and with different but closely related sounds, it should have been imitated with increasing independence of the initial mouth position.

The attempt to move into Stage II produced one success and a number of failures. The one success was the imitation of the sound "a." This was learned, for there was no innate tendency to imitate sounds and the Stage I behavior did not include sustained sequences of "a." Nevertheless, after an input string of "a," the simulation responded with its own "a." This imitation was made possible by the simulation having learned to coordinate and associate certain mouth position variables with certain sounds. The model also tried to imitate "s" but failed to close the velum to stop air from escaping through the nasal cavity.

There were a number of apparent reasons for the failure to learn anything else. When a reflex system had memory elements added to it, its output activities tended to increase and so, together with the memory elements, it dominated attention and prevented elements from other reflex systems from ever reaching reverberation. It would be desirable for the input elements to developed reflexes and those to undeveloped reflexes to compete on an equal basis for space of attention span, and so to be included in new coordinations. Also, output elements tended to have lower activities than input elements and hence failed to be included in new coordinations. In fact a reflex could develop new coordinations only if both its input and output elements were simultaneously reverberating. But typically the input element reverberated first, causing the output element to reverberate in the next cycle, by which time the corresponding mouth variable had changed and the input element was no longer active.

In attempts to overcome some of these difficulties, Cunningham and Gray made various alterations to the model. Most important was the uniting of input and output element into a single element so that there would automatically be simultaneous reverberation. With these changes, they claimed that the simulation had progressed through Stage II and into Stage III (imitation of familiar sounds not behaviorally evident for some time, and coordination of two or more sounds). Although no data was presented to substantiate the claim that Stage II behavior was operative, they did present two outputs which were supposed to exhibit the sound-pair coordination characteristic of Stage III.

References

Alberghina, L. Modeling the control of cell growth. *Simulation*, **31**, 37-41, 1978.

Antonelli, P. L., T. D. Rogers, and M. A. Willard. Geometry and the exchange principle in cell aggregation kinetics. *Journal of Theoretical Biology*, **41**, 1, 1973.

Antonelli, P. L., D. I. McLaren, T. D. Rogers, and M. A. Willard. Transitivity, pattern reversal, engulfment and duality in exchange type cell aggregration kinetics. *Journal of Theoretical Biology*, **49**, 1975.

Bagley, J. D. The behavior of adaptive systems which employ genetic and correlation algorithms. Ph.D. Dissertation, Department of Computer and Communication Sciences, University of Michigan, 1967.

Baker, R. W. and G. T. Herman. Simulation of organisms using a developmental model. *Bio-Medical Computing*, **3**, 201-215 (Part 1), 251-267 (Part 2), 1972.

Baron, R. J. A model for cortical memory. *Journal of Mathematical Psychology*, **7**, 37-59, 1970a.

Baron, R. J. A model for the elementary visual networks of the human brain. *Man-Machine Studies*, **2**, 267-290, 1970b.

Baron, R. J. A theory for the neural basis of language. *Man-Machine Studies*, **6**, 13-48 (Part 1), 155-204 (Part 2), 1974.

Brender, R. F. A programming system for the simulation of cellular spaces. Ph.D. Dissertation, Department of Computer and Communication Sciences, University of Michigan, 1970.

Brindle, A. Genetic algorithms for function optimization. Ph.D. Dissertation, Department of Computing Science, University of Alberta, 1980.

Cassano, W. F. A biochemical simulation system. *Computers and Biomedical Research*, **10**, 383-392, 1977.

Cavicchio, D. J. Adaptive search using simulated evolution. Technical Report 03296-4-T, Department of Computer and Communication Sciences, University of Michigan, 1970.

Covington, A. R., J. R. Sampson, and R. G. Peddicord. A frequency response neuron model: Interactive implementation and further simulation experiments. *Kybernetes*, **7**, 45-60, 1978.

Cunningham, M. *Intelligence: Its organization and development*. New York: Academic Press, 1972.

Cunningham, M. A. and H. J. Gray. Design and test of a cognitive model. *Man-Machine Studies*, **6**, 49-104, 1974.

De Jong, K. A. Analysis of the behavior of a class of genetic adaptive systems. Technical Report 185, Logic of Computers Group, University of Michigan, 1975.

De Jong, K. A. Adaptive systems design: A genetic approach. *IEEE Transactions on Systems, Man, and Cybernetics*, **SMC-10**, 566-574, 1980.

Descheneau, C. Modelling and simulation in developmental genetics. Ph.D. Dissertation, Department of Computing Science, University of Alberta, 1981.

Donaghey, C. E. and B. Drewinko. A computer simulation program for the study of cellular growth kinetics and its application to the analysis of human lymphoma cells in vitro. *Computers and Biomedical Research*, **8**, 118-128, 1975.

Duchting, W. A model of disturbed self-reproducing cell systems. In A. J. Valleron and P. D. M. Macdonald (eds.), *Biomathematics and cell kinetics*, New York: Elsevier, 1978.

Duchting, W. and T. Vogelsaenger. Three-dimensional simulation of tumor growth. *Simulation*, **40**, 163-170, 1983.

Evert, C. F. CELLDYN: A digital program for modeling the dynamics of cells. *Simulation*, **24**, 55-61, 1975.

Finley, M. An experimental study of the formation and development of Hebbian cell assemblies by means of a neural network simulation. Technical Report 08333-1-T, Department of Computer and Communication Sciences, University of Michigan, 1967.

Fishman, G. S. *Concepts and methods in discrete event digital simulation.* New York: Wiley, 1973.

Garfinkel, D. A machine independent language for the simulation of complex chemical and biochemical systems. *Computers and Biomedical Research*, **2**, 31-44, 1968.

Garfinkel, D. Simulating biochemical activity in physiological systems. *Simulation*, **28**, 193-196, 1977.

Garfinkel, D., M. C. Kohn, M. J. Achs, J. Phifer, and G.-K. Roman. Construction of more reliable complex metabolic models without repeated solution of their constituent differential equations. *Mathematics and Computers in Simulation*, **22**, 18-27, 1978.

Gierer, A. and H. Meinhardt. A theory of biological pattern formation. *Kybernetik*, **12**, 30-39, 1972.

Goel, N. S., R. D. Campbell, R. Gorden, R. Rosen, H. Martinez, and M. Ycas. Self-sorting of isotropic cells. *Journal of Theoretical Biology*, **28**, 423, 1970.

Goodman, E. D. Adaptive behavior of simulated bacterial cells subjected to nutritional shifts. Ph.D. Dissertation, Department of Computer and Communication Sciences, University of Michigan, 1972.

Goodman, E. D., R. Weinberg, and R. A. Laing. A cell-space embedding of simulated living cells. *Bio-Medical Computing*, **2**, 121-136, 1971.

Herman, G. T. and W. H. Liu. The daughter of Celia, the French flag, and the firing squad. *Simulation*, **21**, 33-41, 1973.

Hilborn, R. A control system for FORTRAN simulation programming. *Simulation*, **20**, 172-175, 1973.

Hogeweg, P. Simulating the growth of cellular forms. *Simulation*, **31**, 90-96, 1978.

Holland, J. H. *Adaptation in natural and artificial systems.* Ann Arbor: University of Michigan Press, 1975.

Huneycutt, C. W. An interactive biochemical simulation system. M.Sc. Thesis, Department of Computing Science, University of Alberta, 1976.

Hurum, S. and J. R. Sampson. Simulation of cell rearrangements using Pascal. In R. Bryant and B. W. Unger (eds.), *Proceedings of the SCS Conference on Simulation in Strongly Typed Languages*, San Diego, 1984.

Kabrisky, M. *A Proposed model for visual information processing in the human brain.* Urbana: University of Illinois Press, 1966.

Klir, J. and M. Valach. *Cybernetic modelling.* London: Iliffe, 1967.

Leith, A. G. and N. S. Goel. Self-sorting of anisotropic cells. *Journal of Theoretical Biology*, **33**, 171, 1971.

Margolis, S. G. and S. Cooper. Computer simulation of bacterial growth, cell division, and DNA synthesis. Report No. 54, School of Engineering, State University of New York at Buffalo, 1970.

Margolis, S. G. and S. Cooper. Simulation of bacterial growth, cell division, and DNA synthesis. *Computers and Biomedical Research*, **4**, 427-443, 1971.

Matela, R. J. and R. J. Fletterick. A topological model for cell self-sorting. *Journal of Theoretical Biology*, **76**, 403-414, 1979.

Matela, R. J. and R. J. Fletterick. Computer simulation of cellular self-sorting. *Journal of Theoretical Biology*, **84**, 673-690, 1980.

Meinhardt, H. A model of pattern formation in insect embryogenesis. *Journal of Cell Science*, **23**, 111-142, 1978.

Mercer, R. E. and J. R. Sampson. Adaptive search using a reproductive meta-plan. *Kybernetes*, **7**, 215-228, 1978.

Mortimer, J. A. A cellular model for mammalian cerebellar cortex. Technical Report 03296-7-T, Department of Computer and Communication Sciences, University of Michigan, 1970.

Mortimer, J. A. A computer model of mammalian cerebellar cortex. *Computers in Biology and Medicine*, **4**, 59, 1974.

Olinick, M. *An introduction to mathematical models in the social and life sciences.* Reading, Mass.: Addison-Wesley, 1978.

Peddicord, R. G. and J. R. Sampson. A neuron model for simulation of frequency response characteristics. *Bio-Medical Computing*, **5**, 285-299, 1974.

Pellionisz, A. and R. Llinas. A computer model of cerebellar Purkinje cells. *Neuroscience*, **2**, 37-48, 1977.

Pellionisz, A., R. Llinas, and D. H. Perkel. A computer model of the cerebellar cortex of the frog. *Neuroscience*, **2**, 19-35, 1977.

Perkel, D. H. A computer program for simulating a network of interacting neurons. *Computers and Biomedical Research*, **9**, 31-43 (Part I), 45-66 (Part II, with M. S. Smith), 67-73 (Part III), 1976.

Plum, T. W.-S. Simulations of a cell-assembly model. Ph.D. Dissertation, Department of Computer and Communication Sciences, University of Michigan, 1972.

Ransom, R. A computer model of cell clone growth. *Simulation*, **28**, 189-192, 1977a.

Ransom, R. Computer analysis of division patterns in the Drosophila head disc. *Journal of Theoretical Biology*, 1977b.

Ransom, R. *Computers and embryos: Models in developmental biology.* New York: Wiley, 1981.

Rogers, G. and N. S. Goel. Computer simulation of cellular movements: Cell-sorting, cellular migration through a mass of cells and contact inhibition. *Journal of Theoretical Biology*, **71**, 141-166, 1978.

Rogers, T. D. and J. R. Sampson. A random walk model of cellular kinetics. *Bio-Medical Computing*, **8**, 45-60, 1977.

Rosenblatt, F. Comparison of a five-layer perceptron with human visual performance. In H. H. Pattee, E. A. Edelsack, L. Fein, and A. B. Callahan (eds.), *Natural automata and useful simulations.* Washington, D. C.: Spartan Books, 1966.

Sampson, J. R. A neural subassembly model of human learning and memory. Ph.D. Dissertation, Department of Computer and Communication Sciences, University of Michigan, 1969.

Sampson, J. R. Simulation studies of automata networks with applications to neural modelling. In *Proceedings of the Fourth Hawaii International Conference on System Sciences*, 1971.

Sampson, J. R. Modelling and simulation using cellular automata. In *Proceedings of the AMSE International Summer Conference on Modelling and Simulation*, Minneapolis, 1984.

Sampson, J. R. and M. Dubreuil. Design of interactive simulation systems for biological modelling. In B. P. Zeigler (ed.), *Methodology in systems modelling and simulation.* New York: North-Holland, 1979.

Sohnle, R. C., J. Tartar, and J. R. Sampson. Requirements for interactive simulation systems. *Simulation*, **20**, 145-152, 1973.

Weinberg, R. and M. Berkus. Computer simulation of a living cell. *Bio-Medical Computing*, **2**, 95-120 (Part I), 167-188 (Part II), 1971.

Weinberg, R., L. K. Flanigan, and R. A. Laing. Computer simulation of a primitive, evolving, eco-system. Technical Report 03296-6-T, Department of Computer and Communication Sciences, University of Michigan, 1970.

Weinberg, R., B. P. Zeigler, and R. A. Laing. Computer simulation of a living cell: Metabolic control system experiments. *Journal of Cybernetics*, **1**, 34-48, 1971.

Whitehead, B. A. A neural network model of human pattern recognition. Technical Report 209, Logic of Computers Group, University of Michigan, 1978.

Wilby, O. K. and D. H. Ede. A model generating the pattern of skeletal elements in the embryonic chick wing. In *Proceedings of the IEEE Conference on Biologically Motivated Automata Theory*, 1974.

Wilby, O. K. and D. H. Ede. Computer simulation of vertebrate limb development. In A. Lindenmayer and G. Rozenberg (eds.), *Automata, languages, development*, New York: North-Holland, 1976.

Wittie, L. D. Large-scale simulation of brain cortices. *Simulation*, **31**, 73-78, 1978a.

Wittie, L. D. Large network models using the brain organization simulation system (BOSS). *Simulation*, **31**, 117-122, 1978b.

Zeigler, B. P. *Theory of modelling and simulation.* New York: Wiley, 1976.

Zeigler, B. P. *Multifacetted modelling and discrete event simulation.* New York: Academic Press, 1984.

Index